An Introduction to Applied
Electromagnetics and Optics

An Introduction to Applied
Electromagnetics and Optics

Vladimir V. Mitin   Dmitry I. Sementsov

CRC Press
Taylor & Francis Group
Boca Raton London New York

CRC Press is an imprint of the
Taylor & Francis Group, an **informa** business

CRC Press
Taylor & Francis Group
6000 Broken Sound Parkway NW, Suite 300
Boca Raton, FL 33487-2742

First issued in paperback 2019

© 2017 by Taylor & Francis Group, LLC
CRC Press is an imprint of Taylor & Francis Group, an Informa business

No claim to original U.S. Government works

ISBN-13: 978-1-4987-7629-5 (hbk)
ISBN-13: 978-0-367-88961-6 (pbk)

Visit the Taylor & Francis Web site at
http://www.taylorandfrancis.com

and the CRC Press Web site at
http://www.crcpress.com

Contents

SECTION III   Electromagnetic Waves in Periodic and Waveguiding Structures

SECTION IV   Advanced Topics in Electromagnetics and Optics

Preface

Modern technology is developing rapidly, and for this reason, future engineers need to acquire advanced knowledge in science and technology that includes advanced knowledge of electromagnetic phenomena. Hence, we have developed a course in applied electromagnetics that resulted in this book. In contrast to the conventional texts on electromagnetics, substantial attention is paid to the optical part of the electromagnetic spectrum and to electromagnetic waves in different media and structures. Each topic has solved examples. Problems at the end of each chapter are supplemented by either hints or answers to help students to master the material and to solve typical problems of electromagnetics and optics.

The book consists of four sections. Section I has three chapters that review the most important aspects of Maxwell's equations. Students are expected to be familiar with the material included in Section I from their physics courses taken previously. If students are well prepared by these prerequisite courses, an instructor can review this section very fast.

Section II (Chapters 4 through 7) and Section III (Chapters 8 through 10) are the core material of the book. Chapter 4, "Electromagnetic Waves in Homogeneous Media without Absorption," covers the basic equations describing electromagnetic waves in a homogeneous, isotropic, nonabsorbing medium, as well as the basic properties of these waves.

Chapter 5, "Electromagnetic Fields and Waves at the Interface between Two Media," discusses the amplitude and energy, reflection, and transmission coefficients of an electromagnetic wave at an interface, the laws of reflection and refraction, total internal reflection, and reflection from a dielectric layer.

Chapter 6, "Electromagnetic Waves in Anisotropic and Optically Active Media," is devoted to topics that are rarely discussed in the previous studies. An anisotropic medium is one in which physical properties vary along different directions. In an anisotropic medium, there are specific directions (or a direction known as the axis of symmetry) that are associated with the structure of the medium or with an external electric and/or magnetic field. Anisotropy of a medium leads to the fact that the magnitude of a wave vector, the group and phase velocity of the wave, and its polarization parameters depend not only on the frequency but also on the direction of the wave propagation with respect to the axis of symmetry.

Chapter 7, "Electromagnetic Waves in Conducting Media," covers dielectric permittivity and impedance of metals, skin effect, and surface waves at the interface between a dielectric and a conductor. An electromagnetic wave propagating through a conducting medium loses some of its energy that is transformed predominantly into the thermal energy of that material. Superconductivity is also discussed as an example of a conducting medium without losses.

Chapter 8, "Waves in Periodic Structures," covers in detail diffraction phenomena such as diffraction by a slit, by a 1D lattice, and by a 3D lattice. Special attention is paid to an often omitted topic: waves in periodic structures (photonic crystals). Forbidden photonic bands in the reflection and transmission spectra are formed for a wave propagating in the periodic medium along a periodicity axis.

Chapter 9, "Waves in Guiding Structures," in addition to the conventional waveguides (two parallel metal planes, a rectangular waveguide, and two-wire, coaxial, and stripline transmission lines), this chapter also covers optical waveguides (optical fibers).

Chapter 10, "Emission of Electromagnetic Waves," is devoted to emission by accelerated charges, by an electric dipole (Hertz antenna), and by an elementary magnetic dipole. Directional diagrams of different sources of electromagnetic radiation and types of antennas are also discussed. Students are introduced to the near-field zone that extends from a source to distances that have the same order of magnitude with the wavelength of the electromagnetic wave generated by the source and to the far-field zone that is located at a distance much greater than the wavelength of the emitted wave.

Section IV (Chapters 11 through 13), "Advanced Topics in Electromagnetics and Optics," introduces students to recent advances in the field of electromagnetics and optics. In particular:

Chapter 11, "Electromagnetic Waves in Gyrotropic Media," discusses the electromagnetic waves in gyrotropic media, magnetoactive plasma, and ferrites. Such media have the ability to rotate the polarization plane of linearly polarized electromagnetic waves, which propagate through them. The properties of the media are controlled by an external magnetic field.

Chapter 12, "Electromagnetic Waves in Amplyifying Media," addresses the media that can amplify optical radiation. Amplification of waves in a medium is due to the induced radiation emitted by the excited atoms of the medium. The basic principles of amplification are discussed and students are introduced to lasers where amplification occurs due to the induced coherent emission by excited atoms under the influence of the electromagnetic wave field.

Chapter 13, "Electromagnetic Waves in Media with Material Parameters That Are Complex Numbers," is devoted to a detailed discussion of wave propagation in media for which the relative dielectric permittivity or the relative magnetic permeability (or both) is a complex number with a nonzero imaginary part. Electromagnetic tunneling and negative refraction in media with negative permittivity and permeability are also discussed.

As it has been shortly presented earlier, the book covers both conventional material and material that is very rarely discussed in undergraduate textbooks, such as superconductors, surface waves, plasmas, photonic crystals, and negative refraction.

Instructors can use the first 10 chapters for the undergraduate course with a short review of Chapters 11 through 13.

This book can also be used as a review of electromagnetics and optics for graduate students. In this case, the first 3 chapters can be discussed briefly, while the last 10 chapters become the focus of the course.

The authors have many professional colleagues and friends who must be acknowledged. Without their contributions, this work would not have been completed. Special thanks to the Division of Undergraduate Education of the National Science Foundation for their partial support of this work through the TUES Program (Program Director Don Lewis Millard). The authors especially thank Professor Athos Petrou for his editorial efforts in critical reading of this book and for his many valuable comments and suggestions. The authors also thank Yudi Liu, Xiang Zhang, Nizami Vagidov, Yanshu Li, and Tim Yore for their help in the preparation of the figures and some editorial help. The earlier versions of this manuscript were used in teaching EE324, "Applied Electromagnetics," a course in fall 2014 and fall 2015 at the University at Buffalo, the State University of New York. The authors also wish to thank the students in their courses whose valuable feedback helped to substantially improve this book. The earliest version was partially used for the EE324 in fall 2013. The authors also thank the students and their instructor, Dr. Victor Pogrebnyak, for some constructive feedback that they have given. Finally, the authors wish to thank our loved ones for their support and for forgiving us for not devoting more time to them while working on this book.

List of Notations

1D	One dimensional
2D	Two dimensional
3D	Three dimensional
A	Surface area
A_{21}	Einstein coefficient
A	Amplitude of a wave
A^+	Amplitude of a forward wave
A^-	Amplitude of a backward wave
$\mathbf{B}$	Vector of magnetic field
B	Magnetic field
B_c	Critical magnetic field
B_{c0}	Critical magnetic field at zero temperature
B_l	Tangential projection of vector $\mathbf{B}$
b	Slit width
C	Capacitance of a conductor or a capacitor
c	Speed of light in vacuum
$\mathbf{D}$	Vector of electric displacement
D	Angular dispersion
d	Lattice constant
$\mathbf{E}$	Vector of electric field
$\mathbf{E}_\parallel$	Longitudinal component of electric vector $\mathbf{E}$
$\mathbf{E}_\perp$	Transverse component of electric vector $\mathbf{E}$
$\mathbf{E}_e$	Vector of electric field of an extraordinary wave
$\mathbf{E}_o$	Vector of electric field of an ordinary wave
E_0	Electric field amplitude of an incident wave
$\mathbf{e}$	Unit vector
e	Absolute value of an electron charge
$\mathbf{F}$	Force, vector
$\mathbf{F}_A$	Ampere's force
$\mathbf{F}_L$	Lorentz force
$\mathbf{F}_r$	Damping force
$\mathbf{F}_{\text{rad}}$	Radiative friction force
f	Frequency
f_l	Oscillator strength
g	Damping constant
g	Transverse wave number
g	Lande g-factor
$\text{grad}\,\varphi$	Potential gradient
$\mathbf{H}$	Magnetic field intensity vector
H	Magnetic field intensity
H	Henri, SI unit of inductance
h_a	Operating height of a receiving antenna
$\mathbf{i},\mathbf{j},\mathbf{k}$	Unit vectors of axis along the Cartesian coordinate axes
I	Current
I_c	Critical current
I_i	Induced current
I_{macro}	Algebraic sum of macroscopic currents enclosed inside a closed path L

I_{micro}	Algebraic sum of microscopic currents enclosed inside a closed path L
j	Current density, vector
j	Current density
k	Wave vector
k	Wave number, absolute value of wave vector
k_0	Wave vector in vacuum
k_{B}	Bloch wave vector
k_{B}	Boltzmann's constant
k_e	Unit system coefficient
k_m	Proportionality coefficient
k''	Attenuation (amplification) coefficient
κ	Extinction coefficient
L	Inductance
L_{opt}	Optical path length
l	Dipole vector
l	Dipole length
l_{coh}	Coherence length
l_m	Equivalent length of magnetic dipole
M	Magnetization, the magnetic moment per unit volume of a medium
M	Mutual inductances
M_0	Saturation magnetization
m	Mass
m_p	Mass of protons
m_e	Mass of electrons
n	Unit vector
n	Refractive index
n_e	Refractive index for an extraordinary wave
n_o	Refractive index for an ordinary wave
n_s	Density of "superconducting" electrons
P	Polarization vector
p	Dipole moment
P	Pressure
P_A	Power loss density
P_{rad}	Emitted power density
p	Longitudinal index
Q	Charge, distributed charge
q	Charge, point charge
$q\varphi$	Charge potential energy
R	Resistance
R_{rad}	Resistance of radiation
R_s	Surface resistance of a conductor
R	Resolving ability
r	Distance, coordinate
r	Complex reflection coefficient
S	Poynting vector
$\mathbf{S}^i$	Poynting vector of an incident wave
$\mathbf{S}^r$	Poynting vector of a reflected wave
$\mathbf{S}_n$	Normal component of Poynting vector
$\mathbf{S}_t$	Tangential component of Poynting vector
$\langle \mathbf{S} \rangle$	Average over the period energy flux density
T	Absolute temperature

T_c	Critical temperature
t	Complex transmission coefficient
U	Electric potential energy
u_e	Energy density
V	Potential difference
Vol	Volume
v_{gr}	Group velocity
v_{ph}	Phase velocity
v	Velocity of light in a media
v_r	Ray velocity
v_s	Velocity of "superconducting" electrons
W	Work
W_0	Copper pair binding energy
w	Volume energy density of electromagnetic waves
X_C	Capacitive resistance
Z	Impedance
Z_0	Impedance of vacuum
Z_s	Surface impedance of a conductor
α	Coefficient of absorption
α	Attenuation coefficient
α_e	Electric polarizability
α_{em}	Electromagnetic polarizability
α_{ef}	Effective (negative) absorption coefficient
β	Wave propagation constant
β	Longitudinal wave number
$|\Gamma|$	Reflection coefficient module
γ	Gyromagnetic ratio
γ	Damping constant
$\Delta = \nabla^2$	Laplace operator
Δk	Specific Faraday rotation
δ	Thickness of the skin layer
δ_{ij}	Kronecker's symbol
δ_κ	Angles of dielectric losses
δ_{κ_m}	Angles of magnetic losses
∇	Nabla, the vector differential operator
$\mathcal{E}_i$	Induced electromotive force
$\varepsilon = \kappa\varepsilon_0$	Permittivity of material with relative permittivity κ
ε_0	Permittivity of vacuum, electric constant
ζ	Impedance phase
η	Spatial dispersion
η	Chirality parameter
θ_0	Angle of incidence
θ_B	Brewster angle
θ_{cr}	Angle of total internal reflection
κ	Relative dielectric permittivity (dielectric constant)
$\tilde{\kappa}$	Complex permittivity of the conduction medium
$\hat{\kappa}$	Tensor of relative dielectric permittivity
$\kappa_{\parallel}$	Longitudinal dielectric constant
$\kappa_\perp$	Transverse dielectric constant
κ_m	Relative magnetic permeability (magnetic constant)

$\hat{\kappa}_m$	Tensor of relative magnetic permeability
Λ	Path difference
λ	Linear charge density
λ	Wavelength
λ_{cr}	Critical wavelength
μ	Magnetic permeability
μ_0	Magnetic constant, permeability of vacuum
μ_B	Bohr magneton
$\boldsymbol{\mu}_m$	Magnetic moment vector
$\boldsymbol{\mu}_{orb}$	Orbital magnetic moment
$\boldsymbol{\mu}_{spin}$	Spin moment
ρ	Resistivity
ρ	Volume charge density
ρ'	Volume density of bound charge
σ	Surface charge density
σ	Conductivity
τ	Time constant
τ_{coh}	Coherence time
Φ_B	Magnetic flux
Φ_E	Electric field flux
Φ_{net}	Net magnetic flux
φ	Electric potential
φ	Phase angle
φ	Bragg's angle
χ	Magnetic susceptibility
χ	Nonreciprocity parameter
ψ_{rp}	Reflected phase shift
ψ_{tp}	Transmitted phase shift
Ω	Solid angle
ω	Wave angular frequency
ω_c	Angular velocity
ω_C	Cyclotron frequency
ω_{ci}	Ion cyclotron frequency
ω_p	Plasma frequency

Section I

Electric and Magnetic Fields in Isotropic Media

1 Electrostatics

Under certain conditions, macroscopic bodies attract or repel each other. This interaction is of electrostatic nature rather than gravitational. The latter is negligible compared to the electric force. The electric force arises from the presence of net charge of bodies (e.g., charges can be caused by friction or irradiation). An electric charge is a physical quantity that characterizes the ability of a body or a particle to participate in electromagnetic interactions.

Charges can be transferred from one body to another (e.g., through a direct contact). Under different conditions, the same body can acquire different charges. There are two types of charges, which are referred to as *positive* and *negative*. Two charges with the same sign repel each other and charges with the opposite polarity attract. This observation highlights the principal difference between the electromagnetic and gravitational forces.

In this chapter, we review the main laws of *electrostatics* in vacuum and in isotropic media; electrostatics considers properties of system of electric charges that are not moving including forces of interaction of electric charges (*electrostatic forces*) and energy of electric charges (*electrostatic energy*).

1.1 ELECTRIC CHARGES, ELECTRIC CHARGE CONSERVATION LAW, AND COULOMB'S LAW

1. All bodies consist of atoms that contain positively charged **protons**, negatively charged **electrons**, and neutrally charged **neutrons**. Protons and neutrons are parts of the atomic nuclei; electrons are located outside the nuclei. The electric charges of a proton and an electron are equal in absolute value to the elementary charge $e = 1.60 \times 10^{-19}$ C. In a neutral atom, the number of protons in a nucleus is equal to the number of electrons. This number is called *the atomic number*. Atom can lose or acquire one or more electrons. In these cases, the neutral atom becomes a positively or negatively charged **ion**.

Since any charge q consists of a collection of elementary charges, it is multiple of the elementary charge, that is,

$$q = \pm ne, \quad n = 1,2,3,\dots. \tag{1.1}$$

Thus, the electric charge of a body is a discrete quantity. Physical quantities that can have only discrete values are named "quantized" quantities. The elementary charge e is the *quantum* (the smallest quantity) of electric charge.

The conservation of electric charge is a fundamental *law* of nature. This law states that the total charge of any isolated system (a system is electrically isolated if no charges flow through its boundaries) does not change. Thus, the algebraic sum of all the charges inside an isolated system stays constant, that is,

$$q_1 + q_2 + q_3 + \cdots + q_n = \text{const.} \tag{1.2}$$

2. The force between two point charges was investigated by Charles Coulomb in 1785. In his experiments, Coulomb measured the forces of attraction and repulsion between charged spheres using a high precision torsion balance. Experiments performed by Coulomb allowed him to deduce the following law: *the magnitude of the force between two stationary point charges is directly* *proportional to the product of the absolute value of the charges and inversely proportional to the square of the distance r_{12} between them.* These forces are acting along the straight line that connects the two charges. Thus, the expression for the force that the first charge exerts on the second charge can be written in a vector form as

$$\mathbf{F}_{12} = k_e \frac{q_1 \cdot q_2}{r_{12}^3} \, \mathbf{r}_{12} = k_e \frac{q_1 \cdot q_2}{r_{12}^2} \, \mathbf{e}_{12}, \tag{1.3}$$

where $\mathbf{e}_{12} = \mathbf{r}_{12}/r_{12}$ is the unit vector in the direction of the radius vector $\mathbf{r}_{12}$ that points from r_1 to r_2; the proportionality coefficient k_e depends on the unit system. In *the International System of Units* (SI), the unit of charge is the coulomb (C). One coulomb is the charge passing a conductor cross section per 1 second (s) at the current of 1 ampere (A). Ampere, the unit of current, is a fundamental unit like the SI units of length, time, and mass. In the SI, the coefficient k_e is written as

$$k_e = \frac{1}{4\pi\varepsilon_0} = 9 \times 10^9 \ \mathrm{m^2 N/C^2}, \tag{1.4}$$

where $\varepsilon_0 = (1/36\pi) \cdot 10^{-9} = 8.85 \times 10^{-12} \ \mathrm{C^2/N \times m^2}$ is the permittivity of vacuum, meter (m) is the unit of length, and Newton (N) is the unit of force (note that $\mathrm{C^2/N \times m^2 = F/m}$, where farad [F] is the unit of capacitance that we will introduce later).

The Coulomb forces obey Newton's third law, that is, $\mathbf{F}_{12} = -\mathbf{F}_{21}$ (see Figure 1.1). Here, $\mathbf{F}_{21}$ is the force that the second charge exerts on the first charge.

Coulomb's law in the form of Equation 1.3 is valid for point charges. It can be used in cases when the size of the charged bodies is much smaller than the distance between them as those extended charges can be approximated by point charges.

3. Experiments show that the Coulomb force between two charges is valid in the presence of other charges in their surroundings. In a system composed of n charges, the resultant force $\mathbf{F}_i$ that the rest $n-1$ charges exert on the charge q_i is given by

$$\mathbf{F}_i = \sum_{j=1, j \neq i}^{n} \mathbf{F}_{ji}. \tag{1.5}$$

If a charged object interacts with other charged objects, the resultant force on it is given by the vector sum of the forces that all these charged bodies exert on it. Thus, the Coulomb force obeys the principle of **superposition**. In this case, the superposition principle states that if a given charged object is interacting with several charges, then the resulting force acting on the considered object

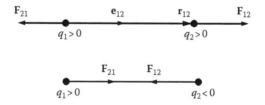

FIGURE 1.1 Two charges with the same sign repel each other (upper part) and charges with the opposite polarity attract (lower part).

is equal to the vector sum of the forces that each individual charge exerts on the object. Figure 1.2 illustrates the superposition principle. Here, the forces $\mathbf{F}_{ji}$ with two indices determine the force that charge j exerts on charge i and the forces $\mathbf{F}_i$ with one index are the resulting force on a given charge i from all other charges.

Equation 1.5 also describes the electrostatic interaction force of nonpoint charges, that is, charges distributed in the bodies of finite size. Considering each body as a set of infinitely small charges dq_1 and dq_2, one can define the force between them as

$$d\mathbf{F}_{12} = k_e \frac{dq_1 \cdot dq_2}{r_{12}^2} \mathbf{e}_{12}, \tag{1.6}$$

where
r_{12} is the distance between these charges
$\mathbf{e}_{12}$ is a unit vector directed from charge 1 to charge 2 (see Figure 1.1)

Generally, the expression for an infinitely small charge is given as

$$dq = \begin{cases} \rho dr^3, \\ \sigma dA, \\ \lambda dl, \end{cases} \tag{1.7}$$

where ρ, σ, and λ are the volume (C/m³), surface (C/m²), and line (C/m) charge densities, respectively. In the most general case, these densities depend on the coordinates. dr^3, dA, and dl are the elements of volume, surface, and length, respectively. To find the resultant Coulomb force, one has to integrate expression (1.6) over the volume (surface, length) occupied by each of the charges. To demonstrate the application of expressions (1.5) through (1.7), consider the example in Exercise 1.1.

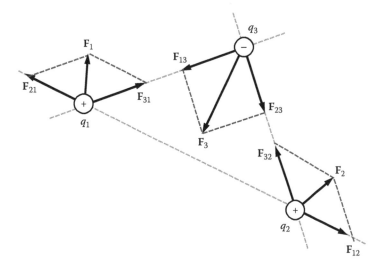

FIGURE 1.2 Illustration of the superposition principle for Coulomb forces.

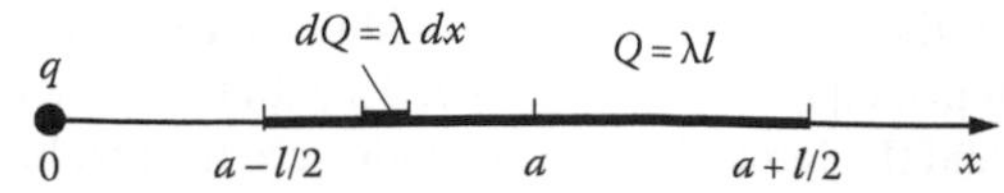

$$dQ = \lambda\, dx \qquad Q = \lambda l$$

$$q$$

$$0 \qquad a - l/2 \qquad a \qquad a + l/2 \qquad x$$

FIGURE 1.3 Interaction of a point charge q with a charged rod of length l.

Exercise 1.1

Find the Coulomb force between a point charge q at the origin and a thin charged rod of length l. The rod has a uniformly distributed charge Q, and it is oriented along the radial direction from the charge q. The geometric center of the rod is at the distance a from the point charge and $a > l/2$.

Solution. Consider an elementary charge $dQ = \lambda\, dx$ on the rod, where the linear charge density $\lambda = Q/l$ (Figure 1.3). The distance between dQ and q is x. The elementary Coulomb force acting on the charge element dQ is oriented along the x-axis, and its magnitude is equal to

$$dF = k_e \frac{q \cdot dQ}{x^2} = k_e \frac{qQ}{l}\frac{dx}{x^2}. \tag{1.8}$$

The total Coulomb force between the charge q and the rod is determined by summing the Coulomb forces between the charge q and all elementary charges dQ in the rod. In the case of a continuous charge distribution along the rod, the sum becomes an integral over the region occupied by the charge:

$$F = k_e \frac{qQ}{l} \int_{a-l/2}^{a+l/2} \frac{dx}{x^2} = -k_e \frac{qQ}{l}\frac{1}{x}\Big|_{a-l/2}^{a+l/2} = k_e \frac{qQ}{a^2 - l^2/4}. \tag{1.9}$$

In the case for which $a \gg l$, expression (1.9) for the Coulomb force becomes the expression for the Coulomb force between two point charges.

1.2 ELECTRIC FIELD VECTOR, PRINCIPLE OF SUPERPOSITION

1. The electric force between charged objects is present without the objects being in direct contact. This can be described by the electric field generated by any charged body in its surroundings. The electric field $\mathbf{E}$ is the vector quantity, which is directly proportional to the force, $\mathbf{F}$, the electric field exerts on a positive charge q at a given point of space and inversely proportional to the charge

$$\mathbf{E} = \frac{\mathbf{F}}{q}. \tag{1.10}$$

The direction of the vector $\mathbf{E}$ at any point of space coincides with the direction of the force acting on a positive charge. The unit of the electric field in the SI is N/C, or V/m, as will be discussed later. Here, volt (V) is the unit of electric potential. Taking into account relation (1.3), the electric field $\mathbf{E}_1$ generated by a point charge q_1 can be written as

$$\mathbf{E}_1 = k_e \frac{q_1}{r^2}\mathbf{e}_r, \tag{1.11}$$

where $\mathbf{e}_r$ is a unit vector directed along vector $\mathbf{r}$ from the position of charge q_1. Electric field lines are commonly used to visualize the electric field. These lines are drawn with the following assumptions: (1) the direction of the vector $\mathbf{E}$ coincides at any point with the direction of tangent to the electric field line, (2) the density of lines in any region is proportional to the magnitude of the electric field vector, and (3) the electric field lines emerge from positive charges and they sink into negative charges. Electric field lines of isolated positive and negative spherically symmetric charges uniformly follow the radial directions. The electric field lines come out of (into) the charge for a positive (negative) charge as it is shown in Figure 1.4.

2. It has been shown earlier that for a system of charges the resultant force acting on a test charge is given by the vector sum of forces exerted by each separate charge on it. Consequently, the electric field produced by a system composed of n charges at a given point is equal to the vector sum of the electric fields produced at the same point by each charge separately (see Figure 1.5):

$$\mathbf{E} = \sum_{i=1}^{n} \mathbf{E}_i. \tag{1.12}$$

This is the principle of superposition, one of the electromagnetic theory fundamentals. It allows calculating the electric field for any system of charges (either discretely or continuously distributed in space). To illustrate the application of the principle of superposition for a discrete charge distribution, consider a system of two charges equal in absolute value but different in sign, q and $-q$, placed

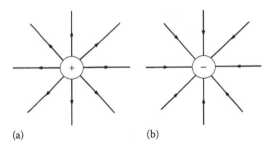

(a)   (b)

FIGURE 1.4   The electric field, $\mathbf{E}$, lines of a positive charge (a) and a negative charge (b).

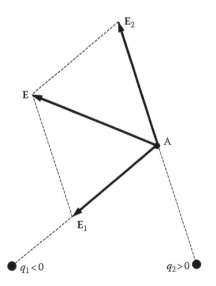

FIGURE 1.5   Illustration of the superposition principle for the electric field.

at a fixed distance l (here $q > 0$). An important parameter in electromagnetic theory is the dipole moment (a vector quantity)

$$\mathbf{p} = q\mathbf{l} \ (C \cdot m), \tag{1.13}$$

where $\mathbf{l}$ is the vector directed from the negative to the positive charge and its magnitude, $|\mathbf{l}| = l$ (see Figure 1.6a). Using the superposition principle given by Equation 1.12, one can show that the electric field on the dipole axis at a point at a distance $b_1 \gg l$ (Figure 1.6b) from the dipole center is given by

$$\mathbf{E}_1 = \frac{2k_e \mathbf{p}}{b_1^3}. \tag{1.14}$$

The electric field along the perpendicular to the dipole axis held at its center at the arbitrary distance b_2 is determined by the equation

$$\mathbf{E}_2 = -\frac{k_e \, \mathbf{p}}{b_2^3}. \tag{1.15}$$

Vector $\mathbf{E}_2$ is directed oppositely to $\mathbf{E}_1$ (Figure 1.6b) and under the condition $b_2 = b_1$, the magnitude of $\mathbf{E}_2$ is two times smaller than the magnitude of $\mathbf{E}_1$.

Many molecules have a nonzero dipole moment. For example, the dipole moment of the water molecule is $p = 6.2 \times 10^{-30} \ C \cdot m$.

Consider n point charges distributed in space points with position vectors $\mathbf{r}_i$. In this case, the electric field vector at a point with the position vector $\mathbf{r}_0$ can be expressed as

$$\mathbf{E} = k_e \sum_{i=1}^{n} \frac{\mathbf{r}_0 - \mathbf{r}_i}{|\mathbf{r}_0 - \mathbf{r}_i|^3} q_i. \tag{1.16}$$

In extended objects, the charge is continuously distributed. To determine the electric field generated by a finite-size object, the object is divided into elements with volume $d(Vol)$ and charge $dQ = \rho d(Vol) = \rho dr^3$. Note, in general case, if ρ depends on coordinates we have $\rho = \rho(\mathbf{r})$.

The field is given by the integral

$$\mathbf{E} = k_e \int_{(Q)} \frac{\mathbf{r}_0 - \mathbf{r}}{|\mathbf{r}_0 - \mathbf{r}|^3} dQ = k_e \int_{(Vol)} \rho(\mathbf{r}) \frac{\mathbf{r}_0 - \mathbf{r}}{|\mathbf{r}_0 - \mathbf{r}|^3} dr^3. \tag{1.17}$$

The integration is carried out over the whole volume, (Vol), over which the charge is distributed.

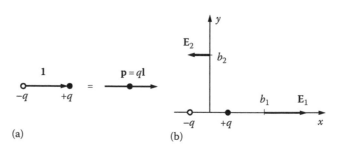

FIGURE 1.6   Definition of a dipole (a) and the electric field of the dipole at points b_1 and b_2 (b).

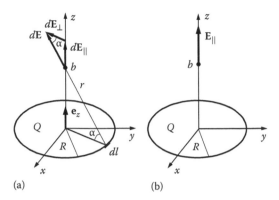

FIGURE 1.7 Electric field of a charged ring: contribution from an element dl (a) and the total field (b).

Exercise 1.2

A uniformly charged ring has a radius R and a total charge Q. Find the electric field vector **E** along the ring axis at a point on the normal to the ring plane at a distance b from the ring center.

Solution. We choose the coordinate system shown in Figure 1.7. We divide the ring into elements of length dl with a charge equal to

$$dQ = \lambda dl = \frac{Q}{2\pi R} dl. \tag{1.18}$$

The vector $d\mathbf{E}$ generated by dQ has two components at the observation point (Figure 1.7a): a longitudinal component $d\mathbf{E}_\parallel$ (along the z-axis) and a transverse component $d\mathbf{E}_\perp$ (perpendicular to the z-axis). Due to the symmetry of the charge distribution, the total transverse component of the electric field is $\mathbf{E}_\perp = 0$ at any point on the z-axis (Figure 1.7b).

The longitudinal component is determined by the following integral:

$$\mathbf{E}_\parallel = E_\parallel \mathbf{e}_z = \mathbf{e}_z \int dE \sin\alpha = \mathbf{e}_z k_e \int_0^{2\pi R} \frac{Qdl}{2\pi R r^2} \frac{b}{r} = \mathbf{e}_z k_e \frac{Qb}{r^3} = \frac{\mathbf{e}_z k_e Qb}{(R^2 + b^2)^{3/2}}. \tag{1.19}$$

From Equation 1.19, the electric field vanishes at the ring center (at $b = 0$). At short distances from the ring center (i.e., for $b \ll R$), the field is $E_\parallel \approx k_e Qb/R^3$, that is, it is linearly dependent on b. At large distances from the ring center (for $b \gg R$), the field becomes the Coulomb field of a point charge: $E_\parallel \approx k_e Q/b^2$. The electric field has the maximum magnitude $E_\parallel^{max} = 2k_e Q/3\sqrt{3}R^2$ at a distance $b = \pm R/\sqrt{2}$.

1.3 ELECTRIC POTENTIAL AND ELECTRIC FIELD ENERGY

1. The elementary work dW that is done by the electric field **E** in moving a charge q by a displacement $d\mathbf{l}$ under the influence of the force $\mathbf{F}_e = q\mathbf{E}$ is equal to

$$dW = \mathbf{F}_e \cdot d\mathbf{l} = qEdl\cos\alpha, \tag{1.20}$$

where α is the angle between the electric field vector **E** and the direction of the charge motion. This equation can be used to find the work performed by the electric field in moving charge q from point

1 to point 2 in the field generated by another stationary charge Q. Since the electric field of a point charge is $\mathbf{E} = (k_eQ/r^2)\mathbf{e}_r$ and $\mathbf{e}_r d\mathbf{l} = dl \cos \alpha = dr$ (Figure 1.8), the total work done by the electric force along the whole path is

$$W_{12} = \int dW = q \int \mathbf{E} \cdot d\mathbf{l} = k_eqQ \int_{r_1}^{r_2} \frac{dr}{r^2} = -k_eqQ \frac{1}{r}\Big|_{r_1}^{r_2} = k_eqQ\left(\frac{1}{r_1} - \frac{1}{r_2}\right). \tag{1.21}$$

From it follows that the work of the electrostatic field W_{12} does not depend on the shape of the path that the charge q follows. This work depends on the difference between the initial, r_1, and final, r_2, positions of the charge q. In such a field, the work performed to move the charge q along any closed contour is equal to zero, that is,

$$q \oint \mathbf{E} \cdot d\mathbf{l} = 0. \tag{1.22}$$

The field is called a conservative field if the work along a closed contour is equal to zero; thus, the electrostatic field is the conservative field. Therefore, the quantities of electric potential and electric potential energy can be introduced for the electrostatic field and charge in this field.

The work W_{12} done by the electric field in moving a charge q is given by the difference between the potential energies at the beginning and at the end of the path, that is,

$$W_{12} = k_eqQ\left(\frac{1}{r_1} - \frac{1}{r_2}\right) = q\left(\frac{k_eQ}{r_1} - \frac{k_eQ}{r_2}\right) = q(\varphi_1 - \varphi_2) = U_1 - U_2. \tag{1.23}$$

Here, we have introduced the electric potential φ of the charge Q: $\varphi = k_eQ/r$. The electric potential is equal to the electric potential energy U of a probe charge q at a given point of the field divided by the value of this charge: $\varphi = U/q$. The electric potential energy can be always defined within an arbitrary constant (only the difference of the potential energies has physical meaning). Therefore, it is convenient to choose the potential energy of a charge at infinity to be zero. If $r_2 \to \infty$, we have

$$W_{1\infty} = q\varphi_1 = U_1 = \frac{k_eqQ}{r_1}. \tag{1.24}$$

This equation determines the work performed by the electric field in moving charge q from point r_1 to infinity. The work is positive if both charges are of the same sign as the potential energy U_∞ of the

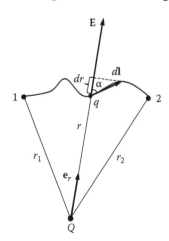

FIGURE 1.8   Work done to move a charge q from point 1 to point 2 in the field of charge Q.

charge q equals zero when charge q is at the infinite distance from the charge Q. It is worth noting here that if we bring charge q from infinity to point r_1, we need to perform work against the electric fields, that is, $W_{\infty 1}$ is negative if both charges are of the same sign: $W_{\infty 1} = -W_{1\infty}$.

To conclude, the potential energy of interaction between charges q and Q at a distance r is equal to

$$U = q\varphi = \frac{k_e q Q}{r}, \tag{1.25}$$

where the electric potential of a point charge Q at a distance r from that charge is $\varphi = k_e Q/r$. The unit of electric potential in the SI is called volt (V).

Exercise 1.3

Find the electric potential of an electric dipole of dipole moment $\mathbf{p}$ at a point A, which is located at a distance r_- and r_+ from the charges $-q$ and $+q$ that constitute the dipole shown in Figure 1.9 (here $q > 0$). Consider the case when the distances r_- and r_+ are much greater than the dipole length l, that is, $r_-, r_+ \gg l$.

Solution. According to the superposition principle, the electric potential of an electric dipole at some point is equal to the sum of the potentials of the two point charges:

$$\varphi_{dip} = \varphi_- + \varphi_+ = -\frac{k_e q}{r_-} + \frac{k_e q}{r_+} = k_e q \left(\frac{r_- - r_+}{r_- r_+} \right). \tag{1.26}$$

As the distances r_- and r_+ are much greater than the dipole length l, we get using the cos relation for two triangulars in Figure 1.9 and taking into account that $\cos(\pi - \alpha) = -\cos\alpha$

$$r_{\mp}^2 = r^2 + \frac{l^2}{4} \pm lr\cos\alpha \approx r^2 \pm lr\cos\alpha. \tag{1.27}$$

Using Taylor series ($\sqrt{1+x} \approx 1 + x/2$ for $x \ll 1$), one can get from Equation 1.27

$$r_{\mp} \approx r \pm \frac{l\cos\alpha}{2} \quad \text{and} \quad r_- - r_+ = l\cos\alpha, \quad r_- r_+ \approx r^2.$$

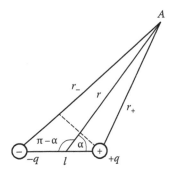

FIGURE 1.9   Electric potential of a dipole at point A.

Introducing these values into Equation 1.26, we have the following expression for the dipole potential:

$$\varphi_{dip} = k_e \frac{ql\cos\alpha}{r^2} = k_e \frac{q\mathbf{l}\cdot\mathbf{r}}{r^3}. \tag{1.28}$$

We note that the dipole potential decreases faster with the distance ($\varphi_{dip} \sim 1/r^2$) than the potential of a point charge ($\varphi_{point} \sim 1/r$).

2. The ***equipotential surface*** is a surface around a charge or a system of charges where the electric potential is the same, that is, φ = const on the equipotential surface. If we move charge q along the equipotential surface, the work against the electrostatic field equals zero (see Equation 1.23). The same follows from Equation 1.21 because $\mathbf{E}$ perpendicular to $d\mathbf{l}$ at each point of the surface φ = const. We will discuss later in more details (see Figure 1.13) the mutual orientation of electric field $\mathbf{E}$ and equipotential surfaces.

Let the surface potential of a conductor be φ. The total charge q of a conductor is the sum of elementary charges dq. Then, the potential energy of a charged conductor equals

$$U_e = \frac{1}{2}\int \varphi dq = \frac{1}{2}\varphi\int dq = \frac{1}{2}\varphi q. \tag{1.29}$$

Note that in Equations 1.23 and 1.24, the energy of charge q is calculated in the external potential φ, but in Equation 1.29, the potential φ is due to charge q of the conductor, and Equation 1.29 has the factor ½ to avoid the double counting of contribution of charge q to the energy. The charge and potential energy of a conductor are proportional to each other:

$$\varphi = \frac{q}{C}, \quad q = C\varphi, \tag{1.30}$$

where the coefficient C is an important characteristic of a conductor defined as the ***capacitance*** of a conductor. Taking into account these relations, the expression for the energy of a charged conductor is given as

$$U_e = \frac{1}{2}\varphi q = \frac{C\varphi^2}{2} = \frac{q^2}{2C}. \tag{1.31}$$

We introduced the capacitance of a conductor first, but more often the capacitance of a capacitor is a starting point. A ***capacitor*** is conventionally defined as a system of two parallel metal plates separated by a dielectric. A charged capacitor has stored energy that is determined by the potential difference $V = \Delta\varphi$ between the plates and by the charge q on the capacitor plates:

$$U_c = \frac{Vq}{2} = \frac{CV^2}{2} = \frac{q^2}{2C}. \tag{1.32}$$

Note that for a parallel plate capacitor, one plate has the charge q and the second plate has the charge $-q$. Capacitance of a parallel plate capacitor can be calculated using the following equation:

$$C = \frac{\varepsilon_0 A}{d},$$

where
   A and d are the area of plates and their separation, respectively
   ε_0 is the permittivity of vacuum between the plates

If it will be a dielectric between the plates, then as we discuss later in Section 1.7, the permittivity of vacuum should be replaced by the permittivity of the dielectric, ε. The permittivity of a dielectric will be introduced and discussed in details in Section 1.7.

The energy of a charged capacitor is stored in the space occupied by the electric field in the gap between the plates. Let us illustrate this for a parallel plate capacitor using the earlier equation for the capacitance. Since the potential difference and electric field in the parallel plate capacitor are related as $E = V/d$, we have

$$U_c = \frac{CV^2}{2} = \frac{CE^2 d^2}{2} = \frac{\varepsilon_0 E^2 Ad}{2} = \frac{\varepsilon_0 E^2}{2} Vol, \tag{1.33}$$

where $Vol = Ad$ is the volume of a gap between plates where the field is localized. If the field is uniform, its energy is distributed in space with the constant volume density:

$$u_e = \frac{U_c}{Vol} = \frac{\varepsilon_0 E^2}{2}. \tag{1.34}$$

The unit of the energy density is J/m³. For a nonuniform field (e.g., point charge field, the field between plates of spherical and cylindrical capacitor), the energy volume density is a function of coordinates—$u(x,y,z)$. In this case, to determine the energy of the electric field localized in volume (Vol), one has to calculate the following integral:

$$U_e = \int_{Vol} u_e(x,y,z) dr^3 = \frac{\varepsilon_0}{2} \int_{Vol} E^2(x,y,z) dr^3. \tag{1.35}$$

It is necessary to stress that this defines the energy inside of the volume of integration; for the parallel plate capacitor, the electric field equals zero outside of the capacitor, so all energy is stored between the plates of the capacitor.

Exercise 1.4

Consider a metal sphere with a radius R, charged with a total charge Q. Determine the energy of the electric field inside and outside the sphere and the capacitance of the sphere.

Solution. The entire charge of the metal sphere is located at the surface of the sphere $(r = R)$ and the charge inside is equal to zero: $Q^{(in)} = 0$. Outside the sphere, the field is equal to the field of the point charge Q placed at the sphere center. Inside the sphere, the field at a point located at a distance r from its center is the field generated by a point charge $Q^{(in)} = 0$ inside of the sphere of radius r, $r < R$. As $Q^{(in)}$ equals zero, the electric field is also equal to zero. So we have

$$E(r) = \begin{cases} \dfrac{k_e Q}{r^2} & \text{for } r \geq R, \\ 0 & \text{for } r < R. \end{cases} \tag{1.36}$$

Figure 1.10 shows the distribution of the electric field given by this relation. Let us introduce this expression into the relation (1.34) for the energy density of the electric field and take into account Equation 1.4:

$$u_e(r) = \frac{k_e Q^2}{8\pi} \begin{cases} \dfrac{1}{r^4} & \text{for } r \geq R, \\ 0 & \text{for } r < R. \end{cases} \tag{1.37}$$

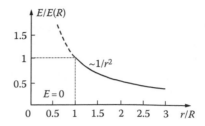

FIGURE 1.10   Electric field of a uniformly charged metal sphere.

Using Equations 1.35 and 1.37, we calculate the electric field energy U outside the sphere as the energy inside the sphere is equal to zero (Equation 1.37). Given the spherical symmetry of the problem, the volume element $d(Vol) = 4\pi r^2 dr$. The following relations give the energy and the capacitance; capacitance is calculated using Equation 1.31:

$$U = \frac{k_e Q^2}{8\pi} \int_R^\infty \frac{4\pi r^2}{r^4} dr = \frac{k_e Q^2}{2R},$$

$$C = \frac{Q^2}{2U} = \frac{R}{k_e} = 4\pi\varepsilon_0 R.$$

(1.38)

The capacitance of a metallic sphere is equal to its radius R multiplied by a universal constant, $4\pi\varepsilon_0$.

1.4   GAUSS'S LAW FOR THE ELECTRIC FIELD

Consider a small surface ΔA penetrated by electric field lines generated by a system of charges. The electric field line direction makes an angle α with the outward normal $\mathbf{n}$ to this surface ($\mathbf{n}$ is a unit vector). Assuming that the magnitude of $\mathbf{E}$ and its direction are approximately constant over surface ΔA, one can define the electric flux through this surface as (Figure 1.11)

$$\Delta\Phi_E = \mathbf{E}\cdot\mathbf{n}\Delta A = E\Delta A\cos\alpha.$$

(1.39)

For an elementary vector surface $d\mathbf{A} = \mathbf{n}dA$, this relation is given as $d\Phi_E = \mathbf{E}\cdot d\mathbf{A}$.

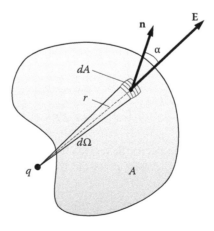

FIGURE 1.11   The calculation of flux of the electric field $\mathbf{E}$ through surface A.

The total flux of the electric field through surface A is calculated as an integral over the whole surface

$$\Phi_E = \int_{(A)} \mathbf{E} \cdot d\mathbf{A}. \tag{1.40}$$

Consider a positive point charge q inside an arbitrary closed surface A. A component $dA_E = dA \cos \alpha$ of a surface element $d\mathbf{A}$ in the direction of vector $\mathbf{E}$ is considered as an element of a spherical surface with radius r and a charge q in its center. dA_E/r^2 is equal to the elementary solid angle $d\Omega$ that subtends element $d\mathbf{A}$. Taking into account the relation for a point charge field, Equation 1.39 becomes

$$d\Phi_E = \frac{q d\Omega}{4\pi\varepsilon_0}. \tag{1.41}$$

Integration over the whole surface surrounding the charge, that is, within the solid angle from 0 to 4π, gives

$$\Phi_E = 4\pi k_e q = \frac{q}{\varepsilon_0}. \tag{1.42}$$

This result is valid for a closed surface of any shape and arbitrary charge system inside this surface. Taking into consideration the principle of superposition, Φ_E can be written as

$$\Phi_E = \oint_{(A)} \mathbf{E} \cdot d\mathbf{A} = \oint_{(A)} \sum_j \mathbf{E}_j \cdot d\mathbf{A} = \sum_j \oint_{(A)} \mathbf{E}_j \cdot d\mathbf{A} = 4\pi k_e \sum_j q_j = \frac{1}{\varepsilon_0} \sum_j q_j. \tag{1.43}$$

Thus, the flux of the electric field through a closed surface of arbitrary shape is proportional to the total charge within this surface. The proportionality coefficient is $4\pi k_e = 1/\varepsilon_0$. Equation 1.43 is known as Gauss's law for the electric field.

If the charge is distributed with a volume density $\rho = dq/d(Vol)$, the total charge inside a closed surface A is equal to

$$Q^{(in)} = \sum q_i = \int_{Vol} \rho dr^3 \tag{1.44}$$

and Gauss's law is expressed as

$$\Phi_E = \oint_A \mathbf{E} \cdot d\mathbf{A} = 4\pi k_e \int_{Vol} \rho dr^3 = \frac{1}{\varepsilon_0} \int_{Vol} \rho dr^3 = \frac{Q^{(in)}}{\varepsilon_0}. \tag{1.45}$$

Note that despite the dependence of the electric field $\mathbf{E}$ at any point on the location of all charges in space, the flux of this vector through an arbitrary closed surface depends only on the charges inside the surface.

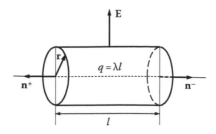

FIGURE 1.12 Electric field in the vicinity of a charged straight line.

Exercise 1.5

Using the integral form of Gauss's law, find the electric field in the vicinity of a charged straight line of infinite length. The linear charge density is uniform and equal to λ.

Solution. Choose a Gaussian surface in the form of a cylinder of length l whose axis coincides with the line of charge (Figure 1.12).

The electric field lines generated by the charge line are perpendicular to the cylinder curved surface. Since the angle between $\mathbf{E}$ and the normal to the cylinder bases is 90°, the flux through both cylinder bases is zero. The total flux through the cylinder surface is $\Phi_E = E\,2\pi r l$, where r is the cylinder radius. The charge inside the cylinder is $q = \lambda l$. According to Gauss's law (Equation 1.42),

$$\Phi_E = E2\pi rl = \frac{\lambda l}{\varepsilon_0}.$$

Consequently, the magnitude of the electric field generated by an infinite uniformly charged line at a distance of r is

$$E = \frac{\lambda}{2\pi r\varepsilon_0}.$$

1.5 RELATION BETWEEN THE ELECTRIC FIELD AND THE ELECTRIC POTENTIAL

1. The elementary work done by the electric field in moving charge q by a distance $d\mathbf{l}$ as it follows from Equations 1.20 and 1.23 is given by

$$dW = dW_{12} = q\mathbf{E}\cdot d\mathbf{l} = q(\varphi_1 - \varphi_2) = -qd\varphi, \tag{1.46}$$

where φ is the electric potential and $d\varphi = \varphi_2 - \varphi_1$. This expression relates the electric potential φ and the electric field vector:

$$\mathbf{E}\cdot d\mathbf{l} = E_x dx + E_y dy + E_z dz = -d\varphi. \tag{1.47}$$

The following three scalar equations are obtained from this equation:

$$E_x = -\frac{\partial\varphi}{\partial x}, \quad E_y = -\frac{\partial\varphi}{\partial y}, \quad E_z = -\frac{\partial\varphi}{\partial z}. \tag{1.48}$$

Thus, vector $\mathbf{E} = \mathbf{i}E_x + \mathbf{j}E_y + \mathbf{k}E_z$ is given by

$$\mathbf{E} = -\left(\mathbf{i}\frac{\partial \varphi}{\partial x} + \mathbf{j}\frac{\partial \varphi}{\partial y} + \mathbf{k}\frac{\partial \varphi}{\partial z} \right). \tag{1.49}$$

Here, the vectors $\mathbf{i}, \mathbf{j}, \mathbf{k}$ are unit vectors along the Cartesian coordinate axes, and the expression in brackets is the *potential gradient*, or grad φ:

$$\operatorname{grad} \varphi = \nabla \varphi = \mathbf{i}\frac{\partial \varphi}{\partial x} + \mathbf{j}\frac{\partial \varphi}{\partial y} + \mathbf{k}\frac{\partial \varphi}{\partial z}, \tag{1.50}$$

where symbol ∇ (nabla) is introduced to denote the vector differential operator referred often to as "del:"

$$\nabla = \mathbf{i}\frac{\partial}{\partial x} + \mathbf{j}\frac{\partial}{\partial y} + \mathbf{k}\frac{\partial}{\partial z}. \tag{1.51}$$

Thus, the potential gradient is given by the product of ∇ "vector" with components $(\partial/\partial x, \partial/\partial y, \partial/\partial z)$ and scalar function φ. Then, grad φ is a vector quantity. The relation between the electric field and the electric potential is given by the expression

$$\mathbf{E} = -\operatorname{grad} \varphi \quad \text{or} \quad \mathbf{E} = -\nabla \varphi. \tag{1.52}$$

This has the following meaning. Consider the equipotential surfaces with the potentials $\varphi_1 < \varphi_2 < \varphi_3$. According to the definition, the direction of vector grad φ is along the direction of the steepest change of φ. This direction coincides with the direction perpendicular to the equipotential surfaces from φ_1 to φ_3. According to Equation 1.52, vector $\mathbf{E}$ at the same point is oppositely directed.

Generalizing this result, one can say that at all points of a continuous surface orthogonal to the electric field lines, the electric field potential is the same (equipotential surface).

Equipotential surfaces are commonly employed to visualize the electric field. Generally, they are drawn so that the potential difference between any two surfaces is the same. Figure 1.13 shows a 2D picture of electric field. Solid lines are electric field lines and equipotential surfaces are dashed lines.

2. According to Gauss's mathematical theorem, the integral over a closed surface can be expressed as a volume integral:

$$\oint_A \mathbf{E} \cdot d\mathbf{A} = \int_{Vol} \operatorname{div} \mathbf{E}\, dr^3, \tag{1.53}$$

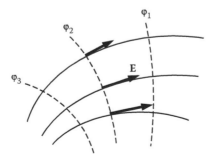

FIGURE 1.13 Electric field is a gradient of the electric potential and equipotential surfaces.

where the scalar div **E** (called the "divergence of **E**") is defined as

$$\operatorname{div} \mathbf{E} = \frac{\partial E_x}{\partial x} + \frac{\partial E_y}{\partial y} + \frac{\partial E_z}{\partial z} = \nabla \cdot \mathbf{E}. \tag{1.54}$$

Since $\nabla \cdot \mathbf{E}$ could be considered as a scalar product of "vector" ∇ and vector **E**, this quantity is a scalar. The left-hand side of Equation 1.53 determines the flux Φ_E over a closed surface that in accordance with Gauss's law (Equation 1.45) is equal to the total charge inside the closed surface (Equation 1.44). Taking these into consideration, Equation 1.53 can be written as follows:

$$\int_{Vol} \operatorname{div} \mathbf{E}\, dr^3 = \frac{1}{\varepsilon_0} \int_{Vol} \rho\, dr^3. \tag{1.55}$$

As the integrals over all ranges are equal, the integrands must be equal as well. Consequently, Gauss's law for the vector of electric field can be written in the following differential form:

$$\operatorname{div} \mathbf{E} = \frac{\rho}{\varepsilon_0}. \tag{1.56}$$

Similar to Equation 1.45, this equation is one (out of four) of Maxwell's equations. Such a field is commonly referred to as a *conservative* or *potential field*, since a scalar potential can be introduced to describe it. Therefore, the curl of vector **E** is zero, that is,

$$\oint_L \mathbf{E} \cdot d\mathbf{l} = 0, \quad \nabla \times \mathbf{E} = 0. \tag{1.57}$$

Here, the vector product of ∇ and **E**, $\nabla \times \mathbf{E}$, denotes the curl of vector **E**, and it is calculated in Cartesian coordinates in accordance with the following rule:

$$\nabla \times \mathbf{E} = \begin{vmatrix} \mathbf{i} & \mathbf{j} & \mathbf{k} \\ \nabla_x & \nabla_y & \nabla_z \\ E_x & E_y & E_z \end{vmatrix} = \mathbf{i}(\nabla_y E_z - \nabla_z E_y) + \mathbf{j}(\nabla_z E_x - \nabla_x E_z) + \mathbf{k}(\nabla_x E_y - \nabla_y E_x)$$

$$= \mathbf{i}\left(\frac{\partial E_z}{\partial y} - \frac{\partial E_y}{\partial z} \right) + \mathbf{j}\left(\frac{\partial E_x}{\partial z} - \frac{\partial E_z}{\partial x} \right) + \mathbf{k}\left(\frac{\partial E_y}{\partial x} - \frac{\partial E_x}{\partial y} \right).$$

Exercise 1.6

Find the electric field vector for the case when the electric potential is $\varphi(r) = \alpha/r$.

Solution. According to Equation 1.49, we can write

$$\mathbf{E} = -\operatorname{grad}\varphi = -\left(\mathbf{i}\frac{\partial \varphi}{\partial x} + \mathbf{j}\frac{\partial \varphi}{\partial y} + \mathbf{k}\frac{\partial \varphi}{\partial z} \right),$$

where

$$\frac{\partial \varphi}{\partial x} = \frac{\partial \varphi}{\partial r}\frac{\partial r}{\partial x}$$

$$\frac{\partial \varphi}{\partial y} = \frac{\partial \varphi}{\partial r}\frac{\partial r}{\partial y}$$

$$\frac{\partial \varphi}{\partial z} = \frac{\partial \varphi}{\partial r}\frac{\partial r}{\partial z}$$

Since $r = \sqrt{x^2 + y^2 + z^2}$, we have $\dfrac{\partial r}{\partial x} = \dfrac{2x}{2\sqrt{x^2 + y^2 + z^2}} = \dfrac{x}{r}, \ \dfrac{\partial r}{\partial y} = \dfrac{y}{r}, \ \dfrac{\partial r}{\partial z} = \dfrac{z}{r}.$

As $\varphi(r) = \dfrac{\alpha}{r}$, then $\dfrac{\partial \varphi}{\partial r} = \dfrac{\partial}{\partial r}\left(\dfrac{\alpha}{r}\right) = -\dfrac{\alpha}{r^2}$. Thus,

$$\mathbf{E} = -\frac{\partial \varphi}{\partial r}\frac{(\mathbf{i}x + \mathbf{j}y + \mathbf{k}z)}{r} = -\frac{\partial \varphi}{\partial r}\frac{\mathbf{r}}{r} = \frac{\alpha \mathbf{r}}{r^3}.$$

1.6 POISSON'S AND LAPLACE EQUATIONS

Equation 1.52 relates the electric potential and the electric field. Let us find the relation between the electric potential and the charge density. For this, we apply the divergence operator to both sides of Equation 1.52:

$$\operatorname{div} \mathbf{E} = -\operatorname{div}(\operatorname{grad} \varphi). \tag{1.58}$$

According to the vector analysis rules,

$$\operatorname{div}(\operatorname{grad} \varphi) = \nabla \cdot (\nabla \varphi) = \nabla^2 \varphi = \left(\frac{\partial^2}{\partial x^2} + \frac{\partial^2}{\partial y^2} + \frac{\partial^2}{\partial z^2}\right)\varphi = \frac{\partial^2 \varphi}{\partial x^2} + \frac{\partial^2 \varphi}{\partial y^2} + \frac{\partial^2 \varphi}{\partial z^2}. \tag{1.59}$$

The differential operator

$$\nabla^2 = \frac{\partial^2}{\partial x^2} + \frac{\partial^2}{\partial y^2} + \frac{\partial^2}{\partial z^2} \tag{1.60}$$

is widely employed in different branches of physics and is referred to as the Laplace operator. Taking into account Equation 1.56, we have the following differential equation:

$$\nabla^2 \varphi = -\frac{\rho}{\varepsilon_0} \quad \text{or} \quad \frac{\partial^2 \varphi}{\partial x^2} + \frac{\partial^2 \varphi}{\partial y^2} + \frac{\partial^2 \varphi}{\partial z^2} = -\frac{\rho}{\varepsilon_0}, \tag{1.61}$$

known as Poisson's equation. For space regions free of electric charges, that is, if $\rho(x, y, z) = 0$, this equation has the following form:

$$\nabla^2 \varphi = 0 \quad \text{or} \quad \frac{\partial^2 \varphi}{\partial x^2} + \frac{\partial^2 \varphi}{\partial y^2} + \frac{\partial^2 \varphi}{\partial z^2} = 0. \tag{1.62}$$

This particular form of Poisson's equation is called Laplace equation. Poisson's equation allows calculating the electric potential of a field generated by volume charges, if their distribution is known for this volume. This equation has to be solved with given boundary conditions.

Let us consider an example of solution of Poisson's equation that is important for the theory of electronic tubes. It is known that metals heated up to high enough temperatures emit electrons from their surface (thermoelectrons) into the surrounding space. If a potential difference is applied between two metal electrodes, the cathode (a negative electrode that is heated to emit electrons) and the anode (a positive electrode), electron flows from cathode to anode. This flux is equivalent to an electric current, which is called thermionic current.

Exercise 1.7

Find the potential distribution between two flat electrodes (cathode and anode) in vacuum for the case of thermionic emission from one of them (cathode). Assume for simplicity that the distance between the cathode and the anode is smaller than the size of those electrodes and that those emitted by the cathode thermoelectrons have zero initial velocity, that is, $v(0) = 0$. The schematic of the thermionic emission in a diode tube is shown in Figure 1.14.

Solution. Let us choose the Cartesian coordinates so that their origin is at the cathode and z-axis is perpendicular to the plane of electrodes and it is directed toward an anode (z-axis is not shown in Figure 1.14). Due to potential difference between the anode and the cathode, the thermoelectrons generated at the cathode flow toward the anode resulting in a current between the cathode and the anode. The electric force performs work on the electrons. This work increases the kinetic energy of each electron:

$$\frac{mv^2(z)}{2} = e\varphi(z), \tag{1.63}$$

where
  m is the electron mass
  $v(z)$ is the electron velocity at a distance z from the cathode
  $\varphi(z)$ is the potential at z

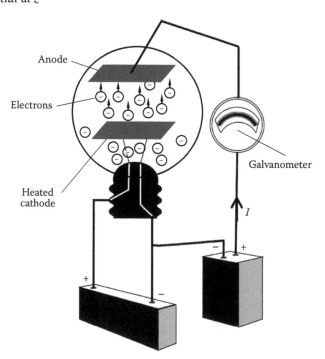

FIGURE 1.14 Schematic of a thermoelectric emission experiment with a diode tube.

For the geometry of the electrodes, the equipotential surfaces are planes parallel to the electrodes. Therefore, in the space between them, the partial derivatives

$$\frac{\partial \varphi}{\partial x} = \frac{\partial \varphi}{\partial y} = 0$$

and Poisson's equation takes the form

$$\frac{d^2\varphi}{dz^2} = -\frac{\rho(z)}{\varepsilon_0}. \tag{1.64}$$

Here, ρ is the charge volume density, and this quantity is related to the density of electrons, $n(z)$, at each point of space between electrodes as follows: $\rho(z) = -en(z)$, where e is the absolute value of the electron charge.

We substitute in Equation 1.64 the charge density by the electric current density $j = en(z)v(z)$, which does not vary with time in the stationary state and independent of the coordinate z (since an equal number of electrons pass per unit time through any plane parallel to electrodes):

$$\frac{d^2\varphi}{dz^2} = -\frac{\rho(z)}{\varepsilon_0} = \frac{en(z)}{\varepsilon_0} = \frac{j}{\varepsilon_0 v(z)}. \tag{1.65}$$

From Equation 1.63, one can express the velocity through the potential and introduce it into Equation 1.65. Since $v = \sqrt{2e\varphi/m}$, we have

$$\frac{d^2\varphi}{dz^2} = \frac{j}{\varepsilon_0}\sqrt{\frac{m}{2e\varphi}} = \frac{j}{\varepsilon_0}\sqrt{\frac{m}{2e}}\varphi^{-1/2} = A^2\varphi^{-1/2}. \tag{1.66}$$

We will solve this equation with the boundary conditions, under which for $z = 0$ equalities

$$\varphi(0) = 0, \quad \left.\frac{d\varphi}{dz}\right|_{z=0} = 0$$

are satisfied, that is, at $z = 0$, both the potential and the electric field are equal to zero. To solve Equation 1.66, let us multiply its right- and left-hand parts by $d\varphi/dz$:

$$\frac{d\varphi}{dz}\frac{d^2\varphi}{dz^2} = A^2\varphi^{-1/2}\frac{d\varphi}{dz}.$$

The earlier equation can be rewritten as

$$\frac{1}{2}\frac{d}{dz}\left(\frac{d\varphi}{dz}\right)^2 = 2A^2\frac{d}{dz}\varphi^{1/2},$$

$$\left(\frac{d\varphi}{dz}\right)^2 = 4A^2\varphi^{1/2}.$$

Taking the square root of the last equation, we obtain the first-order differential equation:

$$\frac{d\varphi}{dz} = 2A\varphi^{1/4}, \quad \frac{d\varphi}{\varphi^{1/4}} = 2A\,dz. \tag{1.67}$$

By integrating this, we obtain the dependence of the potential on z:

$$\frac{4}{3}\varphi^{3/4} = 2Az, \quad \varphi(z) = \left(\frac{3}{2}Az\right)^{4/3}. \tag{1.68}$$

Thus, for these conditions, the potential increases proportionally to $z^{4/3}$.

1.7   ELECTRIC FIELD IN A MEDIUM, ELECTRIC DISPLACEMENT

1. The electric field can exist not only in vacuum but in matter as well. However, there is a crucial difference between the field distribution inside a conductor and in a dielectric. In a conductor, there are free electric charges (electrons) that move due to electric forces. When a conductor is placed in an electric field, uncompensated positive and negative charges appear at the conductor surface due to free charge redistribution. These charges are referred to as induced charges. Induced charges produce their own field $\mathbf{E}'$ that compensates exactly the external field $\mathbf{E}_0$ resulting in a zero net electric field inside the conductor, that is, $\mathbf{E} = \mathbf{E}_0 + \mathbf{E}' = 0$. Here, the potentials at all points inside the conductor are the same and equal to the potential at the conductor surface.

In a dielectric, charges are bound on molecules (and/or atoms), and thus, they are not free to move under the influence of electric forces. When a dielectric is placed in an external electric field, extra uncompensated bound charges appear on the dielectric surface. This process is called polarization of the dielectric. Bound charges produce an electric field $\mathbf{E}'$ inside the dielectric that is oppositely directed to the external field $\mathbf{E}_0$. Thus, the total electric field inside the dielectric is nonzero, and it is lower in magnitude than the external field:

$$\mathbf{E} = \mathbf{E}_0 + \mathbf{E}' \neq 0, \quad |\mathbf{E}| < |\mathbf{E}_0|. \tag{1.69}$$

The ratio of the external electric field magnitude to the magnitude of total field inside an isotropic dielectric is referred to as the relative dielectric permittivity (or dielectric constant) of the material:

$$\kappa = \frac{E_0}{E}. \tag{1.70}$$

If an isotropic dielectric with dielectric permittivity $\varepsilon = \kappa\varepsilon_0$ contains a point charge q, the electric field generated by this charge and the potential are κ times smaller compared to the corresponding values in vacuum:

$$\mathbf{E} = k_e \frac{q}{\kappa r^2}\mathbf{e}_r, \quad \varphi = k_e \frac{q}{\kappa r}. \tag{1.71}$$

2. The presence of bound charges in a dielectric is reflected in the formulation of Gauss's law. If the volume density of free charges ρ and that of bound charges ρ' are introduced, Gauss's law for the vector $\mathbf{E}$ in an integral form has the following form:

$$\oint \mathbf{E} \cdot d\mathbf{A} = 4\pi k_e \int_{(Vol)} (\rho + \rho')dr^3 = \frac{1}{\varepsilon_0}\int_{(Vol)} (\rho + \rho')dr^3 = \frac{1}{\varepsilon_0}(q + q'). \tag{1.72}$$

The differential form of Gauss's law is

$$\text{div } \mathbf{E} = \frac{\rho + \rho'}{\varepsilon_0}. \tag{1.73}$$

The density of bound charges ρ' is related to the polarization via the equation $\rho' = -\nabla \cdot \mathbf{P}$. Here, the polarization vector $\mathbf{P}$ is equal to the total dipole moment of a unit volume of a dielectric, that is,

$$\mathbf{P} = \frac{1}{Vol} \sum_j \mathbf{p}_j \; (\text{C/m}^2). \tag{1.74}$$

The polarization vector is proportional to the electric field vector, that is,

$$\mathbf{P} = \varepsilon_0 \chi \mathbf{E}, \tag{1.75}$$

where the dimensionless parameter χ is the dielectric susceptibility of a substance, and it is related to the dielectric permittivity as follows: $\kappa = 1 + \chi$. Taking into account Equation 1.75 and the earlier relation $\rho' = -\nabla \cdot \mathbf{P}$, Equation 1.73 transforms into

$$\text{div}(\varepsilon_0 \mathbf{E} + \mathbf{P}) = \rho. \tag{1.76}$$

Let us introduce the electric displacement vector

$$\mathbf{D} = \varepsilon_0 \mathbf{E} + \mathbf{P}, \tag{1.77}$$

which takes into account the polarization of a substance and is given in C/m². Taking into consideration Equation 1.75, this relation can be given as

$$\mathbf{D} = \varepsilon_0 \mathbf{E} + \varepsilon_0 \chi \mathbf{E} = \varepsilon_0 (1 + \chi) \mathbf{E} = \varepsilon_0 \kappa \mathbf{E}. \tag{1.78}$$

Consequently, Equation 1.76 is written as

$$\text{div } \mathbf{D} = \rho. \tag{1.79}$$

An integral form of this equation is expressed as follows:

$$\oint \mathbf{D} \cdot d\mathbf{A} = \int_{(Vol)} \rho \, dr^3 = q. \tag{1.80}$$

Equations 1.79 and 1.80 express Gauss's law in the presence of matter. According to Equations 1.79 and 1.80, the field defined by $\mathbf{D}$ is associated with the free charges, where its displacement lines begin and end. As it follows from Equations 1.72 and 1.73, the electric field $\mathbf{E}$ is determined by free and bound charges, that is, by the total charge.

Taking into consideration the relation between vectors $\mathbf{E}$ and $\mathbf{D}$, the volume energy density of electric field of Equation 1.34 can be given by one of the following expressions:

$$u_e = \frac{\kappa \varepsilon_0 E^2}{2} = \frac{D^2}{2\kappa\varepsilon_0} = \frac{\mathbf{E} \cdot \mathbf{D}}{2}. \tag{1.81}$$

This should be used instead of Equation 1.34 if the dielectric with the permittivity $\varepsilon = \varepsilon_0 \kappa$ is placed in the space between the plates of a capacitor. The energy stored in a capacitor is proportional to the relative dielectric permittivity of a dielectric.

Exercise 1.8

A spherical insulating glass shell is uniformly charged with volume density $\rho = 100$ nC/m³. The inner radius of the shell is equal to $R_1 = 5.00$ cm and the outer radius $R_2 = 10.0$ cm. Find the electric field for three points at distances $r = 3.00$, 6.00, and 12.0 cm from the center of the shell.

Solution. Since the charge distribution is spherically symmetric, the electric field points radially outward. This allows us to apply Gauss's law for the solution of the problem. From the symmetry of the problem, it follows that vector $\mathbf{E}$ is directed along $\mathbf{r}$ and its magnitude depends only on the distance r from the center of the shell. We choose the surface of integration to be a sphere of radius r centered at point O, the center of the shell. Let us take into account that the magnitude of the electric field is the same at all points of this surface and $E_n = E_r$. We apply Gauss's law for the electric displacement vector $\mathbf{D}$. The flux of the displacement vector through the selected Gaussian surface (a sphere of radius r) is equal to

$$q = \oint \mathbf{D}\, d\mathbf{A} = \oint D_r dA = D \cdot A = D \cdot 4\pi r^2.$$

Space is divided into three areas: (1) $0 < r < R_1$, (2) $R_1 < r < R_2$, and (3) $r > R_2$. Gauss's law is applied for each area.

Region 1: $0 < r < R_1$.
The free charge within the first region is equal to zero. Consequently, the flux of the displacement vector is also equal to zero. Since the surface area of the Gaussian surface is not equal to zero, then the displacement and the electric field within the first region are equal to zero:

$$D_1 = 0, \quad E_1 = \frac{D_1}{\varepsilon_0} = 0.$$

Region 2: $R_1 < r < R_2$.
The free charge inside the Gaussian surface is given by the following expression:

$$q_{free} = \frac{4\pi}{3}(r^3 - R_1^3)\rho.$$

Applying Gauss's law, we get

$$D_2 4\pi r^2 = \frac{4}{3}\pi(r^3 - R_1^3)\rho \rightarrow E_2 = \frac{D_2}{\varepsilon_0 \kappa} = \frac{\rho}{3\kappa\varepsilon_0}\left(r - \frac{R_1^3}{r^2}\right),$$

where κ is the permittivity of glass. Substituting the numerical values, we get

$$E_2 = \frac{100\times10^{-9}}{3\times8.85\times10^{-12}\times7}\left(0.06 - \frac{0.05^3}{0.06^2}\right) = 13.6 \text{ V/m}.$$

Region 3: $r > R_2$.
The entire charge of the shell is located inside the Gaussian surface:

$$q_{free} = \frac{4\pi}{3}(R_2^3 - R_1^3)\rho.$$

Applying Gauss's law, we obtain

$$D_3 4\pi r^2 = \frac{4}{3}\pi\left(R_2^3 - R_1^3\right)\rho \rightarrow E_3 = \frac{D_3}{\varepsilon_0} = \frac{\rho}{3\varepsilon_0}\left(\frac{R_2^3 - R_1^3}{r^2}\right),$$

$$E_3 = \frac{100\times10^{-9}(0.10^3 - 0.05^3)}{3\times8.85\times10^{-12}\times0.12^2} \approx 229\,\text{V/m}.$$

PROBLEMS

1.1 Four identical point charges q are placed at the corners of a square with side a. Find the net force (magnitude and direction) acting on each of the charges and the potential energy of the system. The charges are located in vacuum. $\left(\textit{Part of the answer:}\,U = \frac{1}{4\pi\varepsilon_0}\cdot\frac{q^2}{a}(4+\sqrt{2}).\right)$

1.2 Charge q is uniformly distributed on a very thin ring of radius R placed in vacuum. Find the electric potential and the electric field at a point on the normal to the ring plane that passes though the ring center (Figure 1.15) as a function of the distance b from the center of the ring.

1.3 A nonpolar molecule is located on the normal to a ring plane (radius R) that passes through the ring center at a distance z from the ring center. The nonpolar molecule dipole moment is proportional to the electric field at the location of the molecule, that is, $\mathbf{p} = \varepsilon_0\alpha\mathbf{E}$, where α is the polarizability of the molecule. Determine the distance z from the center of the ring, where the force $\mathbf{F}$ acting on the molecule is equal to zero. Assume that the system is in vacuum and ignore all other forces acting on the dipole. *Note*: There are two such positions. *Hint*: See Exercise 1.2. (*Part of the answer:* $z_2 = \pm R/\sqrt{2}$.)

1.4 Two metal spheres of radii R_1 and R_2 have charges Q_1 and Q_2 and they are located in vacuum. Find the energy U that will be released if the spheres are connected by a thin conductor.

1.5 Find the work that must be done by an external agent to move a small electric dipole of dipole moment p from the surface of a uniformly charged sphere to infinity. The sphere radius is equal to R and its charge equal to Q. The dipole moment is oriented radially (see Figure 1.16).

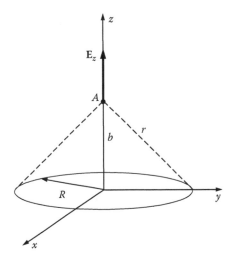

FIGURE 1.15 Electric field of a circular charged ring of R.

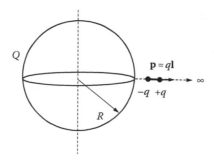

FIGURE 1.16   Interaction of charged sphere with an electric dipole.

1.6   Consider an insulating sphere with a radius R, uniformly charged over its volume with a total charge Q. Determine the energy of the electric field inside and outside the sphere, if the electric field in the sphere and that outside of the sphere are

$$
E(r) = \begin{cases} \dfrac{k_e Q}{r^2}, & r \geq R, \\[3mm] \dfrac{k_e Q}{R^2} \cdot \dfrac{r}{R}, & r \leq R. \end{cases}
$$

1.7   Using the integral form of Gauss's law, find the electric field in the vicinity of an infinite uniformly charged insulating plane with constant surface charge density σ.

1.8   The potential associated with an electric field depends on the Cartesian coordinates according to the equation (i) $\varphi = a(x^2 - y^2)$ and (ii) $\varphi = bxyz$, where a and b are constants. Determine the electric field in each case. (*Part of the answer*: $\mathbf{E} = -b(\mathbf{i}yz + \mathbf{j}xz + \mathbf{k}xy)$.)

1.9   Determine the electric field $\mathbf{E}$ of an electric dipole with a dipole moment $\mathbf{p}$ at a point located at a distance r_- and r_+ from the charges $-q$ and $+q$ as shown in Figure 1.9 (here $q > 0$). Assume that r_- and r_+ are much larger than the charge separation l.

1.10   An infinitely long insulating rod is uniformly charged with linear charge density $\lambda_1 = 3.00 \times 10^{-7}$ C/m. A second insulating rod of finite length $l = 20.0$ cm is uniformly charged with linear density $\lambda_2 = 2.00 \times 10^{-7}$ C/m. The two rods have their axes perpendicular to each other as shown in Figure 1.17. The distance $r_0 = 10.0$ cm. Determine the electric force between the two rods. (*Answer*: $F = 1.19$ mN.)

1.11   A thin insulating rod is bent into a semicircle of radius R. The rod is uniformly charged with a linear density $\lambda = 133$ nC/m. Calculate the work that has to be done by the electric field to move a charge $q = 6.70$ nC from the center of the semicircle to infinity (use Figure 1.18 as a hint). (*Answer*: $W = 2.50 \times 10^{-5}$ J.)

1.12   A metal sphere of radius R is charged with charge Q. The surface of the sphere is covered by an uncharged dielectric shell of thickness h. Determine the polarization of the dielectric shell if the dielectric permittivity of the shell is equal to κ.

1.13   A constant voltage $V = 300$ V is applied to two capacitors, $C_1 = 100$ pF and $C_2 = 200$ pF, connected in series. Determine the energy stored in each capacitor.

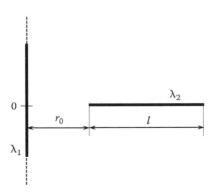

FIGURE 1.17 Interaction of an infinitely long insulating rod with charge density λ_1 and an insulating rod of length l with linear density λ_2.

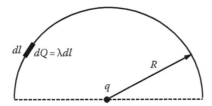

FIGURE 1.18 Semicircle thin insulating rod charged with the linear density λ.

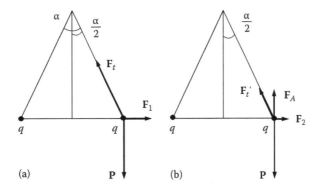

FIGURE 1.19 Interaction of point charges in vacuum (a) and in oil (b).

1.14 Two identical charged spheres of radius a are suspended from a point on insulating threads of equal length. The threads are separated by an angle α. The spheres are immersed in oil of density $\rho_O = 8.00 \times 10^2$ kg/m³. Find the oil permittivity κ_O, if the angle α remains unchanged before and after immersion in the oil. The density of spheres is equal to $\rho_b = 1.6 \times 10^3$ kg/m³. Use Figure 1.19 as a hint.

2 Magnetostatics

In 1820, Ampere has experimentally established the law of interaction of electric currents. His experiments showed that two straight parallel conductors, which carry currents, attract if both currents have the same direction and repel if two currents have opposite directions. The magnitude of the interaction force is directly proportional to the current in each conductor and inversely proportional to the distance between them. It is known that in metals the total charge of the positively charged ions and negatively charged free electrons is equal to zero. Moreover, positive and negative charges are uniformly distributed in the conductor. Therefore, the net electric force between the conductors is zero, and in the absence of current, the conductors do not interact with each other. However, in the presence of current, that is, orderly movement of charge carriers, the conductors interact and therefore, this interaction is ***magnetic***. In the space surrounding the electric current, the magnetic field is generated. The existence of a magnetic field can be detected by the deviation of the magnetic needle near a current-carrying conductor or by the impact of that magnetic field on a moving charged particle, as well as by the impact on other current-carrying conductor.

In analogy with electrostatics, in order to characterize the magnetic field, the ***magnetic field vector*** **B** is introduced. As in the case of the electrostatic field, the magnetic field may be represented using magnetic field lines using the same convention; at each point, **B** is tangential to the local magnetic field line. The magnetic field lines, as well as the electric field lines, do not intersect.

2.1 INTERACTION OF MOVING CHARGES

Electric and magnetic phenomena are closely related. Moving charges are the sources of the magnetic field. No isolated magnetic charges have been discovered so far. A magnetic force is exerted on moving electric charges due to a magnetic field. Current consists of moving charges. Oersted's experiments demonstrated that near a current-carrying conductor, a magnetic needle is deflected by the magnetic force. A force between two parallel current-carrying conductors placed at a distance b was observed by Ampere whose experiments led to the formulation of an equation that describes this force. According to this equation, the magnitude of the force dF_A exerted on an element of length dl of each parallel conductor that carries currents I_1 and I_2 is given by the following expression:

$$dF_A = k_m \frac{2I_1I_2}{b} dl. \tag{2.1}$$

In the SI, the proportionality coefficient is given as $k_m = \mu_0/4\pi$, where the magnetic constant $\mu_0 = 4\pi \times 10^{-7}$ H/m and sometime μ_0 is referred to as the permeability of vacuum (note that H/m = N/A^2, where henry [H] is the unit of inductance that we will introduce later). The force is attractive if I_1 and I_2 are parallel and it is repulsive if I_1 and I_2 flow in opposite directions. The interaction between currents is due to their magnetic fields; the magnetic field of one current exerts a force, which is referred to as ***Amperes' force***, on the other current. In contrast to the electric field, the magnetic field acts upon the moving charges only.

The magnetic field **B** is described by the forces it exerts on moving charges. The magnetic force acting on a charge q traveling with velocity **v** is given by

$$\mathbf{F}_L = q\mathbf{v} \times \mathbf{B}. \tag{2.2}$$

This force is known as ***the Lorentz force***. It is given by the vector product of two vectors, which in Cartesian coordinates is given by the following equation:

$$\mathbf{v}\times\mathbf{B} = \begin{vmatrix} \mathbf{i} & \mathbf{j} & \mathbf{k} \\ v_x & v_y & v_z \\ B_x & B_y & B_z \end{vmatrix} = \mathbf{i}\left(v_yB_z - v_zB_y\right) + \mathbf{j}\left(v_zB_x - v_xB_z\right) + \mathbf{k}\left(v_xB_y - v_yB_x\right). \qquad (2.3)$$

The magnitude of the Lorentz force is given as

$$F_L = qvB\sin\alpha,$$

where α is the angle between **v** and **B**.

Equations 2.2 and 2.3 give the Lorentz force direction that is always perpendicular to vectors **v** and **B**. Therefore, the magnitude of the velocity of a charged particle moving in a magnetic field does not change with time. The right-hand rule is applied to determine the direction of the Lorentz force acting on a positively charged particle. The right-hand rule is applied as follows to determine the direction of $\mathbf{v}\times\mathbf{B}$. Rotate **v** in the plane defined by the vectors **v** and **B** along the shortest angle so that it coincides with **B**. Then curl the fingers of the right hand in the same direction (or rotate palm); the direction of the thumb gives the direction of $\mathbf{v}\times\mathbf{B}$. (Please note that the same rule is applied to any other vector/cross product as well as in determining the right-handed Cartesian coordinate system (x, y, z) with unit vectors **i**, **j**, **k**: $\mathbf{i}\times\mathbf{j} = \mathbf{k}$, $\mathbf{j}\times\mathbf{k} = \mathbf{i}$, $\mathbf{k}\times\mathbf{i} = \mathbf{j}$.)

Vectors **v**, **B**, and $\mathbf{F}_L$ for a positively charged particle moving with the velocity v in a uniform magnetic field in a direction perpendicular to **B** are illustrated in Figure 2.1. In the example, the velocity of the particle **v** lies in the plane perpendicular to the vector **B**. The trajectory is circular with radius $R = mv/qB$. The angular velocity (also known as "cyclotron frequency") of the particle moving along a circular trajectory is given by the equation $\omega_c = qB/m$.

The force that the magnetic field **B** exerts on a length of wire $d\mathbf{l}$ (the wire is assumed to have a very small cross section) that carriers a current I is referred to as Ampere's force as it can be deducted from Equation 2.1 (in some books it is also mentioned as Laplace force as it can be deducted from Equation 2.2). This force is determined by the relation

$$d\mathbf{F}_A = Id\mathbf{l}\times\mathbf{B}, \qquad (2.4)$$

where the direction of the vector $d\mathbf{l}$ coincides with the current direction at the wire segment dl. Ampere's force is the superposition of all the Lorentz forces acting on individual moving charges that form the current I.

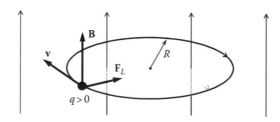

FIGURE 2.1   Relative position of vectors **v**, **B**, and $\mathbf{F}_L$ for a moving positively charged particle in the magnetic field.

Exercise 2.1

A thin semicircular wire (radius R) that carries a current I is placed in a uniform magnetic field $\mathbf{B}$. The wire plane is perpendicular to the magnetic field. What is the force F that the magnetic field exerts on the wire?

Solution. Let us select an element $d\mathbf{l}$ with a current I (Figure 2.2). Ampere's force $d\mathbf{F} = Id\mathbf{l} \times \mathbf{B}$ acts on this current element $d\mathbf{l}$. The right-hand rule discussed earlier is used to determine the direction of this force. The force $d\mathbf{F}$ has two projections along the x- and y-axes.

$$d\mathbf{F} = \mathbf{i}dF_x + \mathbf{j}dF_y.$$

The total force $\mathbf{F}$ acting on the entire wire is determined by the integral

$$\mathbf{F} = \int_L d\mathbf{F} = \mathbf{i}\int_L dF_x + \mathbf{j}\int_L dF_y,$$

where integration is carried out over the semicircle whose length $L = \pi R$. Symmetry arguments give a zero projection of the total force on the x-axis, that is,

$$\mathbf{i}F_x = \mathbf{i}\int_L dF_x = 0, \quad \mathbf{F} = \mathbf{j}F_y = \mathbf{j}\int_L dF_y,$$

$dF_y = dF\cos\alpha$, where α is the angle between $d\mathbf{F}$ and $\mathbf{j}$. Since vector $d\mathbf{l}$ is perpendicular to vector $\mathbf{B}$, $dF = IBdl$, $dl = Rd\alpha$, then $dF = IBRd\alpha$ and $dF_y = IBR\cos\alpha\, d\alpha$.

To find the total force, we integrate $dF_y = IBR\cos\alpha\, d\alpha$ between $-\pi/2$ and $+\pi/2$:

$$\mathbf{F} = \mathbf{j}\int_L dF_y = \mathbf{j}IBR\int_{-\pi/2}^{\pi/2} \cos\alpha\, d\alpha = \mathbf{j}IBR\sin\alpha\Big|_{-\pi/2}^{\pi/2} = 2\mathbf{j}IBR.$$

From this equation, it follows that force $\mathbf{F}$ is oriented along the positive direction of the y-axis (i.e., along vector $\mathbf{j}$).

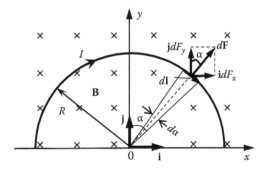

FIGURE 2.2 Calculation of the force acting on a semicircular wire of radius R that carries current I in a uniform magnetic field $\mathbf{B}$ that is perpendicular to the plane of the wire.

2.2   FIELD OF MOVING CHARGES AND CURRENTS, THE BIOT–SAVART LAW

The magnetic field of a point charge q moving with the velocity $\mathbf{v}$ is given by

$$\mathbf{B} = k_m \frac{q\mathbf{v} \times \mathbf{r}}{r^3}, \tag{2.5}$$

where vector $\mathbf{r}$ is directed from the charge to the point where $\mathbf{B}$ is determined by this equation. From this, it follows that vector $\mathbf{B}$ is perpendicular to vectors $\mathbf{v}$ and $\mathbf{r}$; the direction of $\mathbf{B}$ is determined by the right-hand rule.

According to the principle of superposition, the magnetic field of a current flowing in a wire is determined by the combined effect of all free charge carriers that compose the current. However, we must integrate all the contribution from the charge carriers in a wire. It is reasonable to divide the wire into elementary segments of length dl. In this case, substituting $\mathbf{B} \to d\mathbf{B}$ and $q\mathbf{v} \to I d\mathbf{l}$ in Equation 2.5, we arrive at the law of Biot–Savart that determines the magnetic field generated by an elementary segment that carries a current I:

$$d\mathbf{B} = k_m \frac{I d\mathbf{l} \times \mathbf{r}}{r^3}, \tag{2.6}$$

where

$\mathbf{r}$ is the vector directed from the wire segment to the observation point
vector $d\mathbf{l}$ is a vector of magnitude dl that has the direction of the current

The direction of vector $d\mathbf{B}$ coincides with the direction of vector product $d\mathbf{l} \times \mathbf{r}$. The right-hand rule is applied to determine the direction of $d\mathbf{l} \times \mathbf{r}$. As it was already described earlier, we rotate $d\mathbf{l}$ in the plane defined by the vectors $d\mathbf{l}$ and $\mathbf{r}$ along the shortest angle so that it coincides with $\mathbf{r}$. Then we rotate the fingers of the right hand in the same direction (or rotate palm). The direction of the thumb gives the direction of $d\mathbf{l} \times \mathbf{r}$. The magnitude of vector $d\mathbf{B}$ is given by the relation

$$dB = k_m \frac{I dl \sin \alpha}{r^2}, \tag{2.7}$$

where α is an angle between vectors $d\mathbf{l}$ and $\mathbf{r}$.

The superposition principle is valid for the magnetic field. The resultant magnetic field created by the entire conductor at a given observation point is equal to the vector sum of the fields generated individually by each current element. Superposing is provided through integration over the conductor:

$$\mathbf{B} = k_m \int_{(l)} \frac{I d\mathbf{l} \times \mathbf{r}}{r^3}. \tag{2.8}$$

If a conductor is not 1D (i.e., its cross section is not negligible), then taking into account $I d\mathbf{l} = \mathbf{j} dV$, where $\mathbf{j}$ is the current density in the plane perpendicular to $d\mathbf{l}$ and $dV = d^3r$ is an element of conductor volume, we can calculate

$$\mathbf{B} = k_m \int_{(Vol)} \frac{\mathbf{j} \times \mathbf{r}}{r^3} d^3r, \tag{2.9}$$

where Vol is the conductor volume, in which current flows. In general, it is difficult to determine $\mathbf{B}$ from these equations. However, if current distribution exhibits certain symmetry, one can easily calculate $\mathbf{B}$ by applying the Biot–Savart law in connection with the superposition.

Exercise 2.2

A current I flows in a thin conducting ring of radius R. Determine the magnetic field vector at a point A located on the ring axis at a distance b from the ring center.

Solution. Let us select an element dl in a ring and draw a position vector $\mathbf{r}$ from it to a point A (Figure 2.3). We then decompose vector $d\mathbf{B}$, which is perpendicular to the plane defined by $d\mathbf{l} \times \mathbf{r}$ into components $d\mathbf{B}_\parallel$ and $d\mathbf{B}_\perp$ parallel and perpendicular to the ring axis, respectively. According to the superposition principle, the magnitude of each component of the net magnetic field is given by integrals

$$B_\parallel = \oint_L dB_\parallel, \quad B_\perp = \oint_L dB_\perp, \tag{2.10}$$

where $dB_\parallel = dB \cos \beta$, $dB_\perp = dB \sin \beta$. Integration is carried out over all the elements dl into which the ring was divided. Due to the symmetry of the current distribution, the component of the net field $B_\perp$ is zero for a closed ring current.

The elementary components of field $d\mathbf{B}_\parallel$ from different elements of the ring have the same direction. Thus, the component of the net field $B_\parallel$ does not vanish. This component can be determined by using

$$dB_\parallel = dB \cos \beta = k_m \frac{Idl}{r^2} \cos \beta, \tag{2.11}$$

where vector $d\mathbf{l}$ is perpendicular to vector $\mathbf{r}$. Hence,

$$B_\parallel = k_m \frac{I}{r^2} \cos \beta \int_0^{2\pi R} dl = k_m \frac{I \cos \beta \cdot 2\pi R}{r^2}. \tag{2.12}$$

Since $\cos \beta = R/r$, we have

$$B_\parallel = k_m \frac{2\pi IR^2}{r^3} = k_m \frac{2\pi IR^2}{(b^2 + R^2)^{3/2}}. \tag{2.13}$$

At the coil center, that is, at $b = 0$, the field reaches its maximum and it is equal to

$$B_\parallel(0) = \frac{k_m 2\pi I}{R} = \frac{\mu_0 I}{2R}. \tag{2.14}$$

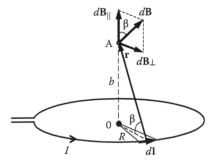

FIGURE 2.3 Calculation of a magnetic field of a circular wire of radius R carrying current I.

Introducing the magnetic moment vector of a coil $\mathbf{\mu}_m$ into Equation 2.13 simplifies the relations. The magnitude of vector $\mathbf{\mu}_m$ is equal to the product of the current and the coil area; its direction is given by the right-hand rule:

$$\mathbf{\mu}_m = IA\mathbf{n} = \pi R^2 I\mathbf{n}. \tag{2.15}$$

Taking into account these relations, the vector $\mathbf{B}$ of a coil is given by

$$\mathbf{B}_{\parallel} = k_m \frac{2IA\mathbf{n}}{r^3} = k_m \frac{2\mathbf{\mu}_m}{r^3}. \tag{2.16}$$

2.3 AMPERE'S LAW

1. For symmetric current configurations, the calculation of the magnetic field is usually simplified. In this case, Ampere's law can be applied quite easily. Let us define the line integral of vector $\mathbf{B}$ along a closed loop. In a space in which a magnetic field is present, we select a closed loop and assign a direction along which we traverse the loop (either clockwise or counterclockwise). For each elementary length dl of this loop, one can define the projection B_l of vector $\mathbf{B}$ along the direction of the tangent to this contour segment (Figure 2.4). The line integral of vector $\mathbf{B}$ is defined as the integral of the scalar product $\mathbf{B} \cdot dl$ over the entire path along the chosen direction:

$$\oint_L \mathbf{B} \cdot d\mathbf{l} = \oint_L B_l dl. \tag{2.17}$$

Some currents can penetrate the area defined by the chosen loop, while others do not. Ampere's law states that the line integral of $\mathbf{B}$ over any closed path is equal to the product of the magnetic permeability μ_0 of vacuum and the sum of all currents that penetrate the area defined by the path, that is,

$$\oint_L \mathbf{B} \cdot d\mathbf{l} = \mu_0 \sum I_j. \tag{2.18}$$

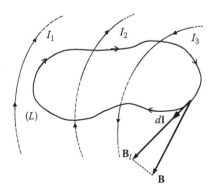

FIGURE 2.4 Explanation of Ampere's law.

The line integral of the vector **B** along a path is nonzero if the path encloses a current. The relation (2.18) is called Ampere's law. Using Stokes' theorem, an integral over a closed path of the vector **B** can be transformed into an integral over the surface that has the path as its border, that is,

$$\oint_L \mathbf{B} \cdot d\mathbf{l} = \int_A \nabla \times \mathbf{B} \cdot d\mathbf{A}. \tag{2.19}$$

Introducing the right-hand part of this relation into Equation 2.18, we arrive at

$$\int_A \nabla \times \mathbf{B} \cdot d\mathbf{A} = \mu_0 \sum I_j = \mu_0 \int_A \mathbf{j} \cdot d\mathbf{A}. \tag{2.20}$$

Comparison of the left- and right-hand parts of this equation gives the following relation:

$$\nabla \times \mathbf{B} = \mu_0 \mathbf{j}. \tag{2.21}$$

The latter two relations contain a differential vector operation *curl* **B** that is determined by a vector product $\nabla \times \mathbf{B}$ and is given in Cartesian coordinates as follows (compare with Equation 2.3):

$$\nabla \times \mathbf{B} = \begin{vmatrix} \mathbf{i} & \mathbf{j} & \mathbf{k} \\ \dfrac{\partial}{\partial x} & \dfrac{\partial}{\partial y} & \dfrac{\partial}{\partial z} \\ B_x & B_y & B_z \end{vmatrix} = \mathbf{i}\left(\dfrac{\partial B_z}{\partial y} - \dfrac{\partial B_y}{\partial z}\right) + \mathbf{j}\left(\dfrac{\partial B_x}{\partial z} - \dfrac{\partial B_z}{\partial x}\right) + \mathbf{k}\left(\dfrac{\partial B_y}{\partial x} - \dfrac{\partial B_x}{\partial y}\right). \tag{2.22}$$

2. Let us introduce the quantity

$$d\Phi_B = \mathbf{B} \cdot d\mathbf{A} \tag{2.23}$$

that defines the element of the magnetic flux (i.e., the flux of the vector **B** through area $d\mathbf{A}$). The unit of this quantity in the SI is *weber* (Wb): $\text{Wb} = \text{T} \cdot \text{m}^2$, where Tesla (T) is the unit of the magnetic fields in the SI. Since the magnetic field lines are always closed, the flux of the vector **B** through any closed surface is zero. Thus, Gauss's theorem for the vector **B** is given as

$$\oint_A \mathbf{B} \cdot d\mathbf{A} = 0. \tag{2.24}$$

This equation is one of the formulations of one of Maxwell's equations known also as "Gauss's law for the magnetic field," and the earlier equation is consistent with Maxwell's equation

$$\text{div } \mathbf{B} = 0. \tag{2.25}$$

A unique potential cannot be assigned for the magnetic field because it would be multivalued. After each traversal of the path, the potential would be incremented by $\mu_0 I$. Such fields are referred to as *solenoidal fields*.

A typical application of Ampere's law is the calculation of the magnetic field of an infinitely long solenoid. A long coil consisting of multiple turns of wire wound in a helical geometry around a cylindrical core is called a *solenoid*. If a current flows through the solenoid, a homogeneous (as long as we stay away from the solenoid ends) magnetic field is generated inside the solenoid. We will consider this field in more details in the next section.

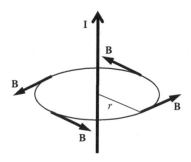

FIGURE 2.5   The calculation of a magnetic field of a thin and infinitely long conductor carrying a current I.

Exercise 2.3

Using Ampere's law, calculate the magnetic field **B** generated by an infinitely long conductor carrying current I.

Solution. Choose an observation point located at a distance r from a conductor. Draw a thin circular integration path through this point (as given in Figure 2.5) so that the plane of the circle is perpendicular to the wire. Since the direction of vector **B** is along the tangent to the circular path at every point, then $\mathbf{B} \cdot d\mathbf{l} = B dl$. The magnitude of **B** is constant at all points of the circular path due to the symmetry of the problem. Thus, from Equation 2.18, we arrive at

$$\oint_L \mathbf{B} d\mathbf{l} = \oint_L B dl = B \oint_L dl = B \cdot 2\pi r = \mu_0 I.$$

Hence, $B = \mu_0 I / 2\pi r = k_m \cdot 2I/r$.

2.4   MAGNETIC FIELD OF A SOLENOID

1. If a direct current is passed through the windings of the solenoid, a static magnetic field is generated inside as well as outside the solenoid. Figure 2.6 shows a schematic diagram of the solenoid and also the magnetic field lines generated by the solenoid.

We will calculate the magnetic field on the axis of a solenoid of finite length L and radius R assuming that the solenoid has N turns. The number of turns per unit length n is equal to N/L: $n = N/L$. It is easy to calculate the magnetic field of the solenoid, if we use Equation 2.12, which gives the magnetic field of a circular coil on its axis:

$$B = k_m \frac{I \cos\beta \cdot 2\pi R}{r^2} = \mu_0 \frac{IR}{2} \cdot \frac{\sin\alpha}{r^2}. \tag{2.26}$$

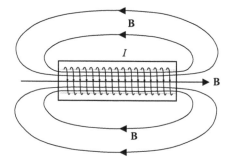

FIGURE 2.6   Schematics of the solenoid and the magnetic field lines of the solenoid carrying current I.

Here, we have introduced the angle α between the axis of the solenoid and the position vector of the point A in Figure 2.7. We want to determine the magnetic field at point A that is included by each circular turn of the coil. It is more convenient to consider angle α that is complementary to angle β (i.e., $\alpha = \pi/2 - \beta$). For each turn in (2.26), there is a corresponding angle α and a position vector **r**. Let us choose an element of the solenoid of length dl containing ndl turns. From Equation 2.26, the magnetic field generated by these turns can be written as

$$dB = \mu_0 \left(\frac{IR}{2} \right) \cdot \left(\frac{\sin \alpha}{r^2} \right) ndl. \tag{2.27}$$

From the definition of angle α, it follows that $\sin \alpha = R/r$. By differentiating this relation, we obtain

$$\cos \alpha \, d\alpha = -\left(\frac{R}{r^2} \right) dr,$$

where $dr = (dl)\cos \alpha$, that is, $d\alpha = -(R/r^2)dl$. By integrating Equation 2.27 over the angle α in the interval $\alpha_1 \le \alpha \le \alpha_2$, we get

$$B = \frac{\mu_0}{2} In(\cos \alpha_2 - \cos \alpha_1). \tag{2.28}$$

Here, α_1 and α_2 are the angles shown in Figure 2.7a. For points lying inside the solenoid, $\pi/2 \le \alpha_1 < \pi$, and therefore, $\cos \alpha_1 < 0$, $0 < \alpha_2 < \pi/2$, and $\cos \alpha_2 > 0$, so B is positive (see Figure 2.7a).

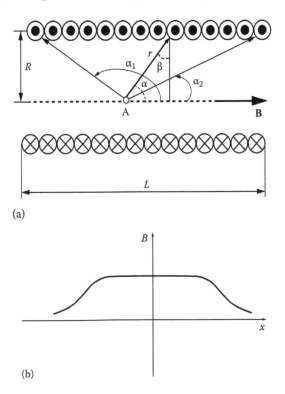

(a)

(b)

FIGURE 2.7 Calculation of the magnetic field of a solenoid (a) and dependence of the magnetic field on the coordinate along the center of the solenoid (b).

Equation 2.28 is an expression for the magnetic field on the axis of a solenoid of finite length. In the case of a very long solenoid ($L \gg 2R$), when the angle $\alpha_1 \to \pi$ and $\cos \alpha_1 \to -1$ and the angle $\alpha_2 \to 0$ and $\cos \alpha_2 \to 1$, the field inside the solenoid is given by the expression

$$B = \mu_0 nI \ [T],\qquad(2.29)$$

where the product nI is called the **number of ampere-turns per meter.**

At the ends of a sufficiently long solenoid, either the angle α_1 or the angle α_2 is equal to $\pi/2$. Therefore, the field in this case is equal to $B(0) = B(L) = \mu_0 nI/2$. Thus, at the ends of the axis of the solenoid, the value of the magnetic field is half the value in the middle.

Next, we will obtain the magnetic field $B(x)$ along the axis of the solenoid. We place the origin $x = 0$ at the center of the solenoid on its axis. An observation point A inside the solenoid has its x-coordinate lying in the interval $-L/2 \le x \le L/2$. In this case, the cosines of the angles α_1 and α_2 are given by the expressions

$$\cos \alpha_1 = -\frac{L/2+x}{\sqrt{(L/2+x)^2 + R^2}}, \quad \cos \alpha_2 = \frac{L/2-x}{\sqrt{(L/2-x)^2 + R^2}}.$$

By substituting these expressions into Equation 2.28, we obtain

$$B(x) = \frac{\mu_0}{2} nI \left(\frac{L/2+x}{\sqrt{(L/2+x)^2 + R^2}} + \frac{L/2-x}{\sqrt{(L/2-x)^2 + R^2}} \right).\qquad(2.30)$$

At the center of the solenoid ($x = 0$), the value of the magnetic field is maximum; at its ends ($x = \pm L/2$), it is minimum; the magnetic field values are given by the relations

$$B(0) = \mu_0 nI \cdot \frac{L}{\sqrt{L^2 + (2R)^2}}, \quad B(\pm L/2) = \frac{\mu_0 nI}{2} \cdot \frac{L}{\sqrt{L^2 + R^2}}.$$

Consider the situation for which the observation point is on the axis of the solenoid outside of the coil. In fact, even in this case, expression (2.30) remains valid, if we take into account that $|x| \ge L/2$. Thus,

$$\cos \alpha_1 = -\frac{x+L/2}{\sqrt{(x+L/2)^2 + R^2}}, \quad \cos \alpha_2 = -\frac{x-L/2}{\sqrt{(x-L/2)^2 + R^2}}$$

and the magnetic field can be written as

$$B(x) = \frac{\mu_0}{2} nI \left(\frac{x+L/2}{\sqrt{(x+L/2)^2 + R^2}} - \frac{x-L/2}{\sqrt{(x-L/2)^2 + R^2}} \right).\qquad(2.31)$$

In the expressions given earlier, we have assumed that the windings are wound tightly on the frame of the solenoid and their planes are perpendicular to the axis of the solenoid. Only in this case, the resulting field of the solenoid has axial symmetry and has only one component parallel to its axis. In a real solenoid, the coil is a spiral with a pitch that is equal to the thickness of the wire. Therefore, a real solenoid always has component of current along the axis that is determined by the angle of inclination of the plane of the turns to the axis of the solenoid. The presence of the longitudinal component of the current leads to a weak magnetic field outside the solenoid. In order for the longitudinal component of the current in the solenoid to be removed, the coil winding is usually placed in two layers—one layer is winding forward and the other layer is winding in the reverse directions. In this case, the transverse components of the current in the layers are added, while the longitudinal components subtracted. Thus, the longitudinal component of one layer offsets the component of the other layer.

Outside the solenoid, the value of B rapidly decreases in either direction. Figure 2.7b shows the distribution of the magnetic field $B(x)$ along the axis of the solenoid, based on expressions (2.30) and (2.31). It is clear that B rapidly tends to zero with increasing distance from the ends of the solenoid.

In order to determine the field outside of the solenoid, let us consider two pairs of current elements equal to each other and arranged symmetrically with respect to cross-sectional plane AA that is perpendicular to the axis of the solenoid (Figure 2.8). Currents Idl_1 are referred to as the "outgoing" currents, while Idl_2 as the "incoming." According to the law of Biot–Savart, the pairs of these current elements will create in each point of the cross-sectional AA magnetic fields whose magnetic fields $d\mathbf{B}_1$ and $d\mathbf{B}_2$ inside the solenoid are directed in one direction and outside the solenoid in the opposite direction. For regions outside of the solenoid, $|d\mathbf{B}_1| > |d\mathbf{B}_2|$, since the elements Idl_1 are closer to the observation point (see distances r_1' and r_1'' in Figure 2.8) than the elements Idl_2 (they are at distances r_2' and r_2''; see Figure 2.8). Therefore, the resulting field $|d\mathbf{B}_1 + d\mathbf{B}_2| = dB_1 - dB_2$ is much smaller than the field generated by these elements inside the solenoid, where the fields of these elements add up. We note that for an infinitely long solenoid, the magnetic field lines outside the solenoid are parallel to the axis of the solenoid and antiparallel to each other. It can be shown that the total field of the selected current elements decreases in proportion $1/b^3$ where the distance b from the axis of the solenoid $b \gg R$. Thus, at points located at a sufficiently large distance from the axis of the solenoid, the magnetic field is practically equal to zero.

2. An infinitely long solenoid is an idealized model of a real, sufficiently long solenoid. The application of Ampere's law to an infinitely long solenoid allows us to get the correct expression for the magnetic field without integration.

Let us select a closed rectangular loop, with two sides parallel and the other two sides perpendicular to the axis of the solenoid (Figure 2.9). Let the loop portion 3→4 be at a distance from the solenoid much greater than its diameter and the section 1→2, which is parallel to the axis of the solenoid, be located in one case inside the solenoid (this particular case is shown in Figure 2.9a) and

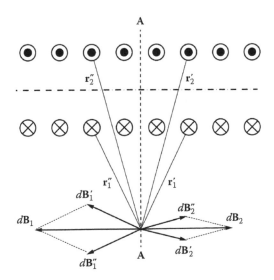

FIGURE 2.8 Calculation of the magnetic field outside of a solenoid.

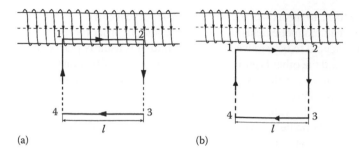

FIGURE 2.9   Calculation of the magnetic field inside (a) and outside of a solenoid (b).

in another case outside the solenoid (this particular case is shown in Figure 2.9b). The line integral of **B** along the closed loop 1→2→3→4 is equal to the sum of the corresponding line integrals for each of the sections:

$$\oint_L \mathbf{B} \cdot d\mathbf{l} = \int_{1\to2} \mathbf{B} \cdot d\mathbf{l} + \int_{2\to3} \mathbf{B} \cdot d\mathbf{l} + \int_{3\to4} \mathbf{B} \cdot d\mathbf{l} + \int_{4\to1} \mathbf{B} \cdot d\mathbf{l}$$

$$= \int_{1\to2} B_l dl + \int_{2\to3} B_l dl + \int_{3\to4} B_l dl + \int_{4\to1} B_l dl. \tag{2.32}$$

Since the magnetic field lines are parallel to the axis, then at all points of section 1→2, B_1 = const. In sections 2→3 and 4→1 of the contour, vector **B** is perpendicular to the contour element $d\mathbf{l}$. Consequently, the projection is $B_l = 0$ at all points on section 2→3 and on section 4→1. Points on section 3→4 are at a distance much greater than the diameter of the solenoid. As noted earlier, at these points we may consider that $B = 0$ with a good degree of accuracy. Thus,

$$\oint_L B_l dl = \int_{1\to2} B_l dl = B \int_{1\to2} dl = Bl, \tag{2.33}$$

where l is the length of section 1→2. Thus, combining Equations 2.18 and 2.33, we can write

$$\oint_L B_l dl = Bl = \mu_0 nlI, \tag{2.34}$$

where n is the density of the windings (number of turns per unit length of the solenoid) and nl is the number of turns per length l. The result of Equation 2.34 does not depend on how far from the axis of the solenoid section 1→2 is as long as it is inside of the solenoid.

    If we move section 1→2 outside of the solenoid, we have no currents enclosed by the loop and $\oint_L B_l dl = 0.$

    Thus, the magnetic field inside an infinitely long solenoid can be considered uniform and equal to $B = \mu_0 nI$, which coincides with (2.29), and outside of it the magnetic field vanishes. A real solenoid, if its length L is much greater than its diameter $2R$, can be approximated by an infinitely long solenoid.

Exercise 2.4

What must the ratio of the length L of a coil to its diameter d be to be able to calculate the magnetic field at the center of the coil using the equation for an infinitely long solenoid so that the resulting error does not exceed 1%?

Solution. For the magnetic field on the axis of the solenoid of finite length, we have from Equation 2.28

$$B = \mu_0 \frac{In}{2}(\cos\alpha_2 - \cos\alpha_1).$$

For an infinitely long solenoid, $B_\infty = \mu_0 In$. For the field intensity at the center of the solenoid due to the symmetry of the problem, we have (consult Figure 2.7)

$$\alpha_1 = \pi - \alpha_2.$$

At the same time, $\cos\alpha_1 = -\cos\alpha_2$ and

$$B = \mu_0 \frac{In}{2}(2\cos\alpha_2) = \mu_0 In\cos\alpha_2.$$

The relative error is

$$\eta = \frac{\Delta B}{B} = \frac{B_\infty - B}{B} = \frac{B_\infty}{B} - 1 = \frac{\mu_0 In}{\mu_0 In\cos\alpha_2} - 1 = \frac{1}{\cos\alpha_2} - 1.$$

Thus, $\cos\alpha_2 = 1/(1 + \eta)$. Since $\cos\alpha_2 = L/\sqrt{L^2 + d^2}$, then

$$\frac{L}{\sqrt{L^2 + d^2}} = \frac{1}{1+\eta}, \quad \sqrt{1 + \frac{d^2}{L^2}} = 1+\eta, \quad \frac{d}{L} = \sqrt{(2+\eta)\eta}.$$

Finally, we get

$$\frac{L}{d} = \frac{1}{\sqrt{(2+\eta)\eta}} = \frac{1}{\sqrt{(2+0.01)\times 0.01}} \simeq 7.00.$$

2.5 MAGNETIC FIELD IN A MEDIUM, MAGNETIC FIELD INTENSITY

1. Experiments show that the magnetic field produced by electric currents in a material differs from the magnetic field generated by the same currents in vacuum. A material placed in a magnetic field is magnetized and itself becomes a source of magnetic field. Materials capable of being magnetized in a magnetic field are called magnetic materials. A magnetized material creates its own magnetic field **B'** generated by *microscopic currents*. This field is added to the magnetic field $\mathbf{B}_0$ generated by currents from external charges (*macroscopic currents*). The resultant field in the material is $\mathbf{B} = \mathbf{B}_0 + \mathbf{B}'$. The absolute value of a ratio between **B** and $\mathbf{B}_0$ is called *the relative magnetic permeability* of the material $\kappa_m = B/B_0$.

The magnetic properties of matter are determined by the magnetic properties of its constituent atoms. The magnetic properties of protons and neutrons are less pronounced than the magnetic properties of electrons by a factor $m_p/m_e = 1836$, where m_p and m_e are mass of protons (neutrons) and electrons. Therefore, the magnetic properties of a material are largely determined by the electrons of its constituent atoms.

In the classic physics picture of an atom, an electron generates a magnetic field by orbiting around the nucleus. This motion can be regarded as a circular current with a corresponding orbital magnetic moment μ_{orb}. In addition, an electron generates its own magnetic field due to its intrinsic (a.k.a. spin) moment μ_{spin}. The magnetic moment of a multielectron atom is the vector sum of the orbital and spin moments from all its electrons. The magnetic fields of electrons (spin and orbital) determine a wide range of magnetic properties. The interaction of the atomic magnetic moments with an external magnetic field accounts for the magnetic properties of the material.

Let us consider a relatively small volume of a particular material. The total magnetic moment of all atoms in this volume is

$$\mu_m = \sum \mu_i. \tag{2.35}$$

The **magnetization** $\mathbf{M}$ of a material is defined as the magnetic moment per unit volume of medium defined as

$$\mathbf{M} = \lim_{Vol \to 0} \left(\frac{1}{Vol} \sum \mu_i \right). \tag{2.36}$$

The magnetization is a characteristic of the material.

2. As mentioned earlier, the net magnetic field $\mathbf{B}$ in a material is generated by all macroscopic and microscopic currents. Ampere's law (see Equation 2.18) for the magnetic field in vacuum can be generalized for a case of magnetic field in matter as follows:

$$\oint_L \mathbf{B} \cdot d\mathbf{l} = \mu_0 \left(I_{macro} + I_{micro} \right), \tag{2.37}$$

where I_{macro} and I_{micro} are an algebraic sum of macroscopic and microscopic currents enclosed inside the closed path L. The sum of the microscopic currents is related to the line integral of the magnetization vector through the relation

$$I_{micro} = \oint_L \mathbf{M} \cdot d\mathbf{l}. \tag{2.38}$$

Taking into account this equation, Ampere's law can be written as

$$\oint_L \left(\frac{\mathbf{B}}{\mu_0} - \mathbf{M} \right) \cdot d\mathbf{l} = I_{macro}. \tag{2.39}$$

Vector $\mathbf{B}/\mu_0 - \mathbf{M} = \mathbf{H}$ is called **the magnetic field intensity**, also terms **magnetic field strength** as well as simply **magnetic field** are used in the literature for $\mathbf{H}$. Thus, Ampere's law for magnetic fields in matter states that the line integral of the magnetic field intensity vector along an arbitrary closed path is equal to the algebraic sum of macroscopic currents enclosed inside the path:

$$\oint_L \mathbf{H} \cdot d\mathbf{l} = I_{macro}. \tag{2.40}$$

This relation expresses Ampere's law in integral form. Its differential form is

$$\nabla \times \mathbf{H} = \mathbf{j}_{macro}. \tag{2.41}$$

The magnetization of an isotropic medium is related to the magnetic field intensity **H** by the expression

$$\mathbf{M} = \chi_m \, \mathbf{H}, \tag{2.42}$$

where χ_m is a dimensionless coefficient that characterizes the magnetic properties of the material, and it is called ***the magnetic susceptibility***. This coefficient is connected to the magnetic permeability of a substance by the relation $\kappa_m = 1 + \chi_m$. Taking into consideration the definition of vector **H**, we arrive at

$$\mathbf{B} = \mu_0(\mathbf{H} + \mathbf{M}) = \mu_0(1 + \chi_m)\mathbf{H} = \mu_0\kappa_m\mathbf{H}. \tag{2.43}$$

Hence,

$$\mathbf{H} = \frac{\mathbf{B}}{\mu_0\kappa_m} \ (\text{A/m}). \tag{2.44}$$

The dimensionless quantity $\kappa_m = 1 + \chi_m$ is the relative magnetic permeability of the material. For diamagnetic materials, χ_m is negative (i.e., the field of the microscopic currents is opposite to the external field). For paramagnetic materials, χ_m is positive (the field of the microscopic currents has the same direction with the external filed). Since the absolute value of the magnetic susceptibility for diamagnetic and paramagnetic materials is very small (some $10^{-4} - 10^{-6}$), for these materials, κ_m differs only slightly from unity. Thus, for diamagnetic materials, $\chi_m < 0$ and $\kappa_m < 1$, and for paramagnetic materials, $\chi_m > 0$ and $\kappa_m > 1$.

Exercise 2.5

A cylindrical infinitely long conductor of radius R carries a current of constant density **j** (Figure 2.10). Find the dependence of the magnetic field B on the distance from the conductor axis.

Solution. The magnetic field lines generated by such a current-carrying conductor are concentric circles enclosing the conductor axis. Using Ampere's law, let us write the following relations for B:

$$\oint_L \mathbf{B} \cdot d\mathbf{l} = \mu_0 I = \mu_0 \int_A \mathbf{j} \cdot d\mathbf{A}$$

where
 j is the current density
 A is the area enclosed by the path L

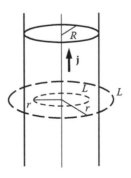

FIGURE 2.10 Calculation of the magnetic field of a cylindrical infinitely long conductor of radius R that carries a current of constant density.

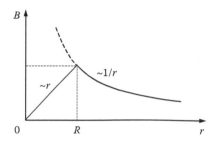

FIGURE 2.11 Dependence of the magnetic field B on the distance from the conductor axis of the wire shown in Figure 2.10.

First, let us consider the case for which $r \leq R$, when the integration path is inside the conductor. The area element vector $d\mathbf{A}$ is parallel with $\mathbf{j}$. The path L coincides with the magnetic field line; hence, the path elementary vector $d\mathbf{l}$ is parallel with $\mathbf{B}$. The surface enclosed by the path has area $A = \pi r^2$, and the total current through this surface is $I_A = jA = j\pi r^2$.

Integration yields $B2\pi r = \mu_0 j\pi r^2$. Hence, the following dependence of the magnetic field on r is obtained:

$$B(r) = \frac{\mu_0 j\pi r^2}{2\pi r} = \frac{\mu_0 jr}{2}, \quad r \leq R.$$

Let us consider the case for which $r > R$, when the integration path is outside the conductor. Since there is no current outside the conductor, the magnetic field is generated by the total current $I = j\pi R^2$. Here, $B2\pi r = \mu_0 j\pi R^2$; hence,

$$B(r) = \frac{\mu_0 j\pi R^2}{2\pi r} = \frac{\mu_0 jR^2}{2r}, \quad r > R.$$

A plot of the $B(r)$ as function of r is shown in Figure 2.11. For $r < R$, the dependence of B on r is linear. For $r > R$, this dependence is hyperbolic. At the boundary $r = R$, the magnetic field is continuous, that is, it does not exhibit a discontinuity due to the absence of surface currents.

PROBLEMS

2.1 Two infinitely long, parallel wires carry currents I_1 and I_2 (Figure 2.12). The distance between the two wires is equal to b. Determine the magnetic field at a point lying on the line connecting the two wires at the distance r_1 from the first wire. Consider two cases: (a) the currents are flowing in opposite directions and (b) the currents are flowing in the same direction.

2.2 A long thin conductor is bent as shown in Figure 2.13. The coil in the x_0z plane consists of three quarters of a full circle of radius R. The conductor carries a current I. Find the vector $\mathbf{B}$ and its magnitude at the center O of the coil. Assume that the coil is in vacuum.

$$\left(Part\ of\ the\ answer: \mathbf{B} = \frac{\mu_0 I}{4\pi R}\left(-\mathbf{i} + \mathbf{j}\frac{3\pi}{2} - \mathbf{k}\right). \right)$$

2.3 A charged particle moves with constant velocity $v = 2.00 \times 10^6$ m/s in a uniform magnetic field $B = 0.50$ T on a circular orbit whose plane is perpendicular to the magnetic field. The radius of the orbit is equal to $R = 2.00$ cm and the kinetic energy of the particle $W = 2.00 \times 10^4$ eV.

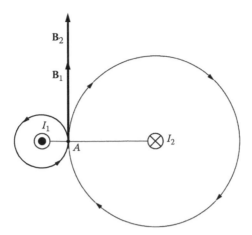

FIGURE 2.12   Magnetic field of two parallel wires.

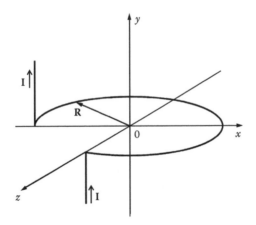

FIGURE 2.13   Magnetic field of three quarters of a circle of radius R.

Determine (a) the charge of the particle, (b) the potential difference that accelerates the particle from rest to $v = 2.00 \times 10^6$ m/s before entering into the magnetic field, and (c) the magnetic moment μ_m of the cyclotron orbit. (*Answer*: (a) $q = 3.20 \times 10^{-19}$ C, (b) $V = 4.00$ kV, and (c) $\mu_m = 6.40 \times 10^{-13}$ A·m².)

2.4   A long cylindrical uniform solenoid is filled by two magnetic materials as shown in Figure 2.14. A current I is flowing in the solenoid's windings. The number of turns per unit length in the winding is equal to n. The magnetic permeability of the inner magnetic material is equal to κ_{m1} and that of the external equal to κ_{m2}. The radius of the inner cylinder is equal to R_1 and that of the external equal to R_2. Determine the density of the microscopic surface currents in the magnetic materials (consult Equations 2.37 and 2.38). (*Answer*: $J_m = (\kappa_{m1} - \kappa_{m2})nI$ at $r = R_1$ and $J_m = (\kappa_{m2} - 1)nI$ at $r = R_2$.)

2.5   Can the magnetic field in vacuum depend on coordinates as (a) $\mathbf{B}(x, y, z) = \alpha(2x\mathbf{i} - y\mathbf{j} + 4z\mathbf{k})$ or (b) $\mathbf{B}(x, y, z) = \alpha(x\mathbf{i} + 2y\mathbf{j} - 3z\mathbf{k})$?

Here, α is a constant with the dimension (T/m) and $\mathbf{i}$, $\mathbf{j}$, and k are the unit vectors of Cartesian system of coordinates. Find the spatial distribution of the current density.

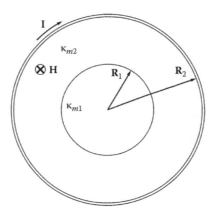

FIGURE 2.14 Solenoid filled by two different magnetic materials with magnetic permeabilities κ_{m1} for the inner material and κ_{m2} for the outer shell.

(*Answer*: (a) the magnetic field cannot have such a dependence on coordinates and (b) the magnetic field can have such a dependence on coordinates. The spatial distribution of current density for case (b) is $\mathbf{j} = 0$.)

2.6 Find the magnetic field generated by an infinitely long (ideal) solenoid with n turns per unit length and current I using Ampere's law. (*Hint*: Consult Figure 2.15) (*Answer*: $B = \mu_0 nI$.)

2.7 A long thin wire is carrying a current I_1. A loop with current I_2 is located in the vicinity of the wire with its plane perpendicular to the wire, as shown in Figure 2.16. The loop consists of two circular arcs 4–1 and 3–2 with radii a and b ($a < b$), connected by straight lines 1–2 and 3–4. Both arcs have a common center located on the wire. The angle between the straight lines is 2φ. Find the torque acting on the loop by the wire. *Reminder*: Torque or moment of force, M, is the tendency of force f to rotate the object to which the force is applied, and magnitude of the torque is equal to $M = fr \sin \varphi$, where $\mathbf{r}$ is the vector from the axis of rotation to the point of force application and φ is the angle between the force vector and the vector $\mathbf{r}$.

$$\left(Answer:\ M = \frac{\mu_0 I_1 I_2}{\pi}(b-a)\sin\varphi. \right)$$

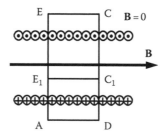

FIGURE 2.15 Calculation of a magnetic field of a solenoid.

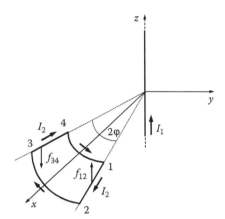

FIGURE 2.16 Calculation of the torque acting on the loop 1–2–3–4 by the magnetic field generated by current I_1.

2.8 A ring ABCD consists of two metal half rings of radius a, joined at points A and C. The diameter of the wire's cross section of the lower half ring ADC is double that of the diameter of the wire's cross section of the upper half ring ABC. The current in the straight sections is equal to I. Find the magnitude of the magnetic field at the center of the ring (point O in the).

$$\left(Answer:\ B(0) = \frac{3\mu_0}{20a} I. \right)$$

2.9 A straight long thin wire carrying a current I is surrounded by a cylinder made of a magnetic material of uniform permeability κ_m. A solenoid that carries a current I is wound onto the outer surface of the cylinder (see). The number of turns per unit length of the solenoid is equal to n. Find the magnitude of the net magnetic field inside and outside the

solenoid. $\left(Part\ of\ the\ answer:\ B = \dfrac{\mu_0 I}{2\pi r}. \right)$

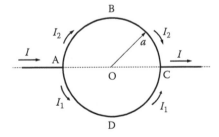

FIGURE 2.17 Calculation of the magnetic field of two semirings.

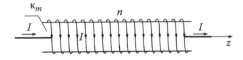

FIGURE 2.18 Calculation of the magnetic field of the straight wire and solenoid.

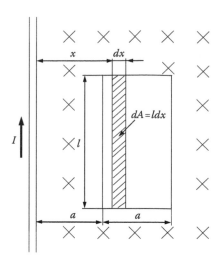

FIGURE 2.19 Calculation of the magnetic flux through the frame of an area $A = al$.

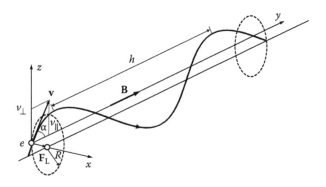

FIGURE 2.20 Helical path of the electron in a magnetic field.

2.10 A rectangular frame with height $l = 65.0$ cm and width a is placed so that its long sides are parallel to an infinitely long, straight wire that carries a current $I = 50.0$ A as shown in. What is the magnetic flux Φ through the frame? (*Answer*: $\Phi = 4.50$ µWb.)

2.11 An electron of velocity $v = 2.00 \times 10^6$ m/s enters a uniform magnetic field of $B = 30.0$ mT at an angle $\alpha = 30°$ to the direction of the magnetic field lines. Determine the radius R and pitch h of the helical path followed by the electron (). (*Answer*: $R = 0.19$ mm and $h = 2.06$ mm.)

3 Maxwell's Equations for Electromagnetic Fields

By adding one additional term in Ampere's law and combining it with Gauss's law for the electric field, Gauss's law for the magnetic field, and Faraday's law, James Clerk Maxwell created a complete theory of electromagnetic fields. Maxwell was able to explain all available experimental facts related to electromagnetism on the basis of these four equations, which were known collectively as *Maxwell's equations*. These equations allow us, using a given spatial distribution of charges and currents and their time dependences, to find the electric and magnetic fields at each point of space at any moment of time.

Maxwell's theory was the greatest contribution to the development of classical physics and allowed to describe a huge range of phenomena, beginning from the electrostatic field of static charges and ending with the electromagnetic nature of light. Maxwell's equations form the basis for wave optics. Propagation of light in vacuum and in media can be described by these equations.

An important property of Maxwell's equations is that they are invariant under the Lorentz transformations. This means that the equations have the same form in all inertial coordinate systems. Nevertheless, like any physical theory, Maxwell's theory has its limits of applicability. Here, we point to two major limitations. First, it can be applied only if the distance between the charges exceeds the interatomic distances in a medium, that is, the distance between charges is larger than 10^{-10} m. Second, the frequency of the electromagnetic field changes should not exceed 10^{15} Hz since at higher frequencies, the quantum properties of radiation are revealed. In this chapter, we will briefly discuss Maxwell's equations and will apply those to several simple problems.

3.1 FARADAY'S LAW

In 1831, the English scientist *Michael Faraday* discovered the important physical phenomenon of *electromagnetic induction*, *in* which a magnetic field B produces an electric current in a closed loop, if the magnetic flux through the surface area of a loop changes with time. An electromotive force (emf) generated by the time-varying magnetic flux and an electric current is induced in a closed loop. Experiments show that the emf generated, referred to as "induced emf," is proportional to the rate, with which the magnetic flux passing through the loop changes with time.

The absolute value of the induced emf in a loop is equal to the rate, with which the magnetic flux passing through the loop changes with time:

$$\mathcal{E}_i = -\frac{d\Phi}{dt}. \tag{3.1}$$

This is known as Faraday's law.

The magnetic flux passing through a surface covering a loop is defined as

$$\Phi = \int_A \mathbf{B} \cdot d\mathbf{A} = \int_A \mathbf{B} \cdot \mathbf{n} \, dA = \int_A B_n \, dA = \int_A B \cos\alpha \, dA, \tag{3.2}$$

where
   $\mathbf{n}$ is the unit vector of the outward normal to the surface element dA
   α is the angle between vectors $\mathbf{B}$ and $\mathbf{n}$ (see Figure 3.1)

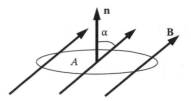

FIGURE 3.1 To the calculation of magnetic flux passing through a surface A of the closed loop.

According to Equation 3.2, a change $d\Phi$ of the magnetic flux can result from changes of the magnitude of $\mathbf{B}$ and/or the surface area A and/or the relative orientation of vectors $\mathbf{B}$ and $\mathbf{n}$. As we already mentioned in Chapter 2, Weber (1 Wb = 1 T $\cdot$ m^2) is a unit of the magnetic flux in the SI units.

The induced emf in a conducting loop results in the appearance of an induced current I_i. The direction of this current is governed by **_Lenz's rule_**: the current in the loop always flows in such a direction as to oppose the change of magnetic flux that produced it. The negative sign in Equation 3.1 means that if the magnetic flux is decreasing ($d\Phi/dt < 0$), then $\mathcal{E}_i > 0$ and the flux generated by the magnetic field of the induced current tries to increase the flux Φ and thus oppose the change. If $d\Phi/dt > 0$, then $\mathcal{E}_i < 0$ and the flux generated by the magnetic field of the induced current tries to decrease the flux Φ and thus also opposes the change. It can be demonstrated that Faraday's law in the form of Equation 3.1 is compatible with the energy conservation law. For this, let us consider a conducting loop with a movable bridge of length l placed in a uniform magnetic field $\mathbf{B}$ (see Figure 3.2). The loop contains an external source of emf, which results in a current I. The magnetic force acting on the bridge moves it by a distance dx during time interval.

In this case, the magnetic force performs work $dW = F_A dx = I\, d\Phi$, where $d\Phi$ is the magnetic flux through the area traversed by the moving bridge during time dt. Let R be the resistance of the loop. According to the energy conservation law, the work done by a current source in time $dt\, dW_{source} = \mathcal{E}I\, dt$ is a sum of work spent for heating the resistor R, $dW_{therm} = I^2R\, dt$, and the work performed by Ampere's force, that is,

$$dW_{source} = dW_{therm} + dW.$$

This relation can be given as

$$\mathcal{E}I\, dt = I^2R\, dt + I\, d\Phi.$$

Hence,

$$I = \frac{1}{R}\left(\mathcal{E} - \frac{d\Phi}{dt}\right) = \frac{\mathcal{E} + \mathcal{E}_i}{R}, \qquad (3.3)$$

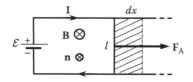

FIGURE 3.2 Illustration of work performed by Ampere's force.

as predicted by Kirchhoff's loop rule for this circuit. The right-hand part of Equation 3.3 in brackets is the sum of the emf's acting in the loop. They are the emf of voltage source $\mathcal{E}$ and the induced emf $\mathcal{E}_i$ given by Equation 3.1. If we have a time-varying magnetic flux through N loops, then

$$\mathcal{E}_i = -\frac{d}{dt}\sum_{i=1}^{N}\Phi_i = -\frac{d\Phi_{net}}{dt}, \tag{3.4}$$

where Φ_{net} is the net magnetic flux. Equation 3.4 is consistent with Faraday's law. If $\Phi_{net} = N\Phi$, Equation 3.4 can be written as

$$\mathcal{E}_i = -\frac{d\Phi_{net}}{dt} = -\frac{d(N\Phi)}{dt} = -N\frac{d\Phi}{dt}, \tag{3.5}$$

that is, the emf is proportional to the number of loops N.

Exercise 3.1

A circular copper wire coil of diameter D is placed in a uniform magnetic field forming an angle α with the coil plane normal (see Figure 3.1). The magnetic field increases at a constant rate b (i.e., $b = dB/dt$). Determine the charge that passes through a cross section of the coil in a time interval τ if the wire cross-sectional area is equal to A_{Cu}.

Solution. When the magnetic field changes by dB, the magnetic flux Φ in a coil changes by

$$\frac{d\Phi}{dt} = \frac{dB}{dt}A\cos\alpha = bA\cos\alpha,$$

where $A = \pi D^2/4$ is the coil area. The change in the magnetic flux results in the appearance of an induced emf equal to

$$\mathcal{E}_i = -\frac{d\Phi}{dt} = -\frac{dB}{dt}A\cos\alpha = -bA\cos\alpha.$$

According to Ohm's law, an induced current is generated in the coil:

$$I_i = \frac{\mathcal{E}_i}{R} = -\frac{b}{R}A\cos\alpha.$$

The charge q that will pass through any cross section of the coil during time Δt is

$$q = |I_i\Delta t| = |I_i\tau| = \frac{b\tau}{R}A\cos\alpha,$$

where we take into account that in our case, $\Delta t = \tau$.

The coil wire resistance has the equation $R = \rho_{Cu}\pi D/A_{Cu}$, where ρ_{Cu} is the resistivity of copper and πD is the length of the circular resistor. Thus, we have

$$q = \frac{b\tau A_{Cu}D}{4\rho_{Cu}}\cos\alpha.$$

Note that the charge per unit area of the wire q/A_{Cu} does not depend on the wire cross-sectional area A_{Cu}.

3.2 SELF-INDUCTANCE AND MUTUAL INDUCTANCE

We have seen that any change in the magnetic flux passing through a loop results in an induced emf. *Self-inductance* is a particular case of this phenomenon: a change in the current in a conducting loop generates an induced emf in that loop.

The electric current in the loop generates a magnetic field through the loop, which is proportional to the current of the loop. The magnetic flux of this magnetic field is proportional to the magnetic field and, hence, to the current in the loop, that is, $\Phi = LI$ (see Figure 3.3). Here, L is known as *the inductance* of the loop. This coefficient depends on a shape and size of the loop and on the magnetic properties of the surrounding medium. The SI unit of inductance is the Henri (H): H = Wb/A. For example, the inductance of a solenoid (see Figure 2.6) of volume *Vol*, length l, cross-sectional area A, and number of turns N in the air is

$$L = \frac{\mu_0 N^2 A}{l} = \mu_0 n^2 (Vol). \tag{3.6}$$

where

Vol is the solenoid volume
$n = N/l$ is a number of turns per unit length (consult Figure 2.6 for a sketch of a solenoid)

For a solenoid whose volume is filled by a core of magnetic permeability κ_m, the inductance is increased according to the following equation:

$$L = \frac{\mu_0 \kappa_m N^2 A}{l} = \mu_0 \kappa_m n^2 (Vol). \tag{3.7}$$

Taking Faraday's law into account, one can derive an expression for the self-induced emf:

$$\mathcal{E}_{si} = -\frac{d\Phi}{dt} = -\frac{d}{dt}(LI) = -\left(\frac{dL}{dt}I + L\frac{dI}{dt}\right). \tag{3.8}$$

If L is constant, then

$$\mathcal{E}_{si} = -L\frac{dI}{dt}. \tag{3.9}$$

According to Lenz's rule, the negative sign indicates that the induced current is flowing in such a direction as to oppose the initial change of the current. The self-induced emf in a loop tends to maintain the current constant: that is, it opposes an increasing current and enhances a decreasing current.

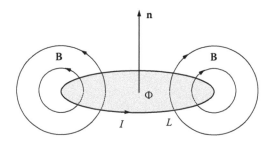

FIGURE 3.3   Illustration of self-inductance phenomenon.

Let us consider two nearby stationary loops labeled 1 and 2. In this case, the current I_1 in loop 1 generates a magnetic flux Φ_2 through loop 2 (see Figure 3.4). The magnetic flux is

$$\Phi_2 = M_{21}I_1. \tag{3.10}$$

Any change in I_1 generates an induced emf in loop 2:

$$\mathcal{E}_{i2} = -\frac{d\Phi_2}{dt} = -M_{21}\frac{dI_1}{dt}. \tag{3.11}$$

Similarly, the current I_2 in loop 2 generates a magnetic flux Φ_1 passing through loop 1. This flux is $\Phi_1 = M_{12}I_2$ (see Figure 3.4). Any change in the current I_2 generates an induced emf in loop 1:

$$\mathcal{E}_{i1} = -\frac{d\Phi_1}{dt} = -M_{12}\frac{dI_2}{dt}. \tag{3.12}$$

The appearance of an induced emf in one of the loops due to a current change in the other is referred to as **mutual inductance**. The coefficients M_{12} and M_{21} are called the mutual inductances of the loops. These coefficients depend on shapes, sizes, mutual position of loops, and on the magnetic permeability of the surrounding medium. In the absence of ferromagnetic materials, these coefficients are equal, that is, $M_{12} = M_{21} = M$. Transformers, which are devices that change the input voltage, are based on mutual induction. In a transformer, an alternating current in one loop (known as the primary) induces an alternating current in another loop (generally a coil known as the "secondary" that has a common core with the primary).

Exercise 3.2

The current in a solenoid of inductance L is changing with time as $I = \beta t - \gamma t^3$. What is the self-induced emf in the solenoid at time τ?

Solution. According to Faraday's law,

$$\mathcal{E}_i = -\frac{d\Phi}{dt}.$$

Since $\Phi = LI$, we have

$$\mathcal{E}_i = -L\frac{dI}{dt} = -L(\beta - 3\gamma t^2) = L(-\beta + 3\gamma \tau^2).$$

Here, we assume that L does not depend on the magnetic field of the solenoid. This assumption is not true for solenoids that contain a ferromagnetic core.

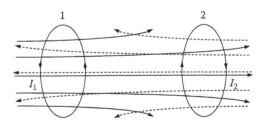

FIGURE 3.4 Illustration of mutual inductance phenomenon: current I_1 in loop 1 generates a magnetic flux Φ_2 through loop 2 and current I_2 in loop 2 generates a magnetic flux Φ_1 passing through loop 1.

3.3   MAGNETIC FIELD ENERGY

By analogy to the electric field energy stored in a charged capacitor, any loop (coil, solenoid) that carries a current flowing in its turns also stores magnetic field energy in the space occupied by the magnetic field. A conductor carrying an electric current is always surrounded by the magnetic field that is generated by that current. The work spent by the current to produce the magnetic field is converted into the magnetic field energy.

We have mentioned earlier that a loop with inductance L carrying a current I is penetrated by its own magnetic flux $\Phi = LI$. For a small change dI of the current in the loop, the magnetic flux changes by $d\Phi = L\,dI$. To change the magnetic flux by $d\Phi$, work,

$$dW = I\,d\Phi = LI\,dI,$$

must be done. The work spent to change the current in the loop from 0 to I is

$$W = \int_0^I LI\,dI = \frac{LI^2}{2} = U_m. \tag{3.13}$$

Here, the magnetic field energy stored in the magnetic field of the loop is given by one of the following equivalent relations:

$$U_m = \frac{LI^2}{2} = \frac{\Phi^2}{2L} = \frac{I\Phi}{2}. \tag{3.14}$$

The magnetic field energy is a function of the parameters characterizing this field in the surrounding medium. For this, let us consider the magnetic field inside a long solenoid. Introducing the relation for inductance of a solenoid defined by Equation 3.7 into Equation 3.14, we obtain

$$U_m = \frac{LI^2}{2} = \frac{\mu_0 \kappa_m N^2 A}{2l} I^2 = \frac{1}{2}\mu_0 \kappa_m n^2 I^2 (Vol), \tag{3.15}$$

where Vol is the solenoid volume. Since $B = \mu_0 \kappa_m nI$ is valid for an ideal solenoid, then

$$U_m = \frac{B^2}{2\mu_0 \kappa_m} Vol. \tag{3.16}$$

Considering the relation of the magnetic field to the magnetic field intensity $B = \mu_0 \kappa_m H$, the expression for the field energy in a solenoid can be written as

$$U_m = \frac{B^2}{2\mu_0 \kappa_m} Vol = \frac{\mu_0 \kappa_m H^2}{2} Vol = \frac{B \cdot H}{2} Vol. \tag{3.17}$$

The magnetic field of a solenoid is uniform and is confined in its interior. Therefore, the energy is enclosed in the solenoid volume and distributed within it with a constant magnetic energy density u_m measured in J/m^3:

$$u_m = \frac{U_m}{Vol} = \frac{B^2}{2\mu_0 \kappa_m} = \frac{\mu_0 \kappa_m H^2}{2} = \frac{B \cdot H}{2}. \tag{3.18}$$

Relation (3.18) for the magnetic energy density has a similar form to the expression for the electric energy density except for the constants. This equation is obtained for a uniform field, but it is valid for nonuniform fields as well. In the last case, B, H, and u_m are functions of coordinates.

Exercise 3.3

Consider a solenoid of length l and cross-sectional area A, containing N turns. The magnetic field inside the solenoid is B. What are the coil current I and its magnetic field energy U_m?

Solution. The magnetic field of a solenoid is given by $B = \mu_0 \kappa_m n I$, where $n = N/l$ is the number of turns per coil length and I is the current in a coil. Hence, the current is

$$I = \frac{B}{\mu_0 \kappa_m n}.$$

The magnetic energy U_m of a coil with current I is given by the relation

$$U_m = \frac{L I^2}{2}.$$

Here, L is the coefficient of inductance in a coil defined by Equation 3.7:

$$L = (\mu_0 \kappa_m n^2) \cdot Vol,$$

where $Vol = lA$ is the volume inside the coil. Taking into consideration the energy relations for a coil, we arrive at

$$U_m = \frac{1}{2} \mu_0 \kappa_m n^2 (Vol) I^2 = \frac{1}{2} \frac{B^2}{\mu_0 \kappa_m} \cdot Vol.$$

3.4 TRANSIENT PROCESSES IN CIRCUITS WITH CAPACITORS AND INDUCTORS

For the case when $\mathcal{E}_i$ is equal to zero, Equation 3.3 reduces to Ohm's law $I = \mathcal{E}/R$ for the simplest circuit that contains just a resistor R and a voltage source with zero internal resistance and emf equal to $\mathcal{E}$. Ohm's law in this form is correct for a direct current conditions when the emf does not depend on time. If $\mathcal{E}$ depends on time and the current will follow the change of $\mathcal{E}$ and Ohm's law can be applied at each instant of time t, then we have $I(t) = \mathcal{E}(t)/R$. In circuits that contain capacitors and inductors and a time-varying $\mathcal{E}$, Ohm's law must be modified. In this section, we discuss the limitations on the relation $I(t) = \mathcal{E}(t)/R$ and introduce criteria when currents can be referred to as slowly varying or quasi stationary. Let us consider the process of establishing a current in a circuit with a resistor and a capacitor and in a circuit with a resistor and a coil.

1. We start with a circuit that contains a capacitor with capacitance C, a resistor R (which includes the internal resistance of the voltage source), a voltage source with emf $\mathcal{E}$, and a switch S (see Figure 3.5). If the switch S is in position 1, the capacitor is being charged by the voltage source. Turning the switch to position 2 leads to the discharging of the capacitor through the resistor. The processes of capacitor charging and discharging are nonstationary electrical processes characterized by a continuous variation of the current in the circuit with time. Let us define the characteristic time τ of capacitor charging and discharging. We start with the process of capacitor charging in the circuit shown in Figure 3.5. Kirchhoff's loop rule, which states that the algebraic sum of voltage

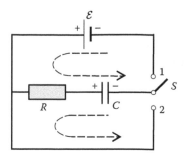

FIGURE 3.5   Circuit for the illustration of charging and discharging of a capacitor.

drops around any closed loop is equal to the algebraic sum of emf in a loop, applied to the upper branch of the loop (Figure 3.5) yields

$$RI(t) + V_C(t) = \mathcal{E}, \tag{3.19}$$

where
  $I(t)$ is the instantaneous current in the circuit
  $V_C(t)$ is the instantaneous voltage across the capacitor

Expressing $I(t)$ and $V_C(t)$ in terms of the instantaneous charge $q(t)$ on the capacitor plates, we have

$$I(t) = \frac{dq(t)}{dt}, \quad V_C = \frac{q(t)}{C}. \tag{3.20}$$

If we substitute Equation 3.20 into Equation 3.19, we have the following differential equation for the capacitor charge:

$$\frac{dq}{dt} + \frac{q}{RC} = \frac{\mathcal{E}}{R}. \tag{3.21}$$

This must satisfy the initial condition at $t = 0$ as we put the switch S in position 1:

$$q(0) = 0. \tag{3.22}$$

The solution of Equation 3.21 that satisfies the initial condition (3.22) is

$$q(t) = C\mathcal{E}\left(1 - e^{-t/\tau}\right), \tag{3.23}$$

Here, $\tau = RC$ is a characteristic time of the circuit known as the "time constant" of the RC circuit.
  To find $I(t)$ and $V_C(t)$, we use Equation 3.23 and relations (3.20):

$$I(t) = \frac{dq(t)}{dt} = \frac{\mathcal{E}}{R} e^{-t/\tau} = I_0 e^{-t/\tau}, \quad V_C(t) = \frac{q(t)}{C} = \mathcal{E}\left(1 - e^{-t/\tau}\right), \tag{3.24}$$

where I_0 is the current in the circuit at $t = 0$. It follows from this equation that the current is maximum at $t = 0$ and it vanishes asymptotically with time; the voltage and the charge are zero at $t = 0$ and they both reach asymptotically their corresponding maximum values $V_C = \mathcal{E}$ and $q = q_0 = C\mathcal{E}$ (Figure 3.6a).

The capacitor discharging takes place when a switch S is turned to position 2 after the capacitor has been charging for a long time. In this case, Equation 3.20 remains the same, but in the right-hand part of Equation 3.21, $\mathcal{E}$ is equal to zero, so we need to solve equation

$$\frac{dq}{dt} + \frac{q}{\tau} = 0. \tag{3.25}$$

The initial condition for this is

$$q_0 = q(t = 0) = C\mathcal{E}. \tag{3.26}$$

The solution of Equation 3.25 that satisfies the initial condition is

$$q(t) = C\mathcal{E}e^{-t/\tau} = q_0 e^{-t/\tau}. \tag{3.27}$$

Taking into account Equation 3.20, one can see that the discharging current maintains the same time dependence as the charging current:

$$I(t) = \frac{dq(t)}{dt} = -\frac{\mathcal{E}}{R}e^{-t/\tau} = -I_0 e^{-t/\tau}. \tag{3.28}$$

The results obtained earlier show that the charging and discharging processes of the capacitor C through the resistor R do not happen instantaneously but instead have specific time dependence. The timescale of establishing equilibrium is given by the parameter $\tau = RC$, referred to as the *time constant* of the RC circuit. The constant τ is the time required for the capacitor voltage and charge to decrease by a factor e after the emf is switched off. Here, e is the basis of natural logarithms.

Note that the current in Equation 3.27 has a sign, which is opposite to that of the current in Equation 3.24 (this is also shown in Figure 3.6b). This indicates that the discharge current is flowing in a direction opposite to the capacitor charging current.

2. Similar processes are observed for a circuit that contains an inductor and a resistor (RL circuit). Let us substitute the capacitor in a circuit in Figure 3.5 with a coil of inductance L (Figure 3.7).

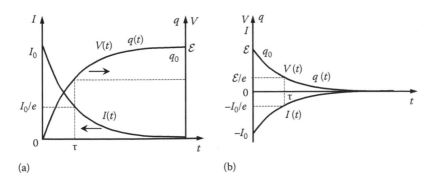

FIGURE 3.6 Dependence of current, charge, and voltage on time for the capacitor charging (a) and discharging (b).

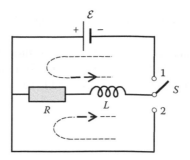

FIGURE 3.7   Circuit for the illustration of transition processes in a circuit with an inductor.

If a switch S is closed, the upper branch of the circuit contains a voltage source that generates the current in the circuit. According to Kirchhoff's loop rule, this process is described by the equation

$$RI(t) = \mathcal{E} + \mathcal{E}_L(t), \tag{3.29}$$

where the second term in the right-hand part of the equation is the self-induced emf across the coil: $\mathcal{E}_L(t) = -L(dI/dt)$. Hence, Equation 3.28 takes the form

$$\frac{dI}{dt} + \frac{R}{L}I(t) = \frac{\mathcal{E}}{L}. \tag{3.30}$$

This equation is similar to Equation 3.21 describing the capacitor charging, except that the current $I(t)$ is the variable here.

The initial condition for Equation 3.29 is like Equation 3.22:

$$I(t = 0) = 0, \tag{3.31}$$

And the solution (compare it with Equation 3.23) is

$$I(t) = \frac{\mathcal{E}}{R}\left(1 - e^{-t/\tau}\right) = I_m\left(1 - e^{-t/\tau}\right), \quad \tau = \frac{L}{R}, \tag{3.32}$$

where the time constant is $\tau = L/R$ and the maximum value of the current is $I_m = \mathcal{E}/R$, which corresponds to the steady-state current I_m that will flow in the circuit after $t \gg \tau$. The dependence of the current on time shown by curve 1 in Figure 3.8 resembles the dependence of charge on time shown in Figure 3.6 for the RC circuit.

Similarly, one can obtain an expression describing the circuit current after the switch is turned to position 2, that is, we need to solve Equation 3.30 with $\mathcal{E} = 0$ (compare with Equation 3.25). Here, the initial current in a circuit is the current established before switching, that is, $I(0) = I_m$ and the solution is

$$I(t) = I_m e^{-t/\tau} = \frac{\mathcal{E}}{R}e^{-t/\tau}. \tag{3.33}$$

In the case of the inductor, switch on and switch off currents are in the same direction; compare Equation 3.32 with Equation 3.33. The dependence of current on time is shown by curve 2 in Figure 3.8.

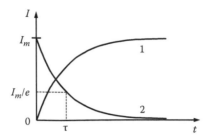

FIGURE 3.8 Dependence of current on time as the switch S on Figure 3.7 is switched in position 1 (curve 1) and in position 2 (curve 2).

The two examples considered earlier demonstrate that the behavior of a circuit with a capacitor (RC circuit) and an inductor (RL circuit) is determined by Equations 3.23 and 3.32. If the emf $\mathcal{E}$ varies with time with an angular frequency ω such that $\omega\tau \ll 1$, the current follows the emf $\mathcal{E}$ (i.e., $I(t) = \mathcal{E}(t)/R$) and such condition is referred to as quasi stationary. In contrast, if $\omega\tau \geq 1$, the current does not follow the variation of the emf. A sinusoidal current with angular frequency ω is known as **alternating current** (AC), which we consider separately.

Exercise 3.4

At what time after the switch is closed (position 1) will the current in an RL circuit be half of the steady-state current?

Solution. According to Equation 3.32, we can write the following time dependence for a current in the circuit:

$$I(t) = I_m\left(1 - e^{-t/\tau}\right).$$

We require that $I(t) = I_m/2$; thus, we can write

$$\frac{I(t)}{I_m} = 1 - e^{-t/\tau} = \frac{1}{2}, \quad e^{-t/\tau} = \frac{1}{2}.$$

Taking the logarithm of this relation yields

$$\frac{t}{\tau} = \ln 2 \quad \text{or} \quad t = \tau \cdot \ln 2 = \frac{L}{R}\ln 2 = 0.69\frac{L}{R}.$$

3.5 DISPLACEMENT CURRENT

In the previous section, we demonstrated that charging a capacitor through a resistor results in a current that depends on time, Equation 3.24. However, the electrons cannot travel from one plate to another, since the capacitor itself is a break in the circuit (there is an insulator or vacuum in the space between the plates). Experiments show that when a current flows in the circuit, a magnetic field is generated between the capacitor plates (Figure 3.9).

To determine the quantitative relations between the time-varying electric field and the magnetic field induced by it, James Clerk Maxwell introduced the concept of **displacement current** and

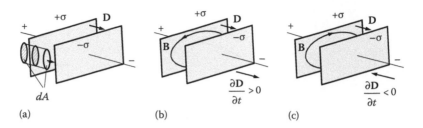

FIGURE 3.9   Calculation of a displacement vector (a) and displacement current (b), (c) in a capacitor.

described correctly the connection between electric and magnetic fields under time-varying conditions. In a circuit with a capacitor, the conduction currents that flow in and out of the connecting wires are replaced by the displacement currents in the space between the plates. A time-varying electric field produces the same magnetic field as generated in the presence of a conduction current between the plates.

As it was discussed in Chapter 1 (see Equation 1.33), the electric field of a capacitor is concentrated between the charged plates of the capacitor and is equal to zero outside of the capacitor. Using Gauss's law of Equation 1.80, we choose the closed surface in the form of the cylinder with the base areas dA parallel to the plate of the capacitor as it is shown in Figure 3.9a. As the displacement vector has only one component D_n perpendicular to the surface of the capacitor and to dA, Equation 1.80 determines that $D_n = \sigma_s$, where σ_s is the surface charge density on the plate of the capacitor. The absolute value of the total charge on each plate $q = \sigma_s A$, where A is the plate area. Hence, the displacement current is

$$I_{ds} = \frac{\partial q}{\partial t} = \frac{\partial}{\partial t}(A\sigma_s) = A\frac{\partial \sigma_s}{\partial t} = A\frac{\partial D}{\partial t},$$
(3.34)

that is, the displacement current is proportional to the rate at which the electric displacement vector changes. Introducing the vector of the displacement current density $j_{ds} = I_{ds}/A$, we arrive at the following vector relation for the displacement current density:

$$\mathbf{j}_{ds} = \frac{\partial \mathbf{D}}{\partial t}.$$
(3.35)

The magnetic field **B** produced by the displacement current is connected to the direction of vector $\partial \mathbf{D}/\partial t$ by the right-hand rule as shown in Figure 3.9b and c.

Exercise 3.5

An alternating current voltage $V(t) = V_0 \cos(\omega t)$ is applied across the plates of a parallel plate capacitor. Find the time dependence of the displacement current density in the space between the capacitor plates if the distance between the plates is equal to d. The space between the plates is completely filled with a dielectric of permittivity $\varepsilon = \varepsilon_0 \kappa$.

Solution. As we indicated earlier, the displacement current I_{ds} is equal to the conduction current I, so the current density $j_{ds} = I/A$, where A is the capacitor area. The current I of a capacitor with a reactance X_C is given by Ohm's law for alternating current:

$$I(t) = \frac{V(t)}{X_C}.$$

The capacitive resistance (reactive) X_C is determined by the equation

$$X_C = \frac{1}{i\omega C} = -\frac{i}{\omega C},$$

where the imaginary unit can be presented as

$$i = \cos\left(\frac{\pi}{2}\right) + i\sin\left(\frac{\pi}{2}\right) = \exp\left(\frac{i\pi}{2}\right).$$

As $V(t) = V_0 \cos(\omega t) = \text{Re}[V_0\exp(i\omega t)]$, we can write down for the current

$$I(t) = \text{Re}\left[\frac{V(t)}{X_C}\right] = V_0\omega C\,\text{Re}\left[\exp(i\omega t)\exp\left(\frac{i\pi}{2}\right)\right]$$

$$= V_0\omega C\,\text{Re}\left[\exp i\left(\omega t + \frac{\pi}{2}\right)\right] = V_0\omega C\cos\left(\omega t + \frac{\pi}{2}\right) = -V_0\omega C\sin(\omega t).$$

As we see from the previous equation, the current and voltage are not in phase but rather shifted by $\Delta\varphi = \pi/2$. Hence,

$$j_{ds}(t) = \frac{I(t)}{A} = -\left(\frac{V_0\omega C}{A}\right)\sin(\omega t).$$

The capacitance is given by

$$C = \frac{\kappa\varepsilon_0 A}{d},$$

Since, in our case, the voltage varies according to $V(t) = V_0 \cos(\omega t)$, the time dependence of the displacement current is expressed as

$$j_{ds}(t) = -\frac{\omega\kappa\varepsilon_0}{d}\cdot V_0\sin(\omega t).$$

3.6 MAXWELL'S EQUATIONS

Maxwell's equations are the equations of classical electrodynamics, and they describe all electromagnetic phenomena. In particular, Maxwell's equations describe mathematically the experimental laws of Coulomb, Ampere, and Faraday. As mentioned earlier, a time-varying magnetic field induces an electric filed and a time-varying electric field induces a magnetic field. Being simultaneously generated, these fields can exist independently of the charges or currents from which they originate. The transformation of one field into another results in the propagation of these fields in space. The displacement current introduced by Maxwell allows creating a unified theory of the electric and magnetic phenomena described by the set of equations known as Maxwell's equations. These equations, as they were formulated, not only provided an explanation of well-known electric and magnetic phenomena but also predicted novel effects, whose existence was confirmed experimentally later.

Maxwell's equations include the following four equations that can be represented either in integral or in differential form:

$$
\begin{aligned}
&1.\ \oint_{S} \mathbf{D} \cdot d\mathbf{A} = \int_{V} \rho \, d(Vol), && \nabla \cdot \mathbf{D} = \rho, \\[2mm]
&2.\ \oint_{L} \mathbf{H} \cdot d\mathbf{l} = \int_{A} \mathbf{j} \cdot d\mathbf{A} + \frac{d}{dt} \int_{A} \mathbf{D} \cdot d\mathbf{A}, && \nabla \times \mathbf{H} = \mathbf{j} + \frac{\partial \mathbf{D}}{\partial t}, \\[2mm]
&3.\ \oint_{L} \mathbf{E} \cdot d\mathbf{l} = -\frac{d}{dt} \int_{A} \mathbf{B} \cdot d\mathbf{A}, && \nabla \times \mathbf{E} = -\frac{\partial \mathbf{B}}{\partial t}, \\[2mm]
&4.\ \oint_{S} \mathbf{B} \cdot d\mathbf{A} = 0, && \nabla \cdot \mathbf{B} = 0.
\end{aligned}
\tag{3.36}
$$

Before discussing Maxwell's equations, we would like to indicate that the fields included in Maxwell's equations are not independent. There is a relationship between them defined by the three constitutive equations that for isotropic media take the form

$$
\mathbf{D} = \kappa \varepsilon_0 \mathbf{E}, \quad \mathbf{B} = \kappa_m \mu_0 \mathbf{H}, \quad \mathbf{j} = \sigma \mathbf{E},
\tag{3.37}
$$

where σ is the conductivity of the medium. Generally, the material parameters κ, κ_m, and σ depend on both the properties of the substance and external parameters such as temperature, frequency, and electric and magnetic fields. In most cases, the material parameters can be considered independent of the magnitude of the external fields. In this case, the constitutive equations (3.37) are linear. Media with linear constitutive equations are called linear media. Nonlinearity in most media occurs only in the presence of very strong fields and in due course, we will define more precisely what we mean by "strong fields."

Maxwell's first equation expresses Gauss's law for the electric field: the electric flux through a closed surface is equal to the net charge inside that surface.

Maxwell's second equation is the generalized Ampere's law. This equation shows that the magnetic field can be generated either by moving charges (electric currents) or by time-varying electric fields.

Maxwell's third equation shows that not only the electric charges but also time-varying magnetic fields can be the source of an electric field. So, the total electric field $\mathbf{E}$ has component $\mathbf{E}_Q$ that can be produced by a system of charges and component $\mathbf{E}_B$ that can be produced by the time-varying magnetic field: $\mathbf{E} = \mathbf{E}_Q + \mathbf{E}_B$.

Maxwell's fourth equation is Gauss's law for the magnetic field. It states that there are no magnetic monopoles in nature or equivalently that the magnetic field lines form closed loops.

Maxwell's equations are not symmetric with respect to the electric and magnetic fields. This is due to the fact that there are electric but no magnetic charges in nature. The differential forms of Maxwell's equations are easily obtained from the equations in integral form. For this, we have to employ Gauss's and Stokes' theorems of vector analysis, respectively:

$$\int_A \mathbf{G} \cdot d\mathbf{A} = \int_{Vol} \nabla \cdot \mathbf{G} d(Vol), \quad \int_L \mathbf{G} \cdot d\mathbf{l} = \int_A \nabla \times \mathbf{G} \cdot d\mathbf{A}. \tag{3.38}$$

where $\mathbf{G}$ is any vector.

Maxwell's equations in differential form describe the electromagnetic field at each point in space. If the charges and currents are continuously distributed in space, the two forms of Maxwell's equations—integral and differential—are equivalent. However, if there is a surface, on which the properties of the medium or fields change abruptly, the integral form of the equations should be used as these equations are more general.

Equations 3.37 is valid for isotropic media, whose properties are identical in all directions, and for which the parameters κ, κ_m, and σ are scalar quantities. In these media, the direction of the vectors $\mathbf{D}$, $\mathbf{B}$, $\mathbf{j}$ coincides with the direction of vectors $\mathbf{E}$, $\mathbf{H}$, $\mathbf{E}$, respectively, while this is not the case in anisotropic media. A medium is homogeneous if the parameters, κ, κ_m, and σ, are the same all over the considered volume. If they are functions of the coordinates, the medium is inhomogeneous.

In what follows, we discuss typical values of material parameters κ, κ_m, and σ in the limit of low frequencies. The dielectric constant for all dielectrics is greater than unity. For air, we have $\kappa \cong 1$, that is, κ is very close to the value of the dielectric permittivity of vacuum. For water, glass, and ceramics, they are $\kappa \approx 81$, $\kappa \approx 4$, and $\kappa \approx 6.5$, respectively. For organic dielectrics, they are $\kappa = 2 - 3$.

The magnetic permeability of free space is $\kappa_m = 1 + \chi = 1$, that is, the magnetic susceptibility of free space is zero: $\chi = 0$. For diamagnetic materials, χ is negative and its absolute value is small: $\chi \sim -10^{-5}$ (e.g., for copper, we have $\kappa_m = 0.99999044$). For paramagnetic materials, χ is positive and its value is small: $\chi \sim 10^{-3} - 10^{-5}$ (e.g., for aluminum, we have $\kappa_m = 1.000022$). For ferromagnets or ferrites, $\chi \gg 1$ and $\kappa_m \gg 1$; moreover, κ_m depends strongly on frequency and on magnetic field intensity.

Metals exhibit high conductivity. The best metal conductors in order of decreasing conductivity with the following values of σ: $(6.17, 5.8, 4.1, \text{ and } 3.72) \times 10^7$ S/m are silver, gold, copper, and aluminum.

For static fields, Maxwell's equations take the form

$$1. \oint_S \mathbf{D} \cdot d\mathbf{A} = \int_V \rho dV, \quad \nabla \cdot \mathbf{D} = \rho,$$

$$2. \oint_L \mathbf{H} \cdot d\mathbf{l} = \int_A \mathbf{j} \cdot d\mathbf{A}, \quad \nabla \times \mathbf{H} = \mathbf{j},$$

$$3. \oint_L \mathbf{E} \cdot d\mathbf{l} = 0, \quad \nabla \times \mathbf{E} = 0,$$

$$4. \oint_A \mathbf{B} \cdot d\mathbf{A} = 0, \quad \nabla \cdot \mathbf{B} = 0. \tag{3.39}$$

This system of equations (in differential form) is used to find the electric and magnetic fields and/or current and/or charge distribution in different problems of electromagnetism that are time independent.

The second of Maxwell's equations in Equations 3.36 is consistent with the charge conservation that relates the charge density $\rho(\mathbf{r})$ with the current density $\mathbf{j}(\mathbf{r})$ at each point of space. To verify this, let us multiply the right- and left-hand parts of the equation

$$\nabla \times \mathbf{H} = \mathbf{j} + \frac{\partial \mathbf{D}}{\partial t}$$

by the operator ∇ and then change the order of time and space derivatives in the right-hand side:

$$\nabla \cdot (\nabla \times \mathbf{H}) = \frac{\partial}{\partial t} \nabla \cdot \mathbf{D} + \nabla \cdot \mathbf{j}. \tag{3.40}$$

In this equation, the left-hand side is equal to zero due to a vector identity:

$$\nabla \cdot (\nabla \times \mathbf{H}) = 0. \tag{3.41}$$

Replacing $\nabla \cdot \mathbf{D}$ by ρ (see first Equation 3.36 in the right-hand part of Equation 3.40), we get

$$\frac{\partial \rho}{\partial t} + \nabla \cdot \mathbf{j} = 0. \tag{3.42}$$

This is referred to as the continuity equation. If we choose a point $\mathbf{r}$ surrounded by a closed surface A and integrate Equation 3.42 over the volume V enclosed by the surface, we get

$$\frac{\partial}{\partial t} \int_{Vol} \rho(\mathbf{r}) \, d\mathbf{r} = -\int_{Vol} \nabla \cdot \mathbf{j} \, d\mathbf{r}. \tag{3.43}$$

The left-hand part is equal to the rate of change of the total charge Q enclosed by the surface A. Using Gauss's theorem, the volume integral in the right-hand part can be converted into an integral over surface A, which encloses that volume, whereby we obtain

$$\frac{dQ}{dt} = -\oint_A \mathbf{j} \cdot d\mathbf{A} = -I_A, \tag{3.44}$$

where I_A is the total current through the surface. This equation states that the rate of change of the charge inside a closed surface is equal to the total current flowing through the surface.

In vacuum, in the absence of charges and currents, the system of equations (3.36) takes the form

$$
\begin{aligned}
&\text{1.} \quad \oint_S \mathbf{D} \cdot d\mathbf{A} = 0, && \nabla \cdot \mathbf{D} = 0, \\[2mm]
&\text{2.} \quad \oint_L \mathbf{H} \cdot d\mathbf{l} = \frac{d}{dt} \int_A \mathbf{D} \cdot d\mathbf{A}, && \nabla \times \mathbf{H} = \frac{\partial \mathbf{D}}{\partial t}, \\[2mm]
&\text{3.} \quad \oint_L \mathbf{E} \cdot d\mathbf{l} = -\frac{d}{dt} \int_A \mathbf{B} \cdot d\mathbf{A}, && \nabla \times \mathbf{E} = -\frac{\partial \mathbf{B}}{\partial t}, \\[2mm]
&\text{4.} \quad \oint_S \mathbf{B} \cdot d\mathbf{A} = 0, && \nabla \cdot \mathbf{B} = 0.
\end{aligned}
\tag{3.45}
$$

As will be shown in the following discussion, the differential form of Maxwell's equations leads to a wave equation that has plane monochromatic waves as a particular solution. The vectors of the electric and magnetic fields are perpendicular to each other and also to the direction of the wave propagation. The oscillations of both fields are in phase. The wave propagates in vacuum with a speed of light $c = 1/\sqrt{\varepsilon_0 \mu_0}$ or $c \cong 2.998 \times 10^8$ m/s.

Exercise 3.6

Write down the differential system of equations (3.36) using Cartesian coordinate system.

Solution. The system of equations (3.36) has two scalar and two vector equations:

$$\nabla \cdot \mathbf{D} = \rho,$$

$$\nabla \cdot \mathbf{B} = 0,$$

$$\nabla \times \mathbf{H} = \mathbf{j} + \frac{\partial \mathbf{D}}{\partial t},$$

$$\nabla \times \mathbf{E} = -\frac{\partial \mathbf{B}}{\partial t}.$$

The first two equations take the form

$$\frac{\partial D_x}{\partial x} + \frac{\partial D_y}{\partial y} + \frac{\partial D_z}{\partial z} = \rho,$$

$$\frac{\partial B_x}{\partial x} + \frac{\partial B_y}{\partial y} + \frac{\partial B_z}{\partial z} = 0.$$

The remaining two equations are vector equations, and therefore, each of them can be presented by three equations that are projections of the vector equations on each of the Cartesian coordinate axis:

$$\frac{\partial H_z}{\partial y} - \frac{\partial H_y}{\partial z} = j_x + \frac{\partial D_x}{\partial t}, \quad \frac{\partial E_z}{\partial y} - \frac{\partial E_y}{\partial z} = -\frac{\partial B_x}{\partial t},$$

$$\frac{\partial H_x}{\partial z} - \frac{\partial H_z}{\partial x} = j_y + \frac{\partial D_y}{\partial t}, \quad \frac{\partial E_x}{\partial z} - \frac{\partial E_z}{\partial x} = -\frac{\partial B_y}{\partial t},$$

$$\frac{\partial H_y}{\partial x} - \frac{\partial H_x}{\partial y} = j_z + \frac{\partial D_z}{\partial t}, \quad \frac{\partial E_y}{\partial x} - \frac{\partial E_x}{\partial y} = -\frac{\partial B_z}{\partial t}.$$

In these equations, $\rho = \rho(t; x, y, z)$ and $j_x = j_x(t; x, y, z)$, $j_y = j_y(t; x, y, z)$ and $j_z = j_z(t; x, y, z)$ are components of vector $\mathbf{j}(t; x, y, z)$.

PROBLEMS

3.1   A coil containing $N = 1000$ turns rotates with constant frequency $f = \omega/2\pi = 10.0\ \text{s}^{-1}$ in a uniform magnetic field $B = 0.10$ T. The area of each turn, A, is equal to 150 cm². Find the instantaneous value of the induced emf, $\mathcal{E}_i$, corresponding to a rotation angle of the coil equal to 30°. (*Answer:* $\mathcal{E}_i = 47.1$ V.)

3.2   When the rate of current change with time in a solenoid is equal to 50 A/s, a self-induced emf is generated across its ends: $\mathcal{E}_i = 0.08$ V. Find the inductance L of the solenoid. (*Answer:* $L = 1.60$ mH.)

3.3   A horizontal conducting rod with a mass m and length l can slide under the action of gravity without friction and without breaking the electrical contact with two vertical conducting rails. A source of emf $\mathcal{E}$ is connected as shown in Figure 3.10. A uniform magnetic field is applied perpendicular to the plane of the rails. Find the ratio of resistors R_1 and R_2 of resistor R_x when the horizontal rod moves with uniform velocity v in opposite directions (up and down, respectively). Neglect the self-inductance of the circuit, resistance of the rod and rails, and the internal resistance of the emf $\mathcal{E}$. $\left(Answer:\ \dfrac{R_1}{R_2} = \dfrac{\mathcal{E} - Blv}{\mathcal{E} + Blv}. \right)$

3.4   The winding of a solenoid consists of a single layer of tightly wound turns of copper wire with a diameter $d = 0.20$ mm. The diameter of the solenoid is $d_1 = 5.00$ cm. The current that flows through a solenoid at $t = 0$ is $I_0 = 1.00$ A. Find the total charge q that flows through the coil of the solenoid if at $t = 0$, the coil is short-circuited. *Hint:* The length of the solenoid, the length of the copper wire, and the number of turns in the solenoid will not enter into the final answer taking into account that the length of the solenoid is equal to a product of the wire diameter and the total number of turns in the solenoid. (*Answer:* $q \approx 145$ μC.)

3.5   A horizontal metal rod of length $l = 50.0$ cm rotates about a vertical axis that passes through one of its ends with a frequency $f = 2.00$ Hz, as shown in Figure 3.11. The vertical component of the magnetic field intensity of the Earth is $H = 40.0$ A/m. Find the potential difference between the ends of the rod. (*Answer:* $\Delta\varphi = 79.0$ mV.)

3.6   Find the magnitude of the displacement current between the square plates of a parallel plate capacitor filled with air (plate side $a = 5.00$ cm) if the electric field between the plates varies with time at the rate $dE/dt = 4.52$ MV/m · s. (*Answer:* Displacement current $I_{ds} = 0.10$ μA.)

3.7   In an ideal oscillatory LC circuit with coil inductance $L = 0.20$ H, the oscillation amplitude of the current is $I_m = 40.0$ mA. Find the energy of the electric field of the capacitor at the moment when the instantaneous value of the current is half the current amplitude. (*Answer:* $W_C = 1.20 \times 10^{-4}$ J.)

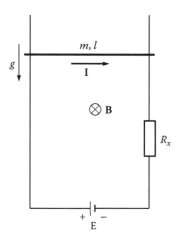

FIGURE 3.10   Calculation of the resistors $R_x = R_1$ and $R_x = R_2$ for the bar moving up and down, respectively.

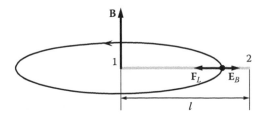

FIGURE 3.11 Calculation of the emf across a rod of length l that rotates about a vertical axis passing through one of its ends in the magnetic field of the Earth.

3.8 A sinusoidal voltage $V(t) = V_m \sin(\omega t)$ with angular frequency ω and amplitude V_m is applied across a circuit that consists of a resistor and an inductor in series as it is shown in Figure 3.12. Find the time dependence of the current in the circuit if the switch is closed at $t = 0$.
$$\left(Answer: I(t) = \frac{V_m}{Z}\left[\sin \varphi \cdot \exp(-t/\tau_L) + \sin(\omega t - \varphi) \right]. \right)$$

3.9 The switch in the *LRC* oscillatory circuit shown in Figure 3.13 is closed at $t = 0$. Before $t = 0$, a voltage V_0 is applied across the capacitor. Find the conditions for the oscillatory and aperiodic modes and the time dependence of the current in the circuit for the oscillatory mode.

3.10 A coil of area $A = 10.0$ cm^2 connected across a capacitor $C = 10.0$ μF. The assembly is placed in a uniform magnetic field. The magnetic field lines are perpendicular to the plane of the coil. The magnetic field varies with t at a rate $\Delta B/\Delta t = 5.00 \times 10^{-3}$ T/s. Find the charge of the capacitor. (*Answer*: $q_C = 5.00 \times 10^{-11}$ C.)

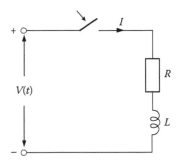

FIGURE 3.12 Circuit with a resistor and an inductor.

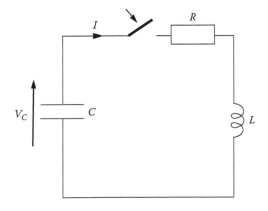

FIGURE 3.13 Circuit with a resistor, a capacitor, and an inductor.

3.11 The wire of a coil of area A is cut at some point and the plates of a capacitor of capacitance C are connected to the cut ends. The coil is placed in a time-varying magnetic field whose field lines are perpendicular to the coil plane. The magnetic field is increasing uniformly with time according to $B(t) = \beta t + \gamma t^2$, where β and γ are constants. Find the time dependence of the charge on the capacitor plates, assuming that the circuit is ideal with the resistance equal to zero. (*Answer:* $q_C(t) = C(\beta + 2\gamma t)A$.)

3.12 Find the ratio of the displacement current density to the conduction current density in seawater (its material parameters are $\kappa = 80.0$, $\kappa_m = 1.00$, $\sigma = 4.00$ S/m) for electromagnetic waves with frequencies $f = 10^3$, 10^7 Hz. (*Part of the answer:* $|j_{ds}/j_{cd}| = 1.26 \times 10^9$. Here, j_{ds} and j_{cd} are displacement and conduction current densities, respectively.)

3.13 Consider a material with magnetic permeability κ_m placed in a magnetic field of a solenoid. The magnetic field B with magnetic energy density u_m is given. Determine (a) the magnetic permeability κ_m and (b) magnetic field intensity H. (*Answer:* $\kappa_m = B^2/2\mu_0 u_m$, $H = 2u_m/B$.)

Section II

Electromagnetic Waves in Homogeneous, Heterogeneous, and Anisotropic Media

4 Electromagnetic Waves in Homogeneous Media without Absorption

Electromagnetic wave processes are extremely widespread in nature. One of the main features that distinguish electromagnetic waves from mechanical waves is that they do not need any material for their propagation because they are able to propagate in vacuum (i.e., in a space free of material). However, electromagnetic waves can also propagate in space filled with matter, altering their behavior to a greater or lesser extent depending on the material. Thus, the characteristics of electromagnetic waves are dependent on the properties of the material in which they are propagating. Despite the significant differences between the electromagnetic waves and mechanical waves, the former in many respects behave like the latter. All types of waves are described quantitatively by the same equations. Electromagnetic fields and waves play an important role in our lives. For example, light is an electromagnetic wave. Radio, television, and various forms of communication are based on the important properties of electromagnetic waves that allow them to transmit information. In this part of the book, we consider the basic equations describing electromagnetic waves in a homogeneous, isotropic, nonabsorbing medium, as well as the basic properties of these waves.

4.1 ELECTROMAGNETIC WAVE SPECTRUM

All electromagnetic waves are time-varying electric and magnetic fields propagating in space and have common properties independently of their wavelength (frequency). However, the methods and devices for electromagnetic wave generation and detection, their range of application, and the effects of the waves on biological systems are strongly wavelength dependent. Therefore, the entire electromagnetic spectrum is divided into several broad ranges. Within each range, the waves share similar properties, methods of generation, applications, detection, and their effects on human beings. We note that there are no sharp boundaries between these ranges.

Let us consider the entire electromagnetic wave spectrum. In order of increasing frequency, this spectrum can be divided into the following seven ranges (Figure 4.1):

1. Radio waves
2. Terahertz (THz) waves
3. Infrared radiation (IR)
4. Visible light
5. Ultraviolet (UV) radiation
6. X-rays
7. Gamma rays

Since the wavelengths within the radio frequency range are too large compared to the size of atoms, propagation of radio waves and their interaction with matter can be described using the laws of classical electrodynamics, that is, Maxwell's equations, without taking into consideration the atomic structure of the matter with which the waves interact.

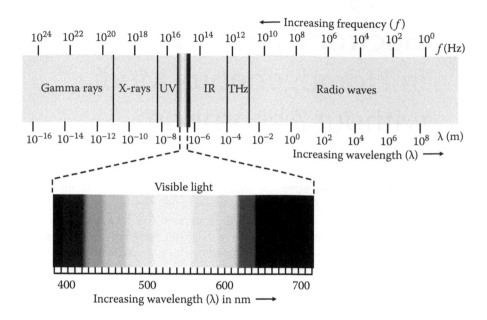

FIGURE 4.1 Electromagnetic spectrum.

THz waves are identified as a unique range that is between the radio frequency range and infrared range. The laws of classical and quantum optics are applicable for infrared, visible, and UV radiation. X-rays and gamma rays are described by the laws and equations of quantum optics only. In the following text, we give more details about the electromagnetic wave spectrum.

During the earlier discoveries, the electromagnetic waves of wavelength $\lambda > 0.1$ mm $= 10^{-4}$ m and frequency $f = c/\lambda < 3 \times 10^{12}$ Hz were considered as **radio waves**. Later after the THz waves start to be treated as a unique spectral range, the range of radio frequencies was reduced to below 3×10^{11} Hz and to wavelengths $\lambda > 1$ mm $= 10^{-3}$ m. The radio wave range is commonly subdivided into several ranges that have different practical applications (note that the terms "superlong waves" and "long waves" are not defined precisely and their definition varies across the literature):

- *Superlong waves* (wavelengths above 10 km, frequencies below 3×10^4 Hz)
- *Long waves* (wavelengths within the range of 1–10 km, frequencies from 3×10^4 to 3×10^5 Hz)
- *Medium waves* (wavelengths within the range of 100 m–1 km, frequencies from 3×10^5 to 3×10^6 Hz)
- *Short waves* (wavelengths range of 10–100 m, frequencies from 3×10^6 to 3×10^7 Hz)
- *Ultrashort waves* (wavelengths range of $10–10^{-3}$ m, frequencies from 3×10^7 to 3×10^{11} Hz).

Ultrashort waves are subdivided into the meter, centimeter, and millimeter waves. The submillimeter waves, as it was already mentioned earlier, are recently considered as "THz waves."

Waves with the wavelength under 1 m (frequency higher than 300 MHz) are also referred to as microwaves.

As it was already mentioned earlier, the THz part of spectrum corresponds to wavelength 1 mm $> \lambda > 30$ µm and frequencies $3 \times 10^{11}–10^{13}$ Hz. This frequency range got a special attention as it was first identified as a window of frequencies that is difficult to generate using either the radio frequency methods (as the frequency is too high) or optical methods (as the frequency is too low). Now the THz gap is practically closed as advanced optical methods allow to generate radiation with frequencies down to about 1 THz = 10^{12} Hz and advanced electrical methods are extended up to about 1 THz.

Infrared, visible, and UV radiation, in a broad sense, compose the optical spectrum of the electromagnetic waves. The joint classification of these three ranges is caused by the similarity in the methods and devices used to study them and by the similarity of their practical applications.

IR corresponds to wavelengths adjacent to the edge of the THz wave range, that is, 3×10^{-5} m $> \lambda > 0.76 \times 10^{-6}$ m and frequencies from 1×10^{13} Hz to 4×10^{14} Hz. We note that the infrared range very often is subdivided into near infrared, infrared, and far infrared, but those details are behind the scope of this course.

Visible light spectrum is restricted to wavelengths within 760 nm $> \lambda > 400$ nm and frequencies from 4×10^{14} to 7.5×10^{14} Hz. The wavelengths $\lambda = 760$ nm and $\lambda = 400$ nm are commonly referred to as the red and violet edges of the visible spectrum, respectively.

The wavelengths adjacent to the edge of short visible light waves, that is, 400 nm $> \lambda > 30$ nm (frequencies from 7.5×10^{14} to 10^{16} Hz), belong to the range of *ultraviolet* (UV) *radiation*.

X-rays correspond to electromagnetic waves of wavelength in the range between 30 nm and 1×10^{-3} nm and frequencies between 1×10^{16} Hz and 3×10^{22} Hz. X-ray radiation is generated during the deceleration of charged particles (electrons, protons), as well as during quantum transitions between inner electron atomic shells.

Gamma radiation corresponds to wavelengths $\lambda < 10^{-5}$ nm and frequencies $f > 3 \times 10^{22}$ Hz. Gamma rays are generated during nuclear processes and nuclear reactions.

Exercise 4.1

An ideal oscillator circuit consists of a coil and a capacitor. The coil's inductance L is equal to 2.00 nH and the capacitance can be varied in the range from $C_{min} = 2.00$ nF to $C_{max} = 50.0$ nF. What is the wave band this oscillator circuit can be tuned to?

Solution. The wavelength of the electromagnetic wave is connected with the propagation speed and frequency through the relationship $\lambda = 2\pi c/\omega$ where the angular frequency is given by $\omega = 2\pi/T$. The period of oscillations of the ideal oscillator circuit is given by $T = 2\pi\sqrt{LC}$. Taking into account these relationships, we get for the wavelength of the electromagnetic radiation the following expression: $\lambda = 2\pi c\sqrt{LC}$. By substituting the given data, we get $\lambda_{min} = 3.77$ m. When the capacitance of capacitor is increased 25 times, the wavelength increases 5 times. Therefore, $\lambda_{max} = 18.8$ m.

4.2 WAVE EQUATION

Maxwell's equations allow us to describe precisely the generation and propagation of electromagnetic waves. *Waves* are oscillations propagating in space and time, which can transport energy but not matter through space. Electromagnetic waves are composed of electric and magnetic fields rapidly varying with time and propagating in space. These waves can propagate through matter as well as through vacuum. The propagation velocity, power per unit area, and polarization are some of the parameters that characterize an electromagnetic wave; these properties can be determined experimentally.

We assume that there are no charges and currents in the region of wave propagation, that is, $\rho = 0$, $j = 0$. Maxwell's equations for a region with no charges and currents are given by Equation 3.45. Substituting the displacement vector $\mathbf{D} = \kappa\varepsilon_0\mathbf{E}$ and the magnetic field $\mathbf{B} = \kappa_m\mu_0\mathbf{H}$ in these equations for the electric field $\mathbf{E}$ and the magnetic intensity vector $\mathbf{H}$, Maxwell's equations in a homogeneous media take the form

$$
\begin{array}{ll}
1.\ \oint_S \mathbf{E} \cdot d\mathbf{A} = 0, & \operatorname{div}\mathbf{E} = 0, \\[2em]
2.\ \oint_L \mathbf{H} \cdot d\mathbf{l} = \kappa\varepsilon_0 \dfrac{d}{dt} \int_A \mathbf{E} \cdot d\mathbf{A}, & \nabla \times \mathbf{H} = \kappa\varepsilon_0 \dfrac{\partial \mathbf{E}}{\partial t}, \\[2em]
3.\ \oint_L \mathbf{E} \cdot d\mathbf{l} = -\kappa_m\mu_0 \dfrac{d}{dt} \int_A \mathbf{H} \cdot d\mathbf{A}, & \nabla \times \mathbf{E} = -\kappa_m\mu_0 \dfrac{\partial \mathbf{H}}{\partial t}, \\[2em]
4.\ \oint_S \mathbf{B} \cdot d\mathbf{A} = 0, & \operatorname{div}\mathbf{H} = 0.
\end{array}
\tag{4.1}
$$

The case of propagation of electromagnetic waves in vacuum is the simplest as the relative dielectric permittivity $\kappa = 1$ and relative magnetic permeability $\kappa_m = 1$. Moreover, for propagation in free space, there is no need to use the integral form of Maxwell's equations, so we will work just with the differential forms that are given in the right column of Equations 4.1. In order to obtain the wave equation, we apply the $\nabla \times$ operation to the right- and left-hand parts of the differential form of the third equation, that is,

$$
\nabla \times (\nabla \times \mathbf{E}) = -\kappa_m\mu_0 \nabla \times \left(\frac{\partial \mathbf{H}}{\partial t} \right).
\tag{4.2}
$$

Transposing the operators $\partial/\partial t$ and ∇, we can write this equation as

$$
\nabla \times (\nabla \times \mathbf{E}) = -\kappa_m\mu_0 \frac{\partial}{\partial t} (\nabla \times \mathbf{H}).
\tag{4.3}
$$

The left-hand part of the equation is transformed using the following equation for the triple vector product $\mathbf{A} \times (\mathbf{B} \times \mathbf{C}) = \mathbf{B}(\mathbf{A} \cdot \mathbf{C}) - \mathbf{C}(\mathbf{A} \cdot \mathbf{B})$; hence,

$$
\nabla \times (\nabla \times \mathbf{E}) = \nabla(\nabla \cdot \mathbf{E}) - \nabla^2 \mathbf{E} = -\nabla^2 \mathbf{E}.
\tag{4.4}
$$

Here, according to the third equation from the system (4.1), $\operatorname{div}\mathbf{E} = \nabla \cdot \mathbf{E} = 0$; thus, we use $\nabla(\nabla \cdot \mathbf{E}) = 0$. Taking into consideration Equation 4.4, we substitute the expression for $\nabla \times \mathbf{H}$ taken from the second equation of system (4.1) into the right-hand part of Equation 4.3. Thus, we arrive at what is known as the wave equation for the electric field vector $\mathbf{E}$:

$$
\nabla^2 \mathbf{E} = \frac{1}{v^2} \frac{\partial^2 \mathbf{E}}{\partial t^2}.
\tag{4.5}
$$

Applying a similar procedure to the second Equation 4.1, one can obtain the wave equation for the magnetic intensity vector $\mathbf{H}$:

$$
\nabla^2 \mathbf{H} = \frac{1}{v^2} \frac{\partial^2 \mathbf{H}}{\partial t^2}.
\tag{4.6}
$$

Here, we introduce the propagation velocity of the electromagnetic wave in a media, which is $v = c/\sqrt{\kappa\kappa_m}$ where c is the speed of light in a vacuum:

$$c = \frac{1}{\sqrt{\varepsilon_0\mu_0}} = 3\times10^8 \text{ m/s.} \tag{4.7}$$

The propagation velocity of any electromagnetic wave in vacuum is constant for all inertial reference frames and thus compatible with the special theory of relativity.

Exercise 4.2

Starting with Maxwell's equations, obtain the wave equation for the 1D case when a plane electromagnetic wave propagates in vacuum along the z-axis.

Solution. The 1D case corresponds to a plane electromagnetic wave. The derivatives $\partial/\partial x$ and $\partial/\partial y$ of both fields are equal to zero. In this case, the second and the third equations of system (4.1) take the form

$$\kappa\varepsilon_0 \frac{\partial E_x}{\partial t} = -\frac{\partial H_y}{\partial z}, \quad \kappa_m\mu_0 \frac{\partial H_x}{\partial t} = \frac{\partial E_y}{\partial z},$$

$$\kappa\varepsilon_0 \frac{\partial E_y}{\partial t} = \frac{\partial H_x}{\partial z}, \quad \kappa_m\mu_0 \frac{\partial H_y}{\partial t} = -\frac{\partial E_x}{\partial z},$$

$$\kappa\varepsilon_0 \frac{\partial E_z}{\partial t} = 0, \quad \kappa_m\mu_0 \frac{\partial H_z}{\partial t} = 0.$$

The last two equations describe time-independent stationary fields. It follows from them that the longitudinal components are constants (i.e., they are time independent): E_z = const, H_z = const. From the first and the fourth equations of system (4.1), it follows that both E_z and H_z do not depend on z so we can choose $E_z = H_z = 0$, that is, electric and magnetic field vectors are perpendicular to the direction of propagation. From the system given earlier, let us consider the equations with the components E_x and H_y:

$$\kappa\varepsilon_0 \frac{\partial E_x}{\partial t} = -\frac{\partial H_y}{\partial z}, \quad \kappa_m\mu_0 \frac{\partial H_y}{\partial t} = -\frac{\partial E_x}{\partial z}.$$

Let us differentiate the first equation with respect to time and the second with respect to z. As a result, we obtain

$$\kappa\varepsilon_0 \frac{\partial^2 E_x}{\partial t^2} = -\frac{\partial^2 H_y}{\partial t\partial z}, \quad \kappa_m\mu_0 \frac{\partial^2 H_y}{\partial t\partial z} = -\frac{\partial^2 E_x}{\partial z^2}.$$

Therefore,

$$\kappa\varepsilon_0\kappa_m\mu_0 \frac{\partial^2 E_x}{\partial t^2} = \frac{\partial^2 E_x}{\partial z^2}.$$

By analogy, we obtain the following equation for the component H_y:

$$\kappa\varepsilon_0\kappa_m\mu_0 \frac{\partial^2 H_y}{\partial t^2} = \frac{\partial^2 H_y}{\partial z^2}.$$

The quantity $1/\varepsilon_0\mu_0 = c^2$, c is the speed of the electromagnetic wave in vacuum. Thus,

$$\frac{\partial^2 E_x}{v^2 \partial t^2} = \frac{\partial^2 E_x}{\partial z^2}, \quad \frac{\partial^2 H_y}{v^2 \partial t^2} = \frac{\partial^2 H_y}{\partial z^2}.$$

These equations describe a wave that propagates along the z-axis in a homogeneous medium. Analogous equations can be easily obtained for components E_y and H_x.

4.3   PLANE MONOCHROMATIC WAVES

Equations 4.5 and 4.6 describe a broad class of electromagnetic waves: from plane to spherical waves and from monochromatic waves to short duration electromagnetic pulses. The simplest and most important type of electromagnetic waves is a monochromatic wave, for which the vectors **E** and **H** have a harmonic dependence on space and time:

$$\mathbf{E}(t,\mathbf{r}) = \mathbf{E}_0 \cos(\omega t - \mathbf{k} \cdot \mathbf{r}),$$

$$\mathbf{H}(t,\mathbf{r}) = \mathbf{H}_0 \cos(\omega t - \mathbf{k} \cdot \mathbf{r}),$$

(4.8)

where
  $\mathbf{E}_0$ and $\mathbf{H}_0$ are the amplitudes of the corresponding fields
  ω is the angular frequency (i.e., the angular frequency of oscillation for vectors **E** and **H**)
  k is the wave vector

The argument of the cosine function determines the vectors **E** and **H** (if their amplitudes $\mathbf{E}_0$ and $\mathbf{H}_0$ are known). Therefore, the argument of the cosine function in the wave equation is called *the wave phase*. The magnitude of the wave vector (the wave number) is related to the wavelength as

$$k = \frac{2\pi}{\lambda} = k_0 \sqrt{\kappa\kappa_m} = \frac{\omega}{c}\sqrt{\kappa\kappa_m},$$

(4.9)

where $k_0 = \omega/c$ is the wave number in vacuum. The wavelength λ in a medium is the distance that the wave travels during one period $T = 2\pi/\omega$ propagating with velocity v:

$$\lambda = v \cdot T = \frac{\lambda_0}{\sqrt{\kappa\kappa_m}}, \quad \lambda_0 = c \cdot T.$$

The propagation of a wave and the wave vector **k** have the same direction:

$$\omega t - \mathbf{k} \cdot \mathbf{r} = \text{const}.$$

(4.10)

The scalar product of two vectors in the Cartesian coordinate system was introduced earlier:

$$\mathbf{k} \cdot \mathbf{r} = k_x x + k_y y + k_z z,$$

(4.11)

where the wave vector components are

$$k_x = k \cos\alpha, \quad k_y = k \cos\beta, \quad k_z = k \cos\gamma.$$

(4.12)

Here, α, β, γ are the angles between the wave vector and the x-, y-, and z-axes, respectively. The quantities $\cos\alpha$, $\cos\beta$, and $\cos\gamma$ are called the direction cosines of the vector **k**, and they are shown in Figure 4.2.

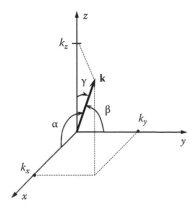

FIGURE 4.2 Vector **k** in Cartesian coordinate system.

Thus, the space variation of the wave phase occurs in the direction of the vector **k**. Therefore, the vector **k** defines the wave propagation direction in space. Solution of Equations 4.5 and 4.6 can be represented by a harmonic function: either by $\cos(\omega t - \mathbf{k} \cdot \mathbf{r})$ or by $\sin(\omega t - \mathbf{k} \cdot \mathbf{r})$.

A linear superposition of these harmonic functions is also a solution of the wave equation. There are four Euler's formulas:

$$
\begin{aligned}
e^{ia} &= \cos a + i \sin a, \\
e^{-ia} &= \cos a - i \sin a, \\
\cos a &= \frac{e^{ia} + e^{-ia}}{2}, \\
\sin a &= \frac{e^{ia} - e^{-ia}}{2i}, \quad i = \sqrt{-1}.
\end{aligned}
\tag{4.13}
$$

Employing the first Euler's formula, we have

$$
\cos(\omega t - \mathbf{k} \cdot \mathbf{r}) + i \sin(\omega t - \mathbf{k} \cdot \mathbf{r}) = \exp\left[i(\omega t - \mathbf{k} \cdot \mathbf{r})\right].
\tag{4.14}
$$

Taking into account Equation 4.14, for convenience, we can express the solution of the wave equation in an exponential (complex) form:

$$
\begin{aligned}
\mathbf{E}(t,\mathbf{r}) &= \mathbf{E}_0 \exp\left[i(\omega t - \mathbf{k} \cdot \mathbf{r})\right] = \left|\mathbf{E}_0\right| \exp\left[i(\omega t - \mathbf{k} \cdot \mathbf{r} + \varphi)\right], \\
\mathbf{H}(t,\mathbf{r}) &= \mathbf{H}_0 \exp\left[i(\omega t - \mathbf{k} \cdot \mathbf{r})\right] = \left|\mathbf{H}_0\right| \exp\left[i(\omega t - \mathbf{k} \cdot \mathbf{r} + \varphi)\right],
\end{aligned}
\tag{4.15}
$$

where the complex amplitudes are given as

$$
\begin{aligned}
\mathbf{E}_0 &= \left|\mathbf{E}_0\right| \exp(i\varphi), \\
\mathbf{H}_0 &= \left|\mathbf{H}_0\right| \exp(i\varphi).
\end{aligned}
\tag{4.16}
$$

Here, the real physical fields correspond to the real parts of solutions (4.15). The exponential form of the wave fields simplifies mathematical manipulations. As an example, we calculate divE and $\nabla \times \mathbf{E}$. The divergence of vector $\mathbf{E}$ in the Cartesian coordinate system is

$$\text{div}\mathbf{E} = \nabla \cdot \mathbf{E} = \frac{\partial E_x}{\partial x} + \frac{\partial E_y}{\partial y} + \frac{\partial E_z}{\partial z}. \tag{4.17}$$

To calculate partial derivatives, we take into account Equation 4.15 for the electric field and Equation 4.11:

$$\frac{\partial}{\partial x}\exp(-i\mathbf{k}\cdot\mathbf{r}) = -i\mathbf{k}\exp(-i\mathbf{k}\cdot\mathbf{r})\cdot\frac{\partial\mathbf{r}}{\partial x} = -i(\mathbf{k}\cdot\mathbf{i})\exp(-i\mathbf{k}\cdot\mathbf{r}) = -ik_x\exp(-i\mathbf{k}\cdot\mathbf{r}),$$

$$\frac{\partial}{\partial y}\exp(-i\mathbf{k}\cdot\mathbf{r}) = -ik_y\exp(-i\mathbf{k}\cdot\mathbf{r}), \tag{4.18}$$

$$\frac{\partial}{\partial z}\exp(-i\mathbf{k}\cdot\mathbf{r}) = -ik_z\exp(-i\mathbf{k}\cdot\mathbf{r}).$$

Hence,

$$\nabla\left[\exp(-i\mathbf{k}\cdot\mathbf{r})\right] = -i\mathbf{k}\exp(-i\mathbf{k}\cdot\mathbf{r}) \quad \text{or} \quad \text{div }\mathbf{E} = \nabla\cdot\mathbf{E} = -i\mathbf{k}\cdot\mathbf{E}. \tag{4.19}$$

Similarly, we can obtain the following expression for $\nabla \times \mathbf{E}$:

$$\nabla\times\mathbf{E} = -i\mathbf{k}\times\mathbf{E}. \tag{4.20}$$

Thus, for monochromatic plane waves represented in the complex form, an application of operator del, ∇, is replaced by a vector $-i\mathbf{k}$. In this case, the differential form of Maxwell's equations (4.1) is

$$\mathbf{k}\cdot\mathbf{E} = 0,$$

$$\mathbf{k}\times\mathbf{H} = -\omega\kappa\varepsilon_0\mathbf{E},$$

$$\mathbf{k}\times\mathbf{E} = \omega\kappa_m\mu_0\mathbf{H}, \tag{4.21}$$

$$\mathbf{k}\cdot\mathbf{H} = 0.$$

The first and last equations state that the electromagnetic waves are transverse waves, that is, $\mathbf{E} \perp \mathbf{k}$ and $\mathbf{H} \perp \mathbf{k}$. We note that in longitudinal waves, the characteristic vector of the wave oscillates along the direction of the wave propagation (longitudinal acoustic waves is one of the examples).

According to the second and third equations and the definition of a vector product, the three vectors $\mathbf{E}$, $\mathbf{H}$, and $\mathbf{k}$ are perpendicular to each other. Consequently, $\mathbf{E} \perp \mathbf{H} \perp \mathbf{k}$, and these vectors are ordered so that they obey the right-hand rule. In addition, the quantitative relations between the amplitudes of the electric and magnetic fields can be obtained from the second and third equations. Taking into consideration the orthogonality of the three vectors $\mathbf{E}$, $\mathbf{H}$, and $\mathbf{k}$, we obtain the following relations for the amplitudes of these vectors:

$$H = \left(\frac{k}{\omega\kappa_m\mu_0}\right)E = \left(\frac{\sqrt{\kappa\kappa_m}}{\kappa_m\mu_0 c}\right)E = \sqrt{\frac{\varepsilon_0\kappa}{\mu_0\kappa_m}}E,$$

$$E = \sqrt{\frac{\mu_0\kappa_m}{\varepsilon_0\kappa}}H = ZH, \tag{4.22}$$

where the quantity equal to the ratio between amplitudes of the electric and magnetic wave fields is referred to as the wave impedance of the medium:

$$Z = \frac{E}{H} = \sqrt{\frac{\mu_0 \kappa_m}{\varepsilon_0 \kappa}}. \tag{4.23}$$

From this equation, we get the impedance of vacuum by substituting $\kappa = \kappa_m = 1$:

$$Z_0 = \sqrt{\frac{\mu_0}{\varepsilon_0}} = \sqrt{\frac{4\pi \times 10^{-7}}{8.85 \times 10^{-12}}} \left(\frac{H}{F}\right)^{-1/2} = 120\pi\,(\Omega) \approx 377\,(\Omega).$$

Here, $\Omega = $ Ohm $= $ V/A is the SI unit of resistance. Harmonic oscillations of the vectors **E** and **H** in a monochromatic wave take place in orthogonal planes. The wave vector lies along the intersection line of these two planes and points toward the wave propagation direction. A snapshot (at fixed time t) of the vectors **E**, **H**, and *k* is shown in Figure 4.3. Vectors **E** and **H** oscillate in the $x0z$ and $y0z$ plane, respectively.

Spherical waves are also solutions of the wave equation. These waves propagate outward from a point source. A spherical wave field generated by a point source of harmonic spherical waves is described by the equation

$$\mathbf{E}(t,r) = \frac{\mathbf{A}}{r} \exp\left[i\omega\left(t - \frac{r}{v}\right) \right]. \tag{4.24}$$

One can see that the amplitude of a spherical wave is inversely proportional to the distance from the source.

Linear sources such as long slits and filaments are often used in optics. Such sources emit what is known as *cylindrical waves*. The amplitude of these waves in the radial direction that are perpendicular to the source axis decreases with increasing distance from the source.

Exercise 4.3

Write a general equation for a plane wave propagating along one of the coordinate axes (e.g., along the x-axis). Assume that a plane with dimensions substantially larger than the wavelength serves as a source of plane waves. The entire surface of the plane radiates harmonic wave along the direction perpendicular to the plane. The electric field vector oscillates in this plane as

$$E(t) = A \sin \omega t.$$

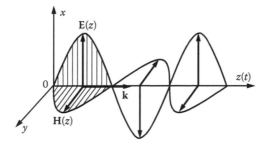

FIGURE 4.3   Propagating electromagnetic wave.

Solution. At a distance x from this plane, the field oscillations do not arrive instantly since the electromagnetic field propagates with the finite speed c. Therefore, the relationship that gives the field oscillations at a distance x from the source can be written as

$$E(t,x) = A \sin \omega \left(t - \frac{x}{c} \right).$$

Here, the oscillations' retardation time is defined by the wave speed c. Taking into account that $\omega = 2\pi/T$, the equation for plane wave can be rewritten in a slightly different form:

$$E(t,x) = A \sin \left(\omega t - \frac{2\pi x}{cT} \right).$$

From the last expression, we see that the field oscillations have temporal as well as spatial periodic dependences. As we already know, the process that is periodic in time and space is called wave process.

Let us introduce the quantity $k = 2\pi/\lambda$, which is called wave number. In this case, the equation for plane wave that propagates in the positive direction of x-axis takes the form that is symmetric with respect to time and space:

$$E(t,x) = A \sin (\omega t - kx).$$

The plane wave equation is usually written in this form where constant A is amplitude of the wave and it is real number.

In the case when the equation for the plane wave is written in the form

$$E(t,x) = A \exp \left(i(\omega t - kx) \right),$$

both constants A and k can be complex numbers:

$$A = A' + iA'' = |A| \exp(i\varphi), \quad k = k' - ik''.$$

In this case, the amplitude of the wave is $|A| = \sqrt{(A')^2 + (A'')^2}$ and $\varphi = \operatorname{arctg}(A''/A')$ determines the phase of the wave, so the equation for the plane wave can be written as

$$E(t,x) = |A| \exp(-k''x) \exp \left(i(\omega t - k'x + \varphi) \right).$$

4.4 POLARIZATION OF ELECTROMAGNETIC WAVES

As mentioned earlier, an important property of electromagnetic waves is the fact that they are transverse, that is, the electric and magnetic field vectors are orthogonal to the wave propagation direction. There is a wide class of electromagnetic waves, in which the electric and magnetic fields oscillate in fixed planes.

An electromagnetic wave is linearly polarized if its vector **E** oscillates along a fixed direction in space. The plane containing vectors **k** and **E** is defined as the polarization plane. During an oscillation period, the tip of the vector **E** (and also **H**) traces a straight line, whose length is equal to twice the amplitudes of **E** (or **H**). A wave is elliptically polarized, if during one period the tip of vector **E** (or **H**) traces an ellipse in the plane perpendicular to the wave propagation (see Figure 4.4b).

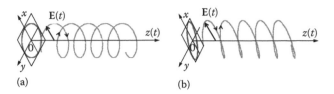

FIGURE 4.4 Circular (a) and elliptically (b) polarized waves.

If a vector traces a circle for the period, the wave is circularly polarized (see Figure 4.4a). Circular and linear polarizations can be considered as special cases of elliptical polarization.

An electromagnetic wave has two independent orthogonal states of polarization. The electromagnetic wave with an arbitrary state of polarization can be obtained as a superposition of two linearly polarized waves with the same frequencies polarized orthogonally to each other and propagating in the same direction (i.e., $\mathbf{E}_1 \perp \mathbf{E}_2$). Generally, these waves have different amplitudes and a constant phase difference. For simplicity, we consider only the electric field components and assume that the waves are propagating along the positive direction of the z-axis. Here, we assume that the electric field vector of the first wave oscillates along the x-axis and that the electric field vector of the second wave along the y-axis. The equations that describe these waves are

$$\mathbf{E}_1(t,z) = \mathbf{i}E_x(t,z) = \mathbf{i}A_x \cos(\omega t - kz),$$

$$\mathbf{E}_2(t,z) = \mathbf{j}E_y(t,z) = \mathbf{j}A_y \cos(\omega t - kz + \varphi),$$

(4.25)

where
A_x and A_y are the amplitudes of the two component waves
φ is the phase difference between the two waves

The net electric field $\mathbf{E} = \mathbf{E}_1 + \mathbf{E}_2$ oscillates in a plane perpendicular to the direction of wave propagation at a fixed coordinate z. The tip of the vector $\mathbf{E}$ traces with time some closed curve in the $x0y$ plane. Let us find an equation for the curve. We transform Equation 4.25 as follows:

$$\frac{E_x}{A_x} = \cos(\omega t - \varphi_x) = \cos \omega t \cdot \cos \varphi_x + \sin \omega t \cdot \sin \varphi_x,$$

$$\frac{E_y}{A_y} = \cos(\omega t - \varphi_y) = \cos \omega t \cdot \cos \varphi_y + \sin \omega t \cdot \sin \varphi_y,$$

(4.26)

where the angles $\varphi_x(z) = kz$, $\varphi_y(z) = kz - \varphi$ are introduced. To obtain the trajectory of the tip of vector $\mathbf{E}$ in the $x0y$ plane, the time must be eliminated in Equation 4.26. We multiply the first equation by $\cos \varphi_y$ and the second by $\cos \varphi_x$, then subtract the second equation from the first one. After that, these equations are multiplied by $\sin \varphi_y$ and $\sin \varphi_x$, respectively, and one equation is subtracted from another. Thus, we arrive to the following set of equations:

$$\left(\frac{E_x}{A_x} \right) \cos \varphi_y - \left(\frac{E_y}{A_y} \right) \cos \varphi_x = \sin \omega t \cdot \sin (\varphi_x - \varphi_y),$$

$$\left(\frac{E_x}{A_x} \right) \sin \varphi_y - \left(\frac{E_y}{A_y} \right) \sin \varphi_x = -\cos \omega t \cdot \sin (\varphi_x - \varphi_y).$$

(4.27)

In this, we take into account that

$$\sin \varphi_x \cos \varphi_y - \sin \varphi_y \cos \varphi_x = \sin(\varphi_x - \varphi_y).$$

Then Equations 4.27 are squared and summed yielding the equation for a curve traced by a tip of the vector **E** for one period in the plane perpendicular to the direction of propagation:

$$\left(\frac{E_x}{A_x}\right)^2 + \left(\frac{E_y}{A_y}\right)^2 - 2\frac{E_x}{A_x}\frac{E_y}{A_y}\cos\varphi = \sin^2\varphi. \qquad (4.28)$$

In this, we take into account that

$$\cos \varphi_y \cos \varphi_x + \sin \varphi_y \sin \varphi_x = \cos(\varphi_x - \varphi_y) = \cos\varphi.$$

If we use E_x and E_y as coordinates, this equation has the general form of an ellipse with its center at the origin ($E_x = E_y = 0$); we note that the ellipse axes do not coincide with the coordinate axes (Figure 4.5). The angle ψ between the x-axis and the major axis of the polarization ellipse is determined by the following expression:

$$tg2\psi = \frac{2A_xA_y}{A_x^2 - A_y^2}\cos\varphi.$$

An ellipse is inscribed into a rectangle with sides $2A_x$ and $2A_y$ and it touches the rectangle sides at points A, A' with coordinates ($\pm A_x, \pm A_y\cos\varphi$) and B, B' ($\pm A_x\cos\varphi, \pm A_y$) (Figure 4.5).

The orientation of the polarization ellipse and its parameters depend on the amplitudes of the constituent waves and the phase difference between them. The rotation direction of vector **E** is determined by the phase difference φ.

Let us consider some particular cases, when Equation 4.28 for the polarization ellipse reduces to a simpler equation describing a simpler trajectory. Thus, for a phase difference $\varphi = \dfrac{m\pi}{2}, m = \pm 1, \pm 3, \pm 5, \ldots$, Equation 4.28 takes a form

$$\left(\frac{E_x}{A_x}\right)^2 + \left(\frac{E_y}{A_y}\right)^2 = 1. \qquad (4.29)$$

This is the equation for an ellipse, whose symmetry axes coincide with the coordinate axes, with semiaxes that are equal to A_1 and A_2. In this case, one of the components of **E** is maximum when the other component is zero.

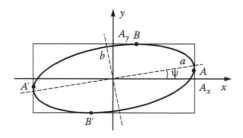

FIGURE 4.5   Polarization ellipse of the elliptically polarized electromagnetic wave.

In the case of *a circularly polarized wave*, the ellipse becomes a circle. The amplitudes of the mutually perpendicular components of the wave electric field are assumed to be equal. Since $A_x = A_y = A_0$, Equation 4.29 becomes

$$(E_x)^2 + (E_y)^2 = A_0^2. \tag{4.30}$$

Tracing a circle, the tip of the vector **E** rotates in the clockwise or counterclockwise direction. To differentiate these two states, the concepts of *left circular polarization* (LCP) and *right circular polarization* (RCP) are introduced. If vector **E** rotates in time in the clockwise direction, for an observer looking in the direction of the wave propagation, the wave will be right circularly polarized. This phenomenon is sometimes called "right-hand circular polarization:" if the observer points out the right thumb in the direction of the propagation, the other four fingers will point in the direction of the rotation of **E**. In the case of the left circularly polarized wave, the vector **E** rotates in the counterclockwise direction for an observer looking in the direction of the wave propagation. That is, if the observer points out the left thumb in the direction of the propagation, the other four fingers will point in the direction of the rotation of **E**. The rotation direction of the vector **E** depends on the sign of the phase difference $\varphi_x - \varphi_y = \varphi$: for right circular polarization $\sin \varphi > 0$, while for left circular polarization $\sin \varphi < 0$.

Another important special case is that of a linearly polarized wave. According to Equation 4.28, at $\varphi = m\pi$, $m = 0, \pm1, \pm2, \pm3, \ldots$, the equation of the polarization ellipse transforms into the equation of a straight line:

$$E_y = (-1)^m \left(\frac{A_y}{A_x} \right) E_x. \tag{4.31}$$

Figure 4.6 shows all polarization types for the two rotation directions of the vector **E** and different values of the phase angle φ.

The polarization types discussed earlier cover all possible polarizations of electromagnetic waves.

Monochromatic radiation is always polarized. Radiation composed of waves with various wavelengths has the vector **E** oscillating either in ordered or in random directions. Radiation with vector **E** that changes randomly is called "unpolarized." An electromagnetic wave can be partially polarized. Partial polarization is characterized by the degree of polarization that is defined as

$$P = \frac{I_{max} - I_{min}}{I_{max} + I_{min}}, \tag{4.32}$$

where I_{max} and I_{min} are the maximal and minimal intensities (flux density of electromagnetic energy) of the radiation passing through a polarizer. Note that unpolarized light becomes polarized after it passes through a polarizer. To find I_{max} and I_{min}, it is necessary to do two measurements: We rotate the polarizer to find the maximum value of the intensity and measure I_{max}. Then we continue to

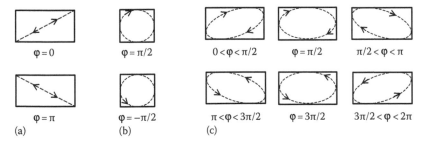

FIGURE 4.6 Possible types of polarization (linear (a), circular (b), and elliptical (c)) for different values of phase angle $\varphi = \varphi_x - \varphi_y$: $A_x \neq A_y$ (a, c), $A_x = A_y$ (b).

rotate the polarizer till we find a minimum value of the intensity and measure I_{min}. Note that intensity is defined by Equation 4.65.

Some sources of electromagnetic radiation generate waves with varying polarizations. Thermal radiation generated by chaotically moving atoms is always unpolarized. Cyclotron radiation generated by electrons moving along circular orbits in a magnetic field is circularly polarized.

Polarizers are used to obtain linearly polarized light. Independently of the principle of their operation, polarizers allow the propagation through them only of electromagnetic wave with the electric field **E** that is parallel to the ***plane of the polarizer***. Figure 4.7 shows that for a wave with vector **E** oriented at an angle α relative to the plane of the polarizer, only the projection E_{tr} of **E** along this plane passes through the polarizer:

$$E_{tr} = E \cos\alpha.$$

Since intensity is determined by the square of the electric field, the intensity of the passed wave is determined by the equation

$$I_{tr} = I_0 \cos^2\alpha,$$

where I_0 is the intensity of the incident wave.

Exercise 4.4

Show that the superposition of a right and a left circularly polarized wave of the same amplitude and frequency produces a linearly polarized wave.

Solution. Using Equation 4.25, we write the equations for a right circularly polarized wave. In this case, $\varphi = \pi/2$ and

$$E_x^R(t,z) = A_0 \cos(\omega t - kz),$$

$$E_y^R(t,z) = A_0 \cos\left(\omega t - kz + \frac{\pi}{2}\right) = -A_0 \sin(\omega t - kz).$$

Analogously, the equations for a left circularly polarized wave are obtained from Equation 4.25 for $\varphi = -\pi/2$:

$$E_x^L(t,z) = A_0 \cos(\omega t - kz),$$

$$E_y^L(t,z) = A_0 \cos(\omega t - kz - \pi/2) = A_0 \sin(\omega t - kz).$$

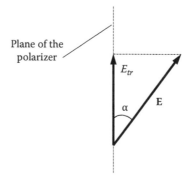

FIGURE 4.7   Polarizer transfer only E_{tr} component of the electric field of waves.

Summing of the corresponding components of these waves yields

$$E_x(t,z) = E_x^R(t,z) + E_x^L(t,z) = 2A_0 \cos(\omega t - kz),$$

$$E_y(t,z) = E_y^R(t,z) + E_y^L(t,z) = 0.$$

Thus, the superposition of a right and a left circularly polarized wave of the same amplitude and frequency produces a linearly polarized wave with polarization along the x-axis, whose electric field vector is

$$\mathbf{E} = \mathbf{E}^R + \mathbf{E}^L = 2iA_0 \cos(\omega t - kz).$$

4.5 SUPERPOSITION OF ELECTROMAGNETIC WAVES

1. Electromagnetic waves obey the principle of superposition. According to this principle, waves generated by different sources do not interact, that is, waves with different frequencies and directions propagate independently in isotropic media. The complex wave field generated by two or more sources is a vector sum of the wave fields generated by the separate sources:

$$\mathbf{E} = \mathbf{E}_1 + \mathbf{E}_2 + \cdots \tag{4.33}$$

The principle of superposition allows not only to combine waves but also to decompose them as sums of independent sine waves. This means that any complex (nonplane, nonmonochromatic) wave can always be represented as a sum of sine waves of different amplitudes, frequencies, phases, and wave vectors. This possibility is widely used in the theory of different (not exclusively electromagnetic) wave processes.

Mathematically, the principle of superposition is a consequence of the linearity of the wave equation that describes the propagation of light waves in vacuum. Indeed, if the fields $\mathbf{E}_1$, $\mathbf{E}_2$, $\mathbf{E}_3$,... are solutions of the wave equation, their sum $\mathbf{E} = \mathbf{E}_1 + \mathbf{E}_2 + \cdots$ is also a solution. You can verify this by substituting into the wave equation the solutions in the form of a sum of waves (e.g., plane waves):

$$\mathbf{E} = \sum_i \mathbf{E}_i = \sum_i \mathbf{A}_{i0} \cos(\omega_i t - \mathbf{k}_i \cdot \mathbf{r}), \tag{4.34}$$

where $\mathbf{k}_i$ are the wave vectors and $k_i = \omega_i/c$ are their magnitudes (wave numbers) in the sum. In this case, the wave equation is a sum of terms describing the individual waves.

Applying the principle of superposition, we can show that two plane monochromatic waves traveling in the same direction and with equal frequencies combine to form a plane monochromatic wave of the same frequency. If the waves have different frequencies or different propagation directions, the resultant traveling wave is not monochromatic.

2. Let us consider the superposition of two plane monochromatic waves traveling in the same direction with similar angular frequencies ω_1, ω_2 and wave numbers k_1, k_2. For simplicity, we assume that the electric field vectors $\mathbf{E}_1$ and $\mathbf{E}_2$ of these waves oscillate along the same direction, for example, along the x-axis. Assume that the direction of propagation is along the z-axis and the amplitudes of both waves are equal:

$$\mathbf{E}_1 = iA_0 \cos(\omega_1 t - k_1 z), \quad \mathbf{E}_2 = iA_0 \cos(\omega_2 t - k_2 z), \tag{4.35}$$

where $|\omega_1 - \omega_2| \ll \omega_1, \omega_2, |k_1 - k_2| \ll k_1, k_2$, and $k_i = \omega_i/c$. Using the principle of superposition, we can express the resulting wave as follows:

$$\mathbf{E} = \mathbf{E}_1 + \mathbf{E}_2 = 2iA_0 \cos\left(\frac{\omega_1 - \omega_2}{2}t - \frac{k_1 - k_2}{2}z\right)\cos\left(\frac{\omega_1 + \omega_2}{2}t - \frac{k_1 + k_2}{2}z\right). \qquad (4.36)$$

Taking into account that $|\omega_1 - \omega_2| \ll \omega_1 + \omega_2$ and $|k_1 - k_2| \ll k_1 + k_2$, the net electric field of the two waves is given by the equation

$$\mathbf{E} = \mathbf{E}_1 + \mathbf{E}_2 = 2iA_0 \cos\left(\frac{\Delta\omega}{2}t - \frac{\Delta k}{2}z\right)\cos\left(\omega t - kz\right), \qquad (4.37)$$

where
$$\Delta\omega = \omega_1 - \omega_2$$
$$\Delta k = k_1 - k_2$$
$$\omega = (\omega_1 + \omega_2)/2$$
$$k = (k_1 + k_2)/2$$

Under the assumptions given earlier, we have the following relations: $|\Delta\omega/2| \ll \omega$ and $|\Delta k/2| \ll k$. Therefore, the argument of the first cosine term varies considerably slower than the second cosine term. Thus, we can assume that Equation 4.37 describes a traveling wave with the variable amplitude:

$$\mathbf{E} = 2iA(t,z)\cos\left(\omega t - kz\right), \qquad (4.38)$$

where the variable amplitude is given by

$$A(t,z) = A_0 \cos\left(\frac{\Delta\omega}{2}t - \frac{\Delta k}{2}z\right). \qquad (4.39)$$

From these relations, one can conclude that the resultant wave with angular frequency ω and wave number k has amplitude that is modulated in space and time by an envelope function that varies with angular frequency $\Delta\omega/2$ and wave number $\Delta k/2$. Figure 4.8 shows the time dependence of the wave electric field at a fixed coordinate $z_0 = \pi/\Delta k$. The solid line indicates oscillations with angular frequency ω, and the dashed line denotes the time-varying amplitude envelope that changes from a maximum value of $2A_0$ to zero. If the amplitudes of the two waves are not equal to each other, the amplitude of the resultant wave changes from $(A_{10} + A_{20})$ to $|A_{10} - A_{20}|$.

The periodic variation of the oscillations amplitude caused by the superposition of two waves with close frequencies is called **beat**. The beat frequency is equal to the frequency difference $\Delta\omega = |\omega_1 - \omega_2|$ between the components. The resulting wave described by Equation 4.37 is modulated not only in

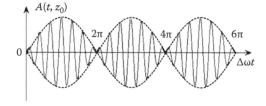

FIGURE 4.8 Time dependence at a fixed coordinate $z_0 = \pi/\Delta k$ of the electric field of the two plane monochromatic waves of the same amplitude traveling in the same direction with similar angular frequencies ω_1, ω_2 and wave numbers k_1, k_2.

time but also in space. Such a traveling wave with varying amplitude is no longer monochromatic. The following equation gives an example of two waves with close frequencies and different amplitudes:

$$E(t,z_0) = 2A_0 \cos\left[5\omega\left(t - \frac{z}{v}\right)\right] + A_0 \cos\left[6\omega\left(t - \frac{z}{v}\right)\right].$$

The electric field is shown in Figure 4.9 as a function of $t - z/v$. This variable is chosen to show dependence on time and space: for the fixed coordinate (e.g., $z = 0$), the plot shows dependence in time, like Figure 4.8, but for fixed time, it shows dependence in space.

3. According to the principle of superposition, any wave can be represented as a sum of sine waves. The result of superposition of such waves with slightly differing frequencies is called a ***wave packet***, because it is localized in space at any time t.

We already considered a simple wave packet comprising two harmonic waves propagating in the same direction, the resultant wave is described by Equations 4.38 and 4.39. The velocity of a wave packet is defined as the velocity of the wave amplitude maximum located in the center of a wave packet. Since the amplitude in Equation 4.39 is constant under the condition

$$\frac{\Delta\omega}{2}t - \frac{\Delta k}{2}z = \text{const}, \tag{4.40}$$

the wave packet velocity is expressed as

$$v_{gr} = \frac{dz}{dt} = \frac{\Delta\omega}{\Delta k}. \tag{4.41}$$

The velocity v_{gr} is called the group velocity. It is defined as the velocity of a group of waves forming the wave packet that is localized in space at each moment of time. Expression (4.41) is obtained for the wave packet comprising two waves.

4. Let us consider now the propagation of a wave packet that is the superposition of monochromatic waves polarized along the x-axis. These waves have different frequencies spread continuously within a narrow spectral range of frequencies $2\Delta\omega \ll \omega_0$ with corresponding wave numbers $2\Delta k \ll k_0$. Here, ω_0 is the central angular frequency of the wave packet and $k_0 = \omega_0/c$ is the corresponding wave number. In this case, the wave packet is represented not as a sum but as an integral of continuously distributed monochromatic waves:

$$\mathbf{E}(t,z) = \mathbf{i} \int_{k_0 - \Delta k}^{k_0 + \Delta k} A(k)\exp\left[i(\omega t - kz)\right]dk, \tag{4.42}$$

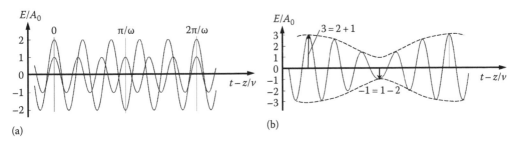

FIGURE 4.9 The electric field of the two plane monochromatic waves with close angular frequencies and different amplitudes (one amplitude is A_0 and the second is $2A_0$) traveling in the same direction is shown as a function of $t - z/v$. (a) Shows electric field E/A_0 of each of those waves and (b) shows the resulting electric field of the sum of these two waves: $E(t)/A_0 = 2\cos(5\omega t') + \cos(6\omega t')$ where $t' = t - z/v$.

where $A(k)$ is the amplitude of a component with wave number k. For the sake of simplicity, the harmonic waves in Equation 4.42 are shown in their exponential form.

Let us expand $\omega(k)$ as a Taylor series over a small parameter $\delta k = k - k_0$ around k_0:

$$\omega(k) = \omega(k_0) + \left(\frac{d\omega}{dk}\right)_{k=k_0} (k-k_0) + \cdots = \omega_0 + v_{gr}\delta k + \cdots \tag{4.43}$$

Here, we introduce a general expression for the group velocity:

$$v_{gr} = \frac{d\omega}{dk}. \tag{4.44}$$

We can represent the phase of the wave in the form

$$\omega t - kz = (\omega_0 + v_{gr}\delta k)t - (k_0 + \delta k)z = (\omega_0 t - k_0 z) - (z - v_{gr}t)\delta k . \tag{4.45}$$

The terms of second order as well as higher orders in δk are omitted. Substituting Equation 4.45 into Equation 4.42, we get

$$\mathbf{E}(t,z) = \mathbf{i}\left\{ \int_{-\Delta k}^{\Delta k} A(k)\exp\left[-i(z-v_{gr}t)\delta k\right]d(\delta k)\right\}\exp(\omega_0 t - k_0 z)$$

$$= \mathbf{i}B(t,z)\exp(\omega_0 t - k_0 z), \tag{4.46}$$

where the function

$$B(t,z) = \int_{-\Delta k}^{\Delta k} A(k)\exp\left[-i(z-v_{gr}t)\delta k\right]d(\delta k) \tag{4.47}$$

is the wave packet envelope. To find the envelope profile, we can simplify Equation 4.47 assuming that the spectral amplitude $A(k)$ (which commonly depends on the wave number k) is a constant. So, we can take it outside the integral in Equation 4.47 and then perform the integration:

$$B(t,z) = A\int_{-\Delta k}^{\Delta k} \exp[-i(z-v_{gr}t)\delta k]d(\delta k)$$

$$= 2A\Delta k \frac{\sin[(z-v_{gr}t)\Delta k]}{(z-v_{gr}t)\Delta k} = 2A\Delta k \frac{\sin\xi}{\xi}, \tag{4.48}$$

where $\xi = (z - v_{gr}t)\Delta k$. Thus, Equation 4.46 describes a plane wave with the frequency ω_0, wave number k_0, and modulated amplitude $B(t, z)$. Figure 4.10a shows the dependence of envelope profile of the wave packet (4.48) on the parameter ξ. The wave packet moves in space with group velocity v_{gr}. The modulated wave packet of Equation 4.46 contains the fast oscillations at ω_0 as shown schematically in Figure 4.10b.

5. In addition to the group velocity, a wave is characterized by a ***phase velocity***, which determines the velocity of propagation of the wave phase. It can be expressed from the condition $\omega t - kz = const$:

$$v_{ph} = \frac{dz}{dt} = \frac{\omega}{k}. \tag{4.49}$$

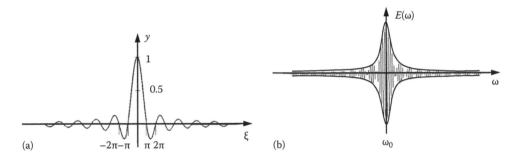

FIGURE 4.10 Dependence of envelope profile of the wave packet of Equation 4.48 on the parameter ξ (a) and the wave packet of Equation 4.46 that moves in space with the group velocity v_{gr} (b).

In vacuum, the phase velocity is always equal to the speed of light, since according to Equation 4.9, $\omega/k_0 = c$. In a transparent optical medium, the phase velocity can be either smaller or greater than the speed of light in vacuum. There is a link between the group and phase velocities that can be found substituting Equation 4.49 into Equation 4.44:

$$v_{gr} = \frac{d\omega}{dk} = \frac{d}{dk}(v_{ph}k) = v_{ph} + k\frac{dv_{ph}}{dk} = v_{ph} - \lambda\frac{dv_{ph}}{d\lambda}, \tag{4.50}$$

where $k = 2\pi/\lambda$.

From Equation 4.50, it follows that the group velocity can be either smaller or larger than the phase velocity depending on the sign of the derivatives dv_{ph}/dk and $dv_{ph}/d\lambda$. In media where these derivatives are zero, the group velocity coincides with the phase velocity. Such a medium is called nondispersive. Vacuum can be considered a special case of nondispersive medium in which phase and group velocities are equal to the speed of light, that is, $v_{gr} = v_{ph} = c$.

4.6 ENERGY AND MOMENTUM OF A WAVE

1. Numerous experiments have demonstrated that electromagnetic waves transfer energy from one point to another in a finite time because of the finite velocity of wave propagation. Examples of such a transfer are heating of a body illuminated by electromagnetic waves, light-induced emission of electrons from metal surfaces, and information transmission with electromagnetic pulses.

The energy of electromagnetic waves propagating in media is stored in the associated electric and magnetic fields. The volume energy density of electromagnetic waves propagating in a medium is determined at each point as the sum of the energies of the electric and magnetic fields:

$$u = \frac{1}{2}(\mathbf{E}\cdot\mathbf{D} + \mathbf{H}\cdot\mathbf{B}) = \frac{\kappa\varepsilon_0 E^2 + \kappa_m\mu_0 H^2}{2}, \tag{4.51}$$

Using the relationship between the electric and magnetic fields $E = \sqrt{\kappa_m\mu_0/\kappa\varepsilon_0}\,H$, we can conclude that the electromagnetic wave energy of the electric field is equal to the energy of the magnetic field.

Let us now define the energy transferred by a wave per unit time and unit area, that is, the energy flux density. We take two equations from four Maxwell's Equations 4.21

$$\mathbf{k}\times\mathbf{H} = -\omega\kappa\varepsilon_0\mathbf{E},$$

$$\mathbf{k}\times\mathbf{E} = \omega\kappa_m\mu_0\mathbf{H}, \tag{4.52}$$

and multiply the first equation by $\mathbf{E}$ and the second equation by $\mathbf{H}$. By subtracting the first from the second, we come up with

$$\mathbf{H} \cdot (\mathbf{k} \times \mathbf{E}) - \mathbf{E} \cdot (\mathbf{k} \times \mathbf{H}) = \omega \left(\kappa \varepsilon_0 E^2 + \kappa \mu_0 H^2 \right). \tag{4.53}$$

The right-hand part expression in brackets is twice the energy density of the electromagnetic field, that is,

$$\kappa \varepsilon_0 E^2 + \kappa_m \mu_0 H^2 = 2(u_e + u_m) = 2u. \tag{4.54}$$

Using the vector identities

$$\mathbf{A} \cdot (\mathbf{B} \times \mathbf{C}) = \mathbf{C} \cdot (\mathbf{A} \times \mathbf{B}) = \mathbf{B} \cdot (\mathbf{C} \times \mathbf{A}) \quad \text{and} \quad \mathbf{A} \times \mathbf{B} = -\mathbf{B} \times \mathbf{A}, \tag{4.55}$$

we get

$$\mathbf{H} \cdot (\mathbf{k} \times \mathbf{E}) - \mathbf{E} \cdot (\mathbf{k} \times \mathbf{H}) = \mathbf{k} \cdot (\mathbf{E} \times \mathbf{H}) - \mathbf{k} \cdot (\mathbf{H} \times \mathbf{E}) = 2\mathbf{k} \cdot (\mathbf{E} \times \mathbf{H}). \tag{4.56}$$

We now introduce the vector

$$\mathbf{S} = \mathbf{E} \times \mathbf{H}, \tag{4.57}$$

which is called **the Poynting vector**. This vector is equal to the energy flux density for the electromagnetic field. Since the vectors $\mathbf{E}$, $\mathbf{H}$, and $\mathbf{k}$ form a right-hand system of vectors, the vector $\mathbf{S}$ is parallel to vector $\mathbf{k}$. Therefore, expression (4.57) can be rewritten as follows:

$$\mathbf{S} = vu \frac{\mathbf{k}}{k}. \tag{4.58}$$

Thus, the energy flux density $\mathbf{S}$ of an electromagnetic wave is a vector quantity with magnitude $S = vu$, $\omega = vk$, $v = c/\sqrt{\kappa_m \kappa}$ and its direction coincident with the direction of the wave propagation.

2. The mean value of the energy flux density of a harmonic wave for one period is physically meaningful. It is convenient to use the complex form of the electric and magnetic fields:

$$\mathbf{E} = \mathbf{E}_0 \exp \left[i(\omega t - \mathbf{k} \cdot \mathbf{r}) \right], \quad \mathbf{H} = \mathbf{H}_0 \exp \left[i(\omega t - \mathbf{k} \cdot \mathbf{r}) \right]. \tag{4.59}$$

Since the energy flux density is a real variable, then the real parts of the fields $(\mathbf{E} + \mathbf{E}^*)/2$ and $(\mathbf{H} + \mathbf{H}^*)/2$ are being used. We substitute these terms into Equation 4.57 and average over time:

$$\langle \mathbf{S} \rangle = \frac{1}{4} \left\langle \left(\mathbf{E} + \mathbf{E}^* \right) \times \left(\mathbf{H} + \mathbf{H}^* \right) \right\rangle = \frac{1}{4} \left\langle \mathbf{E} \times \mathbf{H} + \mathbf{E}^* \times \mathbf{H}^* + \mathbf{E}^* \times \mathbf{H} + \mathbf{E} \times \mathbf{H}^* \right\rangle. \tag{4.60}$$

Here $\mathbf{E}^* = \mathbf{E}_0 \exp \left[-i(\omega t - \mathbf{k} \cdot \mathbf{r}) \right], \quad \mathbf{H}^* = \mathbf{H}_0 \exp \left[-i(\omega t - \mathbf{k} \cdot \mathbf{r}) \right].$

The first two terms in the right-hand part of Equation 4.60 vanish, that is, $\langle \mathbf{E} \times \mathbf{H} \rangle = \langle \mathbf{E}^* \times \mathbf{H}^* \rangle = 0$. Indeed, for the first term,

$$\langle \mathbf{E} \times \mathbf{H} \rangle = (\mathbf{E}_0 \times \mathbf{H}_0) \frac{e^{-2i\mathbf{k}\cdot\mathbf{r}}}{T} \int_0^T e^{2i\omega t} dt = (\mathbf{E}_0 \times \mathbf{H}_0) \frac{e^{-2i\mathbf{k}\cdot\mathbf{r}}}{T} \frac{1}{2i\omega} \left(e^{2i\omega T} - 1 \right) = 0, \qquad (4.61)$$

since $\omega = 2\pi/T$ and $e^{2\pi in} = 1$, where n is an integer. For the second term, the calculations are similar.

As a result, the average energy flux density is expressed as

$$\langle \mathbf{S} \rangle = \frac{1}{4} \langle \mathbf{E}^* \times \mathbf{H} + \mathbf{E} \times \mathbf{H}^* \rangle. \qquad (4.62)$$

We must take the real part of this expression. One can see that the real parts of both terms are equal. Since the electric and magnetic fields given by Equation 4.59 are complex, they can be represented as $\mathbf{E} = \mathbf{E}_1 + i\mathbf{E}_2$ and $\mathbf{H} = \mathbf{H}_1 + i\mathbf{H}_2$, so their complex conjugates are $\mathbf{E}^* = \mathbf{E}_1 - i\mathbf{E}_2$ and $\mathbf{H}^* = \mathbf{H}_1 - i\mathbf{H}_2$. After operation of multiplication for each term in the right-hand side of Equation 4.62, we get for their real parts

$$\mathrm{Re}(\mathbf{E}^* \times \mathbf{H}) = \mathrm{Re}(\mathbf{E} \times \mathbf{H}^*) = \mathbf{E}_1 \times \mathbf{H}_1 + \mathbf{E}_2 \times \mathbf{H}_2. \qquad (4.63)$$

Thus, the average energy flux density can be represented in a form

$$\langle \mathbf{S} \rangle = \frac{1}{2} \mathrm{Re}(\mathbf{E} \times \mathbf{H}^*). \qquad (4.64)$$

3. The amount of energy transported by a wave is often defined as the *wave intensity I*. By definition, it is the energy flux density averaged over one period. Using Equation 4.58 for the electric and magnetic fields and substituting $S = vu$, we get

$$I = \langle S \rangle = \frac{1}{T} \int_0^T S(t) dt = v \kappa \varepsilon_0 E_0^2 \left\langle \cos^2(\omega t - \mathbf{k}\cdot\mathbf{r}) \right\rangle = \sqrt{\frac{\kappa \varepsilon_0}{\kappa_m \mu_0}} \frac{E_0^2}{2} = \frac{E_0^2}{2Z}. \qquad (4.65)$$

Thus, the wave intensity is proportional to the square of the electric field amplitude. The units of the intensity in the SI unit are the same as for the energy flux density, $J/(s \cdot m^2) = W/m^2$.

Electromagnetic waves transport energy as well as momentum. The wave momentum density (i.e., the momentum per unit volume) is given by

$$\pi = \frac{\mathbf{S}}{v^2} = \frac{1}{v^2} \mathbf{E} \times \mathbf{H}. \qquad (4.66)$$

Due to its momentum, an electromagnetic wave exerts pressure on any surface on which the wave is incident. Let a plane wave be incident normally to the surface that completely absorbs the wave. The pressure on the surface is equal to the force acting on the surface divided by its area A, that is,

$$P = \frac{F}{A} = \frac{1}{A} \frac{\Delta p}{\Delta t} = \frac{\Delta z \pi}{\Delta t} = v\pi = u. \qquad (4.67)$$

In the case of a mirror surface that totally reflects the incident wave, the momentum transmitted to the surface is doubled; the pressure $P = 2u$. Let us estimate the value of this pressure for a laser beam. Its electromagnetic wave in vacuum is characterized by an electric field $E_0 = 10^3$ V/m:

$$P = 2\varepsilon_0 E_0^2 = 2 \times 8.85 \times 10^{-12} \times 10^6 = 1.8 \times 10^{-5} \text{ Pa.} \qquad (4.68)$$

This is 10 orders of magnitude smaller than the normal atmospheric pressure ($P_0 \approx 10^5$ Pa) (pascal [Pa] is the SI unit of pressure). Nevertheless, the pressure produced by an electromagnetic wave is easily detected and can be experimentally measured with high accuracy.

Exercise 4.5

Consider a plane electromagnetic wave incident on a plane surface of a weakly conducting nonmagnetic medium. Using Ohm's law (in differential form) and Ampere's force, find the relation between momentum and energy of an electromagnetic wave.

Solution. The wave's electric field generates in the conducting medium a current with density $\mathbf{j} = \sigma \mathbf{E}$. The wave's magnetic field exerts on this current Ampere's force:

$$\mathbf{F} = I(\mathbf{l} \times \mathbf{B}) = \mu_0 I(\mathbf{l} \times \mathbf{H}),$$

where
  $\mathbf{l}$ is a vector along the current (or along vector $\mathbf{E}$) and its magnitude is equal to the length of the medium in the direction of current
  I is current in the medium that is generated by the electric field $\mathbf{E}$ of the wave
  $\mathbf{B} = \mu_0 \mathbf{H}$

The current density is determined by the equation $j = I/A$, where A is the area perpendicular to the vector $\mathbf{l}$ through which the current flows. The force per unit volume can be written as

$$\mathbf{f} = \frac{\mathbf{F}}{Vol} = \frac{\mathbf{F}}{Al} = \mu_0 \frac{I}{A}\left(\frac{\mathbf{l}}{l} \times \mathbf{H}\right) = \mu_0(\mathbf{j} \times \mathbf{H}) = \mu_0 \sigma(\mathbf{E} \times \mathbf{H}),$$

where we took into account that $\mathbf{l}/l$ is the unit vector in the direction of the current $\mathbf{j} = (I/A)(\mathbf{l}/l) = \sigma \mathbf{E}$. The following linear momentum is transferred from the electromagnetic wave to the conducting surface layer during the time Δt:

$$\frac{\Delta \mathbf{p}}{Vol} = \mathbf{f}\Delta t = \mu_0 \sigma(\mathbf{E} \times \mathbf{H})\Delta t.$$

We note that we used Newton's law in this equation and the momentum is written per unit volume. The conducting layer during Δt absorbs energy of electromagnetic wave: $\Delta U/Vol = u = \sigma E^2 \Delta t$. Let us divide these two expressions:

$$\frac{|\Delta \mathbf{p}|}{u} = \frac{\mu_0 \sigma E H \Delta t}{\sigma E^2 \Delta t} = \mu_0 \frac{H}{E} = \sqrt{\varepsilon_0 \mu_0} = \frac{1}{c},$$

where we took into account that $\sqrt{\varepsilon_0} E = \sqrt{\mu_0} H$ for the amplitudes of the wave fields.

4.7 STANDING WAVES

In the traveling waves considered earlier, the electric and magnetic fields at each point oscillate in phase. However, this property of electromagnetic waves is not universal. A type of waves known as "standing waves" is characterized by a time phase shift between the electric field and the magnetic field. Standing waves occur when two counterpropagating monochromatic waves with the same frequency, amplitude, and polarization interfere. For example, this occurs when we have complete reflection of a wave from a surface.

In the following text, we obtain expressions for the electric and magnetic fields in a standing wave. For a wave propagating in the positive z-axis, the fields are given by the equations

$$E_x = E_0 \cos(\omega t - kz),$$
$$H_y = H_0 \cos(\omega t - kz).$$

(4.69)

Here, we have taken into account that $\mathbf{E}$, $\mathbf{H}$, and $\mathbf{k}$ form right-hand system of vectors. Therefore, for the reflected wave traveling along the negative z-axis, vector $\mathbf{E}$ or $\mathbf{H}$ must have different directions to those of the vectors of the incident wave. The field components for a reflected wave can be written as

$$E_x = -E_0 \cos(\omega t + kz), \quad H_y = H_0 \cos(\omega t + kz).$$

(4.70)

When we add the incident and reflected waves, we get

$$E_x = -2E_0 \sin \omega t \sin kz, \quad H_y = 2H_0 \cos \omega t \cos kz.$$

(4.71)

These field equations describe a standing electromagnetic wave. One can see that the wave is not propagating, since the factor $(t \pm z/v)$ typical of a running waves is absent in Equation 4.71. The presence of the factors $\cos \omega t$ and $\sin \omega t$ in Equation 4.71 shows that the electric and magnetic fields at any point are oscillating at the same frequency. The factors $2E_0 \sin kz$ and $2H_0 \cos kz$ describe the amplitudes of the field oscillations at point z. The amplitude dependence on z is harmonic. At the points for which $\sin kz = 0$, the electric field vanishes. These points are called electric field **nodes**. At the points for which $\sin kz = \pm 1$, the amplitude of the electric field oscillations is maximum. These points are called electric field **antinodes**. The distance between neighboring nodes (or between neighboring antinodes) equals to half of the wavelength. In a standing wave, the oscillating electric and magnetic field vectors are shifted in phase by $\pi/2$ in both space and time as it is shown in Figure 4.11.

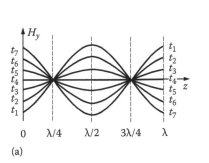

(a)

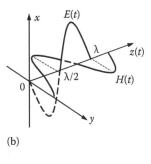

(b)

FIGURE 4.11 Example of a standing wave: change of electric field in time (a) and shift of oscillating electric and magnetic fields in phase by $\pi/2$ (b).

The energy flux density of an electromagnetic wave is described by the Poynting vector $\mathbf{S} = \mathbf{E} \times \mathbf{H}$. Therefore, there is no energy flux at points where either $\mathbf{E}$ or $\mathbf{H}$ is equal to zero. The electric field antinodes coincide with the nodes of the magnetic field and vice versa; hence, the flow of energy in a standing wave through the nodes and antinodes is zero. Therefore, the energy in a standing wave transforms from pure electric energy with $\mathbf{E}$ vector maxima at the electric field anti-nodes into magnetic energy with $\mathbf{H}$ vector maxima in the magnetic field antinodes. These maxima are shifted in space by a quarter of the wavelength ($\lambda/4$). Therefore, in a standing wave, we have no energy transport. Since the oscillations of $\mathbf{E}$ and $\mathbf{H}$ are not synchronized in phase, relation (4.22) is valid only for the maximum amplitudes of fields, that is, $\sqrt{\kappa \varepsilon_0}\, E_0 = \sqrt{\kappa_m \mu_0}\, H_0$.

Exercise 4.6

Two plane waves with electric fields $E_1 = A \sin(\omega t - kx)$ and $E_2 = A \sin(\omega t + kx)$ have the same angular frequencies $\omega = 4.00 \times 10^{15}$ s^{-1} propagate in opposite directions in an isotropic medium with index of refraction $n = 1.50$. The waves interfere with each other and form a standing wave. Find the magnitude of the net electric field vector at the instant $t_0 = (\pi/16) \times 10^{-15}$ s and the amplitude of the standing wave's electric field at all points between nodes separated by a distance $l = \lambda/8$ from each other.

Solution. A standing wave is formed when waves $E_1(t, x)$ and $E_2(t, x)$ are added as a result of interference:

$$E(t,x) = E_1(t,x) + E_2(t,x) = -2A \sin kx \cos \omega t.$$

From this expression for standing wave, we can see that at each point, the electric field oscillates with angular frequency ω. The oscillation amplitude at point x is

$$A(x) = 2A \left| \sin kx \right|.$$

Therefore, at points where $\sin kx = 0$, the oscillation amplitude vanishes. These points are the electric field nodes of standing wave. The coordinate of the first node can be found from the expression $kx_1 = 0$ and the second node at $kx_2 = \pi$. Taking into account that $k = 2\pi/\lambda$, we get $x_2 = \lambda/2$. The distance between two successive nodes is equal to $\Delta x = \lambda/2$, where

$$\lambda = v \times T = \frac{2\pi c}{\omega n} \quad \text{and} \quad \Delta x = \frac{\pi c}{\omega n}.$$

Therefore, in the interval between two successive nodes, we can have five points separated by a distance $l = \lambda/8$. The coordinates of these points are $x_m = m\lambda/8$, $m = 0, 1, 2, 3, 4, \ldots$.

The amplitude of the electric field oscillations at these points is given by the condition

$$A_m = 2A \left| \sin km \frac{\lambda}{8} \right| = 2A \left| \sin\left(m \frac{\pi}{4} \right) \right| = 2A \sin\left(m \frac{\pi}{4} \right),$$

that is,

$$A_0 = 0, \quad A_1 = \sqrt{2}A, \quad A_2 = 2A, \quad A_3 = \sqrt{2}A, \quad A_4 = 0.$$

The magnitude of the standing wave's electric field vector at t_0 can be found from the equation of the standing wave (below here we take into account that $\omega t_0 = \pi/4$ and $\cos(\pi/4) = \sqrt{2}/2$):

$$E(t_0, x_m) = 2A \sin\left(m \frac{\pi}{4} \right) \cos(\omega t_0) = \sqrt{2} A \sin\left(m \frac{\pi}{4} \right).$$

4.8 INTERFERENCE AND COHERENCE OF ELECTROMAGNETIC WAVES

1. When electromagnetic waves generated by a number of sources propagate in the same space, their fields are superposed at each point. If two fields have the same oscillation direction and the same frequency ω, the amplitude of the resultant oscillation depends on their phase difference. The resulting amplitude can vary from a minimum value equal to the difference of the two wave amplitudes to a maximum value equal to their sum. If the amplitudes of two waves are equal and the oscillations have zero phase difference, the amplitude of the resultant wave is doubled, and thus, the wave intensity is quadrupled. In contrast, if the phases of the waves are shifted by π, they cancel out each other. The result of wave superposition varies from point to point in the overlap area. *Interference* is the phenomenon when the superposition of two or more waves leads to a redistribution of the wave energy in space, that is, the appearance of local maxima and minima of the intensity at various points.

An interference pattern cannot be obtained with any type of sources. For example, switching on two lamps instead of one increases the illumination but does not cause a redistribution of energy with the appearance of intensity maxima and minima, because these sources are not coherent. Only coherent waves, that is, the waves with a constant phase difference during the observation period, exhibit interference. For coherent waves, the phase difference is always constant. We note that the phases of all natural light sources vary randomly with a high rate. Therefore, the phase difference of two natural light sources also varies randomly in time. This leads to rapid changes in the interference pattern. Radiation detectors are not able to detect such fast changes but instead are sensitive only to the intensity averaged over time. Thus, using natural light sources, we detect simply the sum of the intensities of the two sources.

Consider two plane electromagnetic waves of the same linear polarization that interfere at some point with position vector **r**:

$$\mathbf{E}_1 = \mathbf{e}E_{10}\cos(\omega_1 t - \mathbf{k}_1 \cdot \mathbf{r} + \varphi_{10}),$$
$$\mathbf{E}_2 = \mathbf{e}E_{20}\cos(\omega_2 t - \mathbf{k}_2 \cdot \mathbf{r} + \varphi_{20}). \tag{4.72}$$

At this point, the electric fields are given by

$$\mathbf{E}_1 = \mathbf{e}E_{10}\cos(\omega_1 t + \varphi_1),$$
$$\mathbf{E}_2 = \mathbf{e}E_{20}\cos(\omega_2 t + \varphi_2), \tag{4.73}$$

where $\varphi_i(\mathbf{r}) = \varphi_{i0} - \mathbf{k}_i \cdot \mathbf{r}$ are the phases at the given point. The resultant net electric field is $\mathbf{E} = \mathbf{E}_1 + \mathbf{E}_2$ and, in accordance with Equation 4.65, the wave intensity is proportional to the square of the electric field averaged over the time:

$$I = \langle S \rangle \sim \langle \mathbf{E}^2 \rangle = \langle (\mathbf{E}_1 + \mathbf{E}_2)^2 \rangle = \langle \mathbf{E}_1^2 \rangle + \langle \mathbf{E}_2^2 \rangle + 2\langle \mathbf{E}_1 \cdot \mathbf{E}_2 \rangle. \tag{4.74}$$

The first two terms on the right-hand side define the intensities of the waves I_1 and I_2. The interference is observed if the third term is not zero. For that to occur, $\mathbf{E}_1$ and $\mathbf{E}_2$ cannot be perpendicular.

Let us consider the interference of two monochromatic plane waves $\mathbf{E}_1$ and $\mathbf{E}_2$ assuming that they are parallel, their angular frequencies $\omega_1 = \omega_2 = \omega$ and their wave vectors $|\mathbf{k}_1| = |\mathbf{k}_2| = \mathbf{k}$.

To be able to observe interference, the phase difference $\Delta\varphi(\mathbf{r}) = \varphi_2(\mathbf{r}) - \varphi_1(\mathbf{r})$ between the two waves at a given point **r** must be time independent.

The magnitude of the resultant electric field $E_0 = |\mathbf{E}_0|$ can be determined from the vector diagram shown in Figure 4.12. One can see that

$$E_0^2 = E_{10}^2 + E_{20}^2 + 2E_{10}E_{20}\cos(\varphi_2 - \varphi_1). \tag{4.75}$$

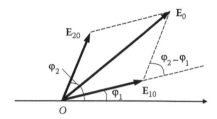

FIGURE 4.12   Vector diagram for sum of vectors $\mathbf{E}_1$ and $\mathbf{E}_2$.

Thus, for the resultant intensity at the given point, we come to

$$I = I_1 + I_2 + 2\sqrt{I_1 I_2}\left\langle\cos(\varphi_2 - \varphi_1)\right\rangle. \qquad (4.76)$$

In sources for which the waves are emitted by individual atoms that are independent from each other, φ_1 and φ_2 change independently. Therefore, the phase difference $\Delta\varphi$ continuously changes taking various values with equal probability. As a result, $\cos(\varphi_2 - \varphi_1)$ averaged over time is equal to zero, that is,

$$\left\langle\cos(\varphi_2 - \varphi_1)\right\rangle = 0. \qquad (4.77)$$

In this case, the total intensity of the resultant wave is just the sum of the intensities of the individual waves, that is, no interference is observed.

In contrast, if the phase difference at each point remains constant over time, the intensity at different points differs from the sum of the individual intensities of the combined waves, that is, interference takes place.

At different points, the superposition of waves depends on the factor $\cos \Delta\varphi$. In particular, when $\cos \Delta\varphi = 1$, the net intensity is maximum:

$$I_{\max} = I_1 + I_2 + 2\sqrt{I_1 I_2} = \left(\sqrt{I_1} + \sqrt{I_2}\right)^2. \qquad (4.78)$$

We get maximum intensity when

$$\Delta\varphi = \varphi_2 - \varphi_1 = 2\pi m, \qquad (4.79)$$

where $m = 0, 1, 2, \ldots$. Integer n is known as the interference order.

At points with $\cos \Delta\varphi = -1$, the intensity is minimum and equal to

$$I_{\min} = I_1 + I_2 - 2\sqrt{I_1 I_2} = \left(\sqrt{I_1} - \sqrt{I_2}\right)^2. \qquad (4.80)$$

Minima in intensity are observed at points for which

$$\Delta\varphi = \varphi_2 - \varphi_1 = (2m + 1)\pi. \qquad (4.81)$$

Equations 4.79 and 4.81 give the conditions for the observation of interference maxima and minima, respectively.

The wave phase at a point depends on the distance traveled by the wave. For coherent waves propagating along the z-axis, the interfering wave phases are given by

$$\varphi_i = \omega t - kz_i = \omega t - \frac{\omega}{c} z_i = \omega t - \frac{2\pi}{\lambda} z_i, \tag{4.82}$$

where λ is the wavelength in vacuum. The value $\Lambda = z_2 - z_1$ is called the path difference. The phase difference and the path difference are related as follows:

$$\Delta \varphi = \frac{2\pi}{\lambda} \Lambda. \tag{4.83}$$

Taking into account Equations 4.79 and 4.81, the conditions for interference maxima and minima can be written as

$$\Lambda = m\lambda \qquad \text{intensity maxima,}$$
$$\Lambda = (2m+1)\lambda/2 \quad \text{intensity minima.} \tag{4.84}$$

To generate coherent waves, various methods based on splitting waves originating from a single source into two or more waves are used. Such waves are coherent and can be used to create an interference pattern. Coherent light waves are emitted by laser sources.

2. In connection with interference of coherent waves, we would like to point out that it is necessary to distinguish **temporal** and **spatial coherence**. The process of interference discussed earlier is an idealization, since monochromatic wave

$$E = E_m \cos(\omega t - kr + \varphi)$$

with constant E_m, ω, k and φ is an abstraction. Any real wave is a result of a superposition of oscillations of all possible frequencies (wavelengths) in a finite range of frequencies $\Delta\omega$ and wave numbers Δk. The amplitude E_m and phase φ undergo random changes with time. If we consider interference of two waves at any particular point in space and take into account that frequencies and phases change with time, the electric field of each of the two waves can be presented as

$$E_1(t) = E_{1m}(t) \cos[\omega_1(t)t + \varphi_1(t)],$$
$$E_2(t) = E_{2m}(t) \cos[\omega_2(t)t + \varphi_2(t)].$$

Further, we assume that the amplitudes E_{1m} and E_{2m} are constants. In this case, the frequency variation and phase change can be reduced to phase variation only. Indeed,

$$E(t) = E_m \cos[\omega_0 t + (\omega(t) - \omega_0)t + \varphi(t)] = E_m \cos[\omega_0 t + \varphi'(t)],$$

where
   ω_0 is the central frequency of the wave
   $\varphi'(t) = (\omega(t) - \omega_0)t + \varphi(t)$

In the resulting function, only the phase oscillations undergo random changes. The time τ_{coh} during which the random variation of the phase $\varphi'(t)$ changes its value by an amount of the order of π is called the **coherence time**. During this time, the oscillation loses its initial phase and ceases to be coherent.

The distance $l_{coh} = c \cdot \tau_{coh}$ traveled by the wave during time τ_{coh} is called the **coherence length**. The coherence length is defined as the distance at which the random change of the phase becomes equal to π. Coherence of the oscillations that interfere at the same point in space but changes with time is called *temporal coherence*.

With the superposition of light from two incoherent sources, the interference is not observed. This means that independent sources are incoherent, even though their emission is monochromatic. The reasons are in the mechanism of light emission by the atoms of the light source. Excited atom emits a single pulse (a *wave train*) for a short period of time $\tau \approx 10^{-8}$ s. Having spent energy through radiation, the atom returns to a lower energy state that is almost always the lowest energy state known as the ground state. After a certain period of time, the atom can be excited again, receiving energy from an outside source and once again radiates. Thus, the observation of the interference of light is possible only when the optical path difference is smaller than the coherence length of the light source. The closer a wave is to a monochromatic wave, the smaller is the width of its frequency spectrum and the longer is its coherence time and therefore the coherence length. The coherence of the oscillations (which is determined by the degree of monochromaticity of the waves) that occur at the same point in space, as we already mentioned earlier, is called temporal coherence.

The frequency spectrum of a real wave includes frequencies from $\omega_0 - \Delta\omega/2$ to $\omega_0 + \Delta\omega/2$. Such a wave over time

$$\Delta t \ll \tau_{coh} = \frac{2\pi}{\Delta\omega}$$

can be approximately considered as a monochromatic wave with frequency ω_0. For example, for the visible sunlight, $\tau_{coh} \approx 10^{-14}$ s, and for lasers of continuous operation, $\tau_{coh} \approx 10^{-5}$ s.

Now let us take into account that the wave number $k = n\omega/c$, where n is the index of refraction. The frequency variation $\Delta\omega$ leads to a spread of values of the wave number $\Delta k = k(\Delta\omega/\omega)$. Thus, the temporal coherence is related to the variation of the magnitude of the wave vector **k** of the wave.

Along with temporal coherence, there is a spatial coherence of the wave too. Spatial coherence is associated with the range of directions of the wave vector **k**. Coherence of oscillations that occur at the same time but at different points in a plane perpendicular to the direction of propagation of the wave is called *spatial coherence*. The two sources (with certain degree of monochromaticity of light) of sizes that allow observing interference are called *spatially coherent*.

Exercise 4.7

A plane monochromatic light wave of wavelength $\lambda_0 = 0.50$ μm is incident normally on an opaque diaphragm with two long narrow slits separated by a distance $d = 2.00$ mm. A system of interference fringes is formed on a screen, which is placed behind the diaphragm at a distance $L = 1.00$ m. Find the spacing of the resulting interference fringes.

Solution. The two narrow slits can be considered as two sources of coherent waves, which interfere at the screen. The width of single interference fringe Δx is equal to the distance between two successive minima at the screen (or between two successive maxima) $\Delta x = x_{n+1} - x_n$ (see Figure 4.13). Using Figure 4.13, we can write

$$x_{n+1} = L \tan\theta_{n+1}, \quad x_n = L \tan\theta_n.$$

If we include the angles θ_{n+1} and θ_n in the equation for interference minima, we get

$$d \sin\theta_{n+1} = \left(2(n+1)+1\right)\frac{\lambda_0}{2}, \quad d \sin\theta_n = (2n+1)\frac{\lambda_0}{2}.$$

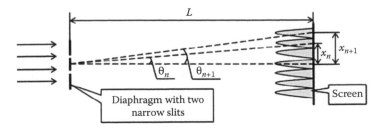

FIGURE 4.13　Light interference from two narrow slits.

According to the conditions of experiment $L \gg d$ and for small angles, the following equation is valid: $\sin \theta \approx \tan \theta$. Taking into account that $\sin \theta \approx \tan \theta$ and replacing $\tan \theta$ by $\sin \theta$, we get

$$\Delta x = L\left(\tan \theta_{n+1} - \tan \theta_n\right) \approx L\left(\sin \theta_{n+1} - \sin \theta_n\right)$$

$$= L\left[2(n+1)+1-(2n+1)\right]\frac{\lambda_0}{2d} = \frac{L\lambda_0}{d}.$$

We note that the spacing of two successive interference maxima (interference fringe) does not depend on n. The calculation gives us

$$\Delta x = \frac{L\lambda}{d} = \frac{1 \times 0.50 \times 10^{-6}}{2.50 \times 10^{-3}}\ \text{m} = 0.20\ \text{mm}.$$

PROBLEMS

4.1　The operating wavelength λ of a radar is 0.30 m. The radar emits 250 pulses per second with pulse duration $\tau = 5.00\ \mu s$. How many oscillations are contained in each pulse and what is the maximum range of the radar? (*Answer*: $n = 5.00 \times 10^3$ oscillations, $L = 6.00 \times 10^5$ m.)

4.2　A plane monochromatic electromagnetic wave of frequency $f = 200$ MHz propagates along the z-axis. At point z_0, the electric field is given by $E(t, z_0) = E_m \sin(2\pi ft - kz_0)$ where the amplitude E_m is equal to 75.0 mV/m. Write expressions for the electric field for points that are located at a distance of 2.00 m from the given point z_0 along the positive and negative directions of the z-axis. (*Answer*: $E(t, z_0 \pm 2) = 75.0 \times 10^{-3} \sin[4\pi(10^8 t - ((z_0 \pm 2)/3))]$ V/m.)

4.3　An electromagnetic plane wave with electric field vector $\mathbf{E}(t, \mathbf{r}) = \mathbf{E}_m \sin(\omega t - \mathbf{k} \cdot \mathbf{r})$ propagates in vacuum. Find the magnetic field vector at $t = T/4$ at a point with position vector whose magnitude $r = (5/4)\lambda$ (a) if the direction of the position vector coincides with the direction of wave vector and (b) the position vector forms an angle $\beta = \pi/6$ with the direction of the wave vector. Here, T and λ are the period and the wavelength of the wave, respectively. $\left(\ Answer\text{: (a) } \mathbf{H}(T/4, \mathbf{r}) = 0,\ \text{(b) } \mathbf{H}(T/4, \mathbf{r}) = \dfrac{\mathbf{k} \times \mathbf{E}_m}{k}\sqrt{\dfrac{\varepsilon_0}{\mu_0}}\dfrac{\sqrt{3}}{2}.\ \right)$

4.4　Two plane electromagnetic waves propagate in vacuum. One of the waves propagates along the x-axis, while the other propagates along the y-axis. The corresponding electric fields are given by

$$\mathbf{E}_1(t, \mathbf{r}) = \mathbf{E}_m \sin\left(\omega t - k_{\bar{x}}x\right), \quad \mathbf{E}_2(t, \mathbf{r}) = \mathbf{E}_m \sin\left(\omega t - k_y y\right)$$

where $k_x = k_y = k$. Both waves are linearly polarized along the z-axis. Find the average value of the energy density flux in the plane $x = y$. (Note that both waves have electric field amplitude E_m in the same direction, namely, in z direction. *Answer*: $\langle S \rangle = \sqrt{2\varepsilon_0/\mu_0}\,(E_m)^2$.)

4.5　Write general expressions for the electric and magnetic fields of a plane wave propagating along one of the coordinate axes (e.g., along the z-axis) as well as for the Poynting vector of the wave and the average flux density. Assume that a plane with dimensions substantially larger than the wavelength serves as the source of plane waves. The entire surface of the plane radiates harmonic wave along the direction perpendicular to the plane. The electric field vector oscillates in this plane as $E(t) = A \sin \omega t$. (*Part of the answer*: Average flux density $S_z = (1/2)\sqrt{\varepsilon_0/\mu_0}\, A^2$.)

4.6 When natural light passes through a system of two liner polarizers that have parallel polarization axes, its intensity is reduced by 50%. The intensity decreases by an additional factor of two when a quartz plate is placed in the beam path between the polarizers. At what angle has the polarization axis of the beam turned inside the quartz plate? Neglect absorption. (*Answer*: $\alpha = \pm\,(\pi/4) \pm n\pi$, where $n = 0, 1, 2, \ldots$.)

4.7 A plane monochromatic wave is incident normally on a plane with three identical parallel slits. Using the principle of superposition, determine the angular distribution of the intensity of the light transmitted through the slits if all the slits have a width equal to b, and the distance between the centers of consecutive slits is d (here $a + b = d$), and the wavelength is equal to λ. Determine the angular position of minima from single slit and additional minima due to interference from three slits. *Part of the answer*: The angular distribution of intensity has the following form:

$$I = I_0 \frac{\sin^2\!\left(\dfrac{\pi b}{\lambda}\sin\varphi\right)}{\left(\dfrac{\pi b}{\lambda}\sin\varphi\right)^2}\left[2\cos\!\left(\dfrac{2\pi d}{\lambda}\sin\varphi\right)+1\right]^2 = I_0 \frac{\sin^2\!\left(\dfrac{\pi b}{\lambda}\sin\varphi\right)}{\left(\dfrac{\pi b}{\lambda}\sin\varphi\right)^2}\frac{\sin^2\!\left(\dfrac{3\pi d}{\lambda}\sin\varphi\right)}{\sin^2\!\left(\dfrac{\pi d}{\lambda}\sin\varphi\right)}.$$

4.8 Determine the pressure exerted by a monochromatic light of frequency f that is incident onto the surface with area A with the energy reflectance coefficient $R = 0.50$. The angle of incidence of light $\alpha = 60°$ and the incident light intensity $I = 2.00$ kW/m^2. (*Answer*: $P = 5.00 \times 10^{-6}$ Pa.)

4.9 In natural light, the vibrations of the associated electric field at different directions have the same amplitude and are distributed with the same probability. Show that natural light can be represented as a sum of two noncoherent waves polarized in mutually perpendicular planes and have the same intensity.

4.10 A plane wave of frequency ω and electric field amplitude E_0 is incident normally on the reflective surface of a mirror. Determine the position of nodes and antinodes of the electric and magnetic fields of the standing wave, which is formed by the superposition of the incident and the reflected waves.

4.11 A plane monochromatic wave of angular frequency ω is incident on a thin film with a refractive index n at an angle α with respect to the film normal. Determine the minimum (nonzero) film thickness at which the intensity maxima or minima in the reflected light are observed. $\left(Answer: d_{\min} = \lambda/\!\left(4\sqrt{n^2 - \sin^2\alpha}\right)\right.$ to observe intensity maxima; $d'_{\min} = \lambda/\!\left(2\sqrt{n^2 - \sin^2\alpha}\right)$ to observe intensity minima.$\Big)$

4.12 Using the principle of superposition, determine the intensity of a wave comprising two linearly polarized plane waves propagating along the same direction in a transparent nonmagnetic medium ($\sigma = 0$, $\mu = 1$):

$$\mathbf{E}_1 = \mathbf{e}_1 E_0 \cos(\omega t - kz),$$

$$\mathbf{E}_2 = \mathbf{e}_2 E_0 \cos(\omega t - kz).$$

Unit vectors $\mathbf{e}_1$ and $\mathbf{e}_2$ indicate the directions of the wave polarization. In general, these vectors are different.

5 Electromagnetic Fields and Waves at the Interface between Two Media

In solving many problems in electrodynamics, we would like to know the electromagnetic field in a given spatial region. Such problems include, for example, the design of generator of high-frequency electromagnetic waves and the development of various wireless devices and power transmission lines. For the calculation of the electromagnetic field in each case, it is required to solve the proper electrodynamic problem taking place at the interfaces between different media. The practical problems of determining the fields or their sources are usually quite challenging as the interfaces may have complicated geometry.

The solution of such problems can be often obtained only by introducing a number of simplifying assumptions. Therefore, in practice, instead of the real problem, some model problem, which approximates the real situation, is considered. Examples of such problems are the use of the point-charge model and the line currents in determining the interaction forces between charged objects and current-carrying conductors, the calculation of the electric field in a parallel-plate capacitor and the magnetic field inside a solenoid neglecting the edge effects, and the study of transients in circuits with capacitances or inductances neglecting the finite velocity of propagation of an electromagnetic signal. In this chapter, in order to determine the amplitude and energy reflection and transmission coefficients of an electromagnetic wave at an interface, we use one of the simplifications—plane monochromatic wave.

5.1 BOUNDARY CONDITIONS AND INVERSE BOUNDARY VALUE PROBLEMS IN ELECTROMAGNETISM

The subject of this book is classical electromagnetism (also referred to as "classical electrodynamics") that deals with electromagnetic phenomena—whenever the relevant length scales and field strengths are large enough, the quantum mechanical effects can be ignored. For the solution of many problems in electromagnetism, it is necessary to know the electromagnetic field in a specific region of space. This task could involve the design of radiating systems (antennas), the maintenance of electromagnetic compatibility of radio devices, and the development of various power transmission lines. For the calculation of the electromagnetic field in each specific case, it is necessary to solve the corresponding problem. There are two wide types of problems in electromagnetism: direct and inverse problems. In direct problems, we determine the field that is created in a given region of space for a known spatial distribution of the field sources. In inverse problems, on the other hand, we determine the source system that creates an electromagnetic field with a given structure.

Direct problems in electromagnetism are often formulated as boundary value problems consisting of finding the electromagnetic field that satisfies certain conditions at the boundary of the considered medium. The problems of determining the fields or their sources are usually rather complicated. The solution of such problems can be obtained either from complicated numerical simulations or with the introduction of a number of simplifying assumptions. Therefore, in practice, instead of the real problem, we consider a related model problem that in its essence reflects the real situation. Examples of such problems are the use of point charges and line currents in determining the interaction forces between charged bodies and current-carrying conductors; the determination of the electric field in a parallel-plate capacitor; the determination of the magnetic field in a solenoid, where we neglect edge effects and thus the finite propagation velocity of electromagnetic signals; the assumption that the current in all parts of a closed circuit is the same; and the use of the model of a plane monochromatic wave in determining the reflection and transmission coefficients across the boundary between different media.

Another important idealization is a model that represents solids, liquids, and gases as continuous media. This does not take into account that a real medium has a specific discrete structure and atomic (or molecular) composition. The continuum model becomes inconsistent in the study of electromagnetic phenomena in a matter, when the wavelength of an electromagnetic wave is close to the average distance between the particles of a medium. The use of the continuum model, when applicable, allows the use of continuous functions for the mathematical description of the electromagnetic phenomena under study. In this case, the electric and magnetic fields in these media are assumed to be continuous but locally averaged. The material parameters of the medium—dielectric constant, conductivity, and permittivity—are also averaged parameters.

Equations 3.36, written in integral and differential forms, are the equations of the electromagnetic field in a medium. Representation of these equations in a differential form assumes that all quantities in space and time change continuously. The differential form of these equations is supplemented with the boundary conditions for the tangential and the normal components of electromagnetic fields at the interface between two media. For simplicity, we restrict ourselves to the derivation of the boundary conditions for static electric and magnetic fields. Generalization of the derivation for the case of time-varying fields can be carried out using the integral form of Maxwell's equations.

Exercise 5.1

Find the steplike change of the electric field E on an infinitely extended charged plane with surface charge density σ. Plot the dependence of the magnitude of the field on the distance from the charged plane.

Solution. In order to determine the electric field, we will use Gauss's law. We choose the surface of integration A to be cylinder of arbitrary height whose axis is perpendicular to the charged plane (see Figure 5.1). Let us divide the surface of integration into one side surface (curved) and two end surfaces (planar). The total flux of the electric field through surface A is

$$\Phi_E = \oint_A \mathbf{E} \cdot d\mathbf{A} = \int_{A_{side}} \mathbf{E} \cdot d\mathbf{A} + 2 \int_{A_{base}} \mathbf{E} \cdot d\mathbf{A} = \int_{A_{side}} EdA \cos \alpha_{side} + 2 \int_{A_{base}} EdA \cos \alpha_{base}$$

$$= 2 \int_{A_{base}} EdA \cos \alpha_{base} = 2EA_{base}.$$

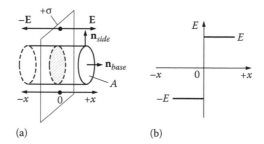

(a)   (b)

FIGURE 5.1 Calculation of the field of a charged plane (a) and (b) the x-component of the electric field versus x.

Here, we have taken into account the fact that the flux through the side surface is equal to zero since on this surface $\mathbf{E} \perp \mathbf{n}_{side}$ and $\cos \alpha_{side} = 0$. Here, $\mathbf{n}_{side}$ is the unit normal vector.

Because of the symmetry of the problem, the electric field lines for an infinitely extended and uniformly charged plane are normal to the plane and point out of the plane (for $\sigma > 0$). Thus, $\mathbf{E} \parallel \mathbf{n}_{base}$ and $\cos \alpha_{base} = 1$. The plane segment inside the cylinder is A_{base}. The charge enclosed by the cylinder is $q^{(in)} = \sigma A_{base}$. According to Gauss's law,

$$\Phi_E = 2EA_{base} = \frac{q^{(in)}}{\varepsilon_0} = \frac{\sigma A_{base}}{\varepsilon_0}.$$

Then, the coordinate dependence of the extended plane field can be written as

$$E(x) = \frac{\sigma}{2\varepsilon_0} \frac{x}{|x|}.$$

Since $x/|x|$ depends only on sign x and does not depend on the absolute value of x, from the obtained expression, it follows that the electric field of an extended plane is homogeneous, that is, it does not depend on the distance and its magnitude is $E = \sigma/2\varepsilon_0$. The field distribution of a uniformly charged plane along the coordinate x is shown in Figure 5.1b. The steplike change of the field direction when passing through the charged surface results in a change of electric field, that is,

$$\Delta E = E(x) - E(-x) = \frac{\sigma}{\varepsilon_0}.$$

Since the electric field does not depend on x, the steplike change of the field takes place directly at the charged surface.

5.2 BOUNDARY CONDITIONS FOR THE ELECTRIC FIELD OF AN ELECTROMAGNETIC WAVE

We will use Gauss's law to obtain the boundary conditions for the electric displacement vector $\mathbf{D}$. We assume that an interface between two media has a surface charge density σ_s. We select a closed surface in the form of a cylinder whose axis is perpendicular to the interface and whose bases are placed at equal distances from the interface (Figure 5.2).

According to Gauss's law (see Equation 1.80), the flux of the electric displacement through any closed surface is equal to the total charge within this surface.

Separating the fluxes through cylinder bases ($A_1 = A_2 = A$) and the curved surface (A_{lat}), we get

$$\oint_A \mathbf{D} \cdot d\mathbf{A} = D_{2n}A - D_{1n}A + \langle D_\tau \rangle A_{lat} = \sigma_s A, \tag{5.1}$$

where
$A_{lat} = 2\pi rh$ is the lateral surface of the cylinder, h and r are the height and radius of the cylinder
$\langle D_\tau \rangle$ is the value of a tangential component of the displacement averaged over a lateral surface D_τ

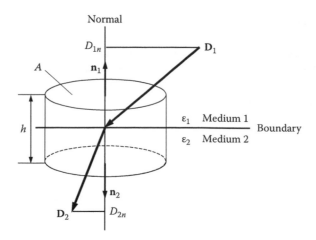

FIGURE 5.2 Boundary conditions for the electric displacement vector **D** based on Gauss's law.

In the limit $h \to 0$, A_{lat} goes to zero. Therefore, from Equation 5.1, it follows that for a charged boundary the normal component of the dielectric displacement vector has a discontinuity described by

$$D_{2n} - D_{1n} = \sigma_s. \tag{5.2}$$

If the boundary is not charged, the normal components of the vector **D** are continuous:

$$D_{2n} = D_{1n}. \tag{5.3}$$

Since media "1" and "2" have different dielectric permittivities, ε_1 and ε_2,

$$\mathbf{D}_i = \varepsilon_i \mathbf{E}_i = \kappa_i \varepsilon_0 \mathbf{E}_i$$

(here $i = 1, 2$), the normal component of the vector **E** has a discontinuity at the interface. From Equation 5.3, we get

$$\varepsilon_2 E_{2n} = \varepsilon_1 E_{1n}, \quad E_{2n} - E_{1n} = \left(\frac{\varepsilon_1}{\varepsilon_2} - 1 \right) E_{1n} = \left(\frac{\kappa_1}{\kappa_2} - 1 \right) E_{1n}. \tag{5.4}$$

It is important to stress here that $\varepsilon_1 / \varepsilon_2 = \kappa_1 \varepsilon_0 / \kappa_2 \varepsilon_0 = \kappa_1 / \kappa_2$, so the relations (5.4) for the electric field are written in the literature in two equivalent forms, using either dielectric permittivities ε_i or dimensionless dielectric permittivities κ_i. To avoid any confusion, we will use ε_i in all relations (like the first parts in Equation 5.4 with the ratio of permittivities), where ε_i can be replaced by κ_i.

The boundary conditions for the tangential components of the vectors **D** and **E** follow from the line integral of the electric field along a closed path (Figure 5.3).

Consider near the interface a closed rectangular path of length l and height h.
Since the line integral for the electric field vector is equal to zero, we have

$$\oint_L \mathbf{E} \cdot d\mathbf{l} = E_{1\tau} l - E_{2\tau} l + \langle E_n \rangle h = 0, \tag{5.5}$$

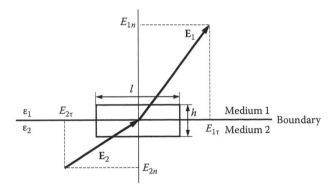

FIGURE 5.3 Derivation of boundary conditions for the electric field vector **E**.

where $\langle E_n \rangle$ is the averaged value of E_n on the sides of the rectangle. Going to the limit $h \to 0$, we get

$$E_{2\tau} = E_{1\tau}. \tag{5.6}$$

For the tangential components of the displacement, the boundary condition has the form

$$\frac{D_{2\tau}}{\varepsilon_2} = \frac{D_{1\tau}}{\varepsilon_1}, \quad D_{2\tau} - D_{1\tau} = \left(\frac{\varepsilon_2}{\varepsilon_1} - 1\right) D_{1\tau}. \tag{5.7}$$

Thus, at the interface that separates two dielectric media, the tangential component of the vector **E** is continuous, while the tangential component of **D** has a discontinuity.

Relations (5.2) through (5.7) allow to determine the orientation change of the electric field and the displacement at the interface. Let us introduce angles α_1 and α_2 that define the orientation of $\mathbf{D}_1$ and $\mathbf{D}_2$ as shown in Figure 5.4 (angles α_1 and α_2 are measured from the vectors to the interface). Using Equations 5.2 through 5.7, it is possible to write the following relation:

$$\frac{\tan \alpha_1}{\tan \alpha_2} = \frac{D_{1n} D_{2\tau}}{D_{1\tau} D_{2n}} = \frac{D_{2\tau}}{D_{1\tau}} = \frac{\varepsilon_2}{\varepsilon_1}.$$

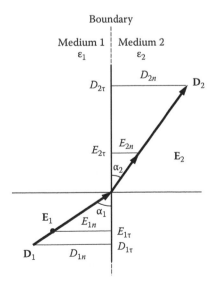

FIGURE 5.4 Vectors **D** and **E** at the interface of two dielectrics with $\varepsilon_1 < \varepsilon_2$.

A similar ratio can be written for the electric field vector:

$$\frac{\tan \alpha_1}{\tan \alpha_2} = \frac{E_{1n}E_{2\tau}}{E_{1\tau}E_{2n}} = \frac{E_{1n}}{E_{2n}} = \frac{\varepsilon_2}{\varepsilon_1}.$$

Figure 5.4 illustrates the change of direction of vectors $\mathbf{D}$ and $\mathbf{E}$ at the interface of two media. The figure shows that at the boundary of two isotropic media, both the displacement and the electric field are refracted to the same degree because vectors $\mathbf{D}$ and $\mathbf{E}$ are collinear: $\mathbf{D} = \varepsilon_0 \kappa \mathbf{E}$. Entering the dielectric with the higher dielectric permittivity, both vectors $\mathbf{D}$ and $\mathbf{E}$ move out of the normal to the interface.

Exercise 5.2

The magnitude of the electric field in the empty space between the plates of a charged parallel-plate capacitor is equal to $\mathbf{E}_0$. Find the relation between the electric field and the electric displacement if the space between the plates is partially filled by two layers of isotropic dielectrics with relative dielectric permittivities $\kappa_1 = 3$ and $\kappa_2 = 2$ (see Figure 5.5).

Solution. If placed in a homogeneous field, dielectrics are polarized in such a way that bound charges are formed at their surfaces. These charges create their own electric field that is directed in the opposite direction to the direction of field $\mathbf{E}_0$. As a result, the net electric field inside the dielectrics decreases and becomes $\mathbf{E}_1 = \mathbf{E}_0/\kappa_1 = \mathbf{E}_0/3$ in the first dielectric and $\mathbf{E}_2 = \mathbf{E}_0/\kappa_2 = \mathbf{E}_0/2$ in the second. In the space between plates, where there is no dielectric, the electric field remains $\mathbf{E}_0$.

The electric field that lines in a parallel-plate capacitor is perpendicular to the capacitor's plates and the dielectric interfaces. Thus, the normal component of the electric field at the interface of two dielectrics abruptly changes its value. The distribution of electric field lines is shown for this case in Figure 5.5a.

The electric displacement $\mathbf{D}$ can be written as

$$\mathbf{D} = \varepsilon_0 \kappa \mathbf{E}.$$

The electric displacement lines are also perpendicular to the dielectric interfaces. In contrast to the electric field, the normal component of vector $\mathbf{D}$ does not change its value at the interface of two

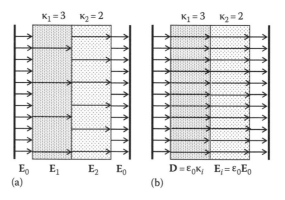

$\kappa_1 = 3$ $\kappa_2 = 2$ $\kappa_1 = 3$ $\kappa_2 = 2$

E_0 E_1 E_2 E_0 $\mathbf{D} = \varepsilon_0\kappa_i$ $\mathbf{E}_i = \varepsilon_0 E_0$

(a)   (b)

FIGURE 5.5   Schematics of the distribution of electric field $\mathbf{E}$ (a) and the displacement $\mathbf{D}$ (b) in a capacitor.

dielectrics, that is, it is continuous. Therefore, at any point between capacitor's plates $\mathbf{D} = \varepsilon_0\mathbf{E}_0$. Indeed,

$$\mathbf{D} = \varepsilon_0\kappa_1\mathbf{E}_1 = \frac{\varepsilon_0\kappa_1\mathbf{E}_0}{\kappa_1} = \varepsilon_0\mathbf{E}_0, \quad \mathbf{D} = \varepsilon_0\kappa_2\mathbf{E}_2 = \frac{\varepsilon_0\kappa_2\mathbf{E}_0}{\kappa_2} = \varepsilon_0\mathbf{E}_0.$$

The distribution of electric displacement lines is shown in Figure 5.5b.

5.3 BOUNDARY CONDITIONS FOR THE MAGNETIC FIELD OF AN ELECTROMAGNETIC WAVE

We now discuss the boundary conditions for the normal and tangential components of the magnetic field vector.

For the normal components of the magnetic field vector $\mathbf{B}$ (Figure 5.6), Gauss's law for the magnetic field gives

$$\oint_A \mathbf{B} \cdot d\mathbf{A} = B_{2n}A - B_{1n}A + \langle B_\tau \rangle A_{lat} = 0. \tag{5.8}$$

The flux of the magnetic field through the curved surface of the cylinder as $h \to 0$ becomes vanishingly small and can be neglected. Hence, at the interface between two homogeneous magnetic materials, the normal components of a magnetic field are continuous, that is,

$$B_{2n} = B_{1n}. \tag{5.9}$$

As $\mathbf{B}_i = \mu_i\mathbf{H}_i = \kappa_{mi}\mu_0\mathbf{H}_i$, the normal component of the magnetic intensity at the interface exhibits a discontinuity

$$\mu_2 H_{2n} = \mu_1 H_{1n}, \quad H_{2n} - H_{1n} = \left(\frac{\mu_1}{\mu_2} - 1\right)H_{1n} = \left(\frac{\kappa_{m1}}{\kappa_{m2}} - 1\right)H_{1n}. \tag{5.10}$$

By analogy to the dielectric permittivities (see discussion after Equation 5.4), we have $\mu_1/\mu_2 = \kappa_{m1}\mu_0/\kappa_{m2}\mu_0 = \kappa_{m1}/\kappa_{m2}$, so relations (5.10) for the magnetic field are also written in the literature in two

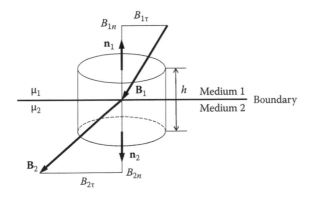

FIGURE 5.6 Boundary conditions for the magnetic field vector $\mathbf{B}$.

equivalent forms, using either magnetic permeabilities μ_i or dimensionless magnetic permeabilities κ_{mi}. To avoid any confusion, we will use μ_i in all relations (like the first parts in Equation 5.4 with the ratio of permeabilities), where μ_i can be replaced by κ_{mi}.

If we have a surface current along the interface then, following Ampere's law, we have

$$\oint_L \mathbf{H} \cdot d\mathbf{l} = H_{1\tau} l - H_{2\tau} l + \langle H_n \rangle h = I_s. \tag{5.11}$$

Here, I_s is the surface current perpendicular to the rectangular $l \cdot h$ shown in Figure 5.7.

The components of the line integral of the field on the shorter sides of the chosen contour of integration in the limit $h \to 0$ vanish (Figure 5.7). Thus, at the interface of two homogeneous magnetic materials, the tangential component of the magnetic intensity exhibits discontinuity:

$$H_{2\tau} - H_{1\tau} = j_s, \tag{5.12}$$

where $j_s = I_s/l$ is the surface current density perpendicular to the rectangular $l \cdot h$ (Figure 5.7).

If the surface current at the interface is zero, the tangential components of the magnetic intensity are continuous, that is,

$$H_{2\tau} = H_{1\tau}. \tag{5.13}$$

Thus, the tangential components of the magnetic field exhibit a discontinuity in the presence of a surface current (Equation 5.12). The following conditions at the interface are satisfied if $j_s = 0$:

$$\frac{B_{1\tau}}{\mu_1} = \frac{B_{2\tau}}{\mu_2}, \quad B_{2\tau} - B_{1\tau} = \left(\frac{\mu_2}{\mu_1} - 1 \right) B_{1\tau}. \tag{5.14}$$

By analogy to the case of electric field (see Figure 5.4), it is possible to determine the change of the orientation of the magnetic field at the interface using Equations 5.9 through 5.14.

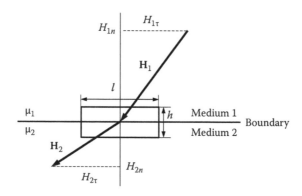

FIGURE 5.7   Boundary conditions for the magnetic intensity **H** based on Ampere's law.

Exercise 5.3

Calculate the magnetic intensity in the air gap of a ferromagnetic core with relative magnetic permeability, $\mu = \mu_0 \kappa_m$ inside a toroidal coil. Assume that the width of the gap d is much smaller than the average radius of the coil R (Figure 5.8).

Solution. According to Ampere's law,

$$\oint_L \mathbf{H} \cdot d\mathbf{l} = \sum_i I_i = NI,$$

where
  I_i is the current in each loop of the coil
  N is the total number of loops

The line integral of $\mathbf{H}$ along the closed path shown in Figure 5.8 is equal to

$$\oint_L \mathbf{H} \cdot d\mathbf{l} = H_1(2\pi R - d) + H_2 d = NI,$$

where
  H_1 is the magnetic intensity in the core
  H_2 is the magnetic intensity in the gap

The magnetic field lines in the toroidal coil are concentric circles, that is, the magnetic field is perpendicular to the core–gap interface. Since the normal component of magnetic intensity at the interface of two media is not continuous, we have from Equations 5.9 and 5.10

$$H_2 = \kappa_m H_1.$$

Let us substitute this expression into the preceding expression for the total current

$$H_1(2\pi R - d) + \kappa_m H_1 d = NI$$

and obtain

$$H_1 = \frac{NI}{2\pi R + d(\kappa_m - 1)}, \quad H_2 = \frac{\kappa_m NI}{2\pi R + d(\kappa_m - 1)}.$$

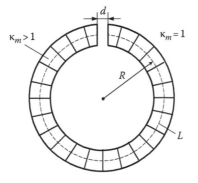

FIGURE 5.8 Ferromagnetic core with relative magnetic permeability $\mu = \mu_0 \kappa_m$ and air gap d inside a toroidal coil.

5.4 LAWS OF REFLECTION AND REFRACTION OF WAVES

Reflection of an electromagnetic wave is observed only in the presence of an inhomogeneity, and the simplest and the best pronounced example of the inhomogeneity is interface between two homogeneous media. To study the basic laws of reflection, we consider the simplest model of a plane wave reflected from a plane boundary between two media (Figure 5.9).

At the interface between two different media, in addition to reflection, we also have refraction of the incident electromagnetic wave. The laws of reflection and refraction follow from the boundary conditions for the electromagnetic field vectors. An analysis of the processes of reflection and refraction for monochromatic waves is carried out in the following.

Suppose that a linearly polarized plane wave with the wave vector $\mathbf{k}_1$ is incident from medium "1" at an angle θ_0 with respect to the interface normal at the plane boundary between two non-absorbing media with real permittivities and permeabilities ε_1, μ_1 and ε_2, μ_2. Here, $k_1 = \omega_1/v_1$, where ω_1 is the frequency and $v_1 = 1/\sqrt{\varepsilon_1 \mu_1} = c/\sqrt{\kappa_1 \kappa_{1m}}$ is the phase velocity of the wave in the medium "1."

In addition to the incident wave, in medium "1," the reflected wave will propagate at the angle θ', returning part of the incident wave energy to medium "1." We will assume that the frequency of this wave is ω' and that its wave vector is $\mathbf{k}_1'$ with magnitude $k_1' = \omega'/v_1$. Here, v_1 is the phase velocity of the wave in medium "1."

Thus, the net field in medium "1" is a superposition of the fields of the incident and the reflected waves.

The field in the second medium exists only in the form of *a refracted* wave that is transmitted through the interface. The refracted wave transfers part of the energy of the incident wave into medium "2." The refracted wave of frequency ω_2 will propagate at an angle of refraction θ_2 in the direction of wave vector $\mathbf{k}_2$, whose magnitude $k_2 = \omega_2/v_2$, where v_2 is the phase velocity of the wave in medium "2."

Using the principle of superposition, the fields in medium "1" can be represented as

$$\mathbf{E}_1 = \mathbf{E}_m^i e^{i(\omega_1 t - \mathbf{k}_1 \cdot \mathbf{r})} + \mathbf{E}_m^r e^{i(\omega_1' t - \mathbf{k}_1' \cdot \mathbf{r})},$$

$$\mathbf{H}_1 = \mathbf{H}_m^i e^{i(\omega_1 t - \mathbf{k}_1 \cdot \mathbf{r})} + \mathbf{H}_m^r e^{i(\omega_1' t - \mathbf{k}_1' \cdot \mathbf{r})},$$

$$(5.15)$$

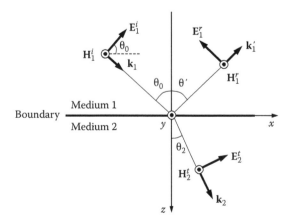

FIGURE 5.9   Orientation of magnetic and electric fields near the interface of two dielectric media.

and in medium "2" in the form

$$\mathbf{E}_2 = \mathbf{E}_m^t e^{i(\omega_2 t - \mathbf{k}_2 \cdot \mathbf{r})}, \quad \mathbf{H}_2 = \mathbf{H}_m^t e^{i(\omega_2 t - \mathbf{k}_2 \cdot \mathbf{r})}. \tag{5.16}$$

In the general case, the complex amplitudes of the electric and magnetic field vectors satisfy the following relationships:

$$E_m^i = Z_1 H_m^i, \quad E_m^r = Z_1 H_m^r, \quad E_m^t = Z_2 H_m^t, \tag{5.17}$$

where $Z_j = \sqrt{\mu_j / \varepsilon_j} = \sqrt{\mu_0 \kappa_{mj} / \varepsilon_0 \kappa_j}$ are the wave resistances (impedances) of the media. These equations together with the boundary conditions allow us to solve the problem of determining the amplitudes of the reflected and transmitted waves.

The wave vectors of all three waves have the same direction as the respective Poynting vectors in the case of isotropic media and lie in the same plane called the ***plane of incidence*** defined by the interface and its normal. It is necessary to distinguish two cases depending on the orientation of the electric field vector, that is, the polarization of the incident wave with respect to the plane of incidence. First, let us consider the case where the electric field vector **E** of the incident wave lies in the plane of incidence; it follows that the magnetic field vector **H** is perpendicular to this plane (Figure 5.9). Often, this type of orientation of the wave vectors of the fields is called ***parallel polarization*** and denoted by the index "*p*." In optics, this polarization is called the ***p-polarization***; in the following text we will use this terminology. For *p*-polarization, the vector **H** for each of the three waves is perpendicular to the plane of incidence with component H_y (see coordinate system in Figure 5.9). The electric field vector **E** has two components (with respect to the interface)—tangential E_x and normal E_z:

$$E_{1x}^i = E_1^i \cos\theta_0, \quad E_{1x}^r = -E_1^r \cos\theta', \quad E_{2x}^t = E_2^t \cos\theta_2,$$

$$E_{1z}^i = -E_1^i \sin\theta_0, \quad E_{1z}^r = -E_{1z}^r \sin\theta', \quad E_{2z}^t = -E_2^t \sin\theta_2. \tag{5.18}$$

We note that the scalar products of the wave vectors and the position vector are represented as

$$\mathbf{k}_1 \cdot \mathbf{r} = k_{1x}x + k_{1z}z = k_1 x \sin\theta_0 + k_1 z \cos\theta_0,$$

$$\mathbf{k}_1' \cdot \mathbf{r} = k_{1x}'x + k_{1z}'z = k_1' x \sin\theta' - k_1' z \cos\theta', \tag{5.19}$$

$$\mathbf{k}_2 \cdot \mathbf{r} = k_{2x}x + k_{2z}z = k_2 x \sin\theta_2 + k_2 z \cos\theta_2.$$

As a result, it is possible to write the expressions for the corresponding projections of the electric and magnetic fields as follows:

$$H_{1y} = H_{1y}^i + H_{1y}^r = H_0 e^{i(\omega_1 t - k_1 \sin\theta_0 \cdot x - k_1 \cos\theta_0 \cdot z)} + H_m^r e^{i(\omega_1 t - k_1' \sin\theta' \cdot x + k_1' \cos\theta' \cdot z)},$$

$$E_{1x} = E_{1x}^i + E_{1x}^r = Z_1 H_{1y}^i \cos\theta_0 - Z_1 H_{1y}^r \cos\theta',$$

$$E_{1z} = E_{1z}^i + E_{1z}^r = -Z_1 H_{1y}^i \sin\theta_0 - Z_1 H_{1y}^r \sin\theta', \tag{5.20}$$

$$H_{2y} = H_{2y}^t = H_m^t e^{i(\omega_2 t - k_2 \sin\theta_2 \cdot x - k_2 \cos\theta_2 \cdot z)},$$

$$E_{2x} = E_{2x}^t = Z_2 H_{2y}^t \cos\theta_2, \quad E_{2z} = E_{2z}^t = -Z_2 H_{2y}^t \sin\theta_2,$$

where H_0, H_m^r, and H_m^t are the amplitudes of the magnetic field of the incident, reflected, and transmitted waves. At the interface (i.e., at $z = 0$), the boundary conditions (5.6) and (5.13) for the tangential components of electric and magnetic wave fields must be satisfied:

$$H_{1y} = H_{2y}, \quad E_{1x} = E_{2x}$$

or

$$H_{1y}^i + H_{1y}^r = H_{2y}^t, \quad E_{1x}^i + E_{1x}^r = E_{2x}^t. \tag{5.21}$$

By combining Equation 5.21 with Equation 5.20 and taking into account that $z = 0$, we obtain

$$e^{i(\omega_1 t - k_1 \sin\theta_0 \cdot x)} + r_p e^{i(\omega_1' t - k_1' \sin\theta' \cdot x)} = t_p e^{i(\omega_2 t - k_2 \sin\theta_2 \cdot x)},$$

$$Z_1 \cos\theta_0 \cdot e^{i(\omega_1 t - k_1 \sin\theta_0 \cdot x)} - Z_1 r_p \cos\theta' e^{i(\omega_1' t - k_1' \sin\theta' \cdot x)}$$

$$= Z_2 t_p \cos\theta_2 e^{i(\omega_2 t - k_2 \sin\theta_2 \cdot x)}, \tag{5.22}$$

$$r_p = \frac{H_m^r}{H_0}, \quad t_p = \frac{H_m^t}{H_0}.$$

These relations are satisfied at any time and at any point of the interface irrespective of the coordinate x, so the time-dependent factors and coordinate-dependent factors in all terms should be the same. Thus, we arrive at the equations

$$\omega_1 = \omega_1' = \omega_2, \quad k_1 \sin\theta_0 = k_1 \sin\theta' = k_2 \sin\theta_2. \tag{5.23}$$

From here the important laws follow: (1) *the frequencies of the reflected and transmitted waves are equal to the frequency of the incident wave*, (2) *the angle of reflection is equal to the angle of incidence*,

$$\theta' = \theta_0 \tag{5.24}$$

and (3) *Snell's law that relates the angle of refraction with the angle of incidence*

$$\frac{\sin\theta_0}{\sin\theta_2} = \frac{k_2}{k_1} = \frac{v_1}{v_2} = \sqrt{\frac{\varepsilon_2}{\varepsilon_1}} = \frac{n_2}{n_1}. \tag{5.25}$$

Here, we introduce the refractive indices of media "1" and "2," $n_1 = \sqrt{\kappa_1}$ and $n_2 = \sqrt{\kappa_2}$, respectively. The laws of reflection and refraction are valid for waves of other polarizations.

Exercise 5.4

A p-polarized wave is incident at an angle θ_0 on the surface of medium "2" with refractive index $n(y) = n_0 + by$, where b is positive constant, y is the coordinate along the normal to the interface, and n_0 is the refractive index in medium "1." Find the trajectory of the refracted light beam in the medium "2." Note that medium "2" exhibits optical inhomogeneity.

Solution. As the refractive index depends on coordinate y, the light beam will undergo refraction at each point y along its trajectory. Thus, it is possible to introduce the refraction angle $\theta_2(y)$ for any point A that lies on the beam's trajectory (see Figure 5.10):

$$\frac{dy}{dx} = \cot \theta_2(y).$$

Let us split the medium 2 on thin slices that are parallel to the surface between two media and have thickness Δy (in the limit it is dy). Snell's law for two consecutive slices is

$$\frac{\sin\theta_0}{\sin\theta_{21}} = \frac{n(y_1)}{n_0}, \quad \frac{\sin\theta_{21}}{\sin\theta_{22}} = \frac{n(y_2)}{n(y_1)}, \quad \frac{\sin\theta_{22}}{\sin\theta_{23}} = \frac{n(y_3)}{n(y_2)}, \quad \frac{\sin\theta_{23}}{\sin\theta_{24}} = \frac{n(y_4)}{n(y_3)}, \ldots,$$

where the first index in θ_{2j} indicates the medium and the second—the slices number. Multiplying the left- and right-hand parts of the preceding equations, we get that at each point y, the following is true:

$$\frac{\sin\theta_0}{\sin\theta_2(y)} = \frac{n(y)}{n_0} = \tilde{n}(y), \quad \cos\theta_2 = \sqrt{1 - \sin^2\theta_2} = \frac{\sqrt{\tilde{n}^2 - \sin^2\theta_0}}{\tilde{n}}.$$

Here, $\tilde{n} = n/n_0$. Taking into account these expressions, we can write

$$\frac{dy}{dx} = \frac{\sqrt{\tilde{n}^2 - \sin^2\theta_0}}{\sin\theta_0} \quad \text{or} \quad dx = \frac{\sin\theta_0}{\sqrt{\tilde{n}^2(y) - \sin^2\theta_0}} \, dy.$$

By integrating the last equation, we find the beam's trajectory:

$$x = \int_0^y \frac{\sin\theta_0}{\sqrt{(1+\tilde{b}y)^2 - \sin^2\theta_0}} \, dy = \frac{\sin\theta_0}{\tilde{b}} \ln \left[\left(\frac{1+\tilde{b}y}{\sin\theta_0} \right) + \sqrt{\left(\frac{1+\tilde{b}y}{\sin\theta_0} \right)^2 - 1} \, \right] \Bigg|_0^y.$$

Here, $\tilde{b} = b/n_0$. After we put values of ln on the upper limit, y, and the lower limit, 0, we get

$$x(y) = \frac{\sin\theta_0}{\tilde{b}} \cdot \ln \frac{(1+\tilde{b}y) + \sqrt{(1+\tilde{b}y)^2 - \sin^2\theta_0}}{1 + \sqrt{1 - \sin^2\theta_0}} =$$

$$= \frac{n_0 \sin\theta_0}{b} \cdot \ln \frac{(n_0 + by) + \sqrt{(n_0 + by)^2 - n_0^2 \sin^2\theta_0}}{n_0(1 + \cos\theta_0)}.$$

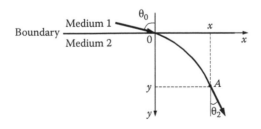

FIGURE 5.10 Refraction in a medium with variable refractive index: $n(y) = n_0 + by$.

5.5 REFLECTION AND TRANSMISSION COEFFICIENTS OF WAVES

In Equations 5.22, we introduced the *coefficients of reflection and transmission* for the *p*-polarized wave:

$$r_p = \frac{H_m^r}{H_0}, \quad t_p = \frac{H_m^t}{H_0}, \tag{5.26}$$

which generally are complex quantities. Taking into account Equations 5.23 and 5.24, the system of Equations 5.22 can be written as

$$1 + r_p = t_p, \quad Z_1(1 - r_p)\cos\theta_0 = Z_2 t_p \cos\theta_2. \tag{5.27}$$

Solving this system, we obtain the following expressions for the reflection and transmission coefficients, known as *Fresnel equations*:

$$r_p = \frac{Z_1 \cos\theta_0 - Z_2 \cos\theta_2}{Z_1 \cos\theta_0 + Z_2 \cos\theta_2} = \left|r_p\right| e^{i\psi_{rp}}$$

$$t_p = \frac{2Z_2 \cos\theta_0}{Z_1 \cos\theta_0 + Z_2 \cos\theta_2} = \left|t_p\right| e^{i\psi_{tp}} \tag{5.28}$$

where

$|r_p|$ and $|t_p|$ determine the real parts of the reflection and transmission coefficients

ψ_{rp} and ψ_{tp} are the phase shifts of the reflected and transmitted waves with respect to the incident wave, respectively

If media "1" and "2" are nonabsorbing dielectrics, that is, we have no loss of electromagnetic energy, the quantities r_p and t_p are real, exception is for the special case of *total internal reflection* (**TIR**), which will be discussed in the following text. This means that the phase shifts of the reflected and transmitted waves are either zero or π to get $e^{i\psi_{rp}}$ and $e^{i\psi_{tp}}$ either +1 or −1. Since always $t_p > 0$, then $\psi_{tp} = 0$, that is, the phase of a transmitted wave is equal to the phase of an incident wave. The phase shift ψ_{rp} of the reflected wave is determined by the sign of the numerator in Equation 5.28 for the coefficient r_p.

Let us now consider the case in which the magnetic field vector **H** of the incident wave lies in the plane of incidence, and thus, the electric field vector **E** is perpendicular to this plane. It is accepted to call this type of polarization of a wave field *perpendicular* and to denote by the index "⊥." It is known as the *s-polarization* following the terminology in optics.

By performing an analysis along the same lines as in the case of the *p*-polarization, we arrive at the following expressions for the reflection and transmission coefficients:

$$r_s = \frac{E_m^r}{E_0} = \frac{Z_2 \cos\theta_0 - Z_1 \cos\theta_2}{Z_2 \cos\theta_0 + Z_1 \cos\theta_2} = \left|r_s\right| e^{i\psi_{rs}},$$

$$t_s = \frac{E_m^t}{E_0} = \frac{2Z_2 \cos\theta_0}{Z_2 \cos\theta_0 + Z_1 \cos\theta_2} = \left|t_s\right| e^{i\psi_{ts}}, \tag{5.29}$$

where E_0 is the electric field amplitude of the incident wave. Equations 5.28 and 5.29 for the reflection and transmission coefficients of waves of two orthogonal polarizations are collectively called

Fresnel equations. In the general case, a plane wave with arbitrary polarization can be represented as the sum of waves with *p*- and *s*-polarizations; then using Fresnel equations, it is possible to find the electric and magnetic fields of the reflected and transmitted waves for an arbitrary polarization of the incident wave.

Let us now explore Fresnel equations for some important special cases. In the case of normal incidence, the angle $\theta_0 = 0$ and the relations (5.28) and (5.29) take the form

$$r_s = \frac{Z_2 - Z_1}{Z_2 + Z_1} = \frac{\sqrt{\varepsilon_1 \mu_2} - \sqrt{\varepsilon_2 \mu_1}}{\sqrt{\varepsilon_1 \mu_2} + \sqrt{\varepsilon_2 \mu_1}}, \quad r_p = -r_s,$$

$$t_s = \frac{2 Z_2}{Z_2 + Z_1} = \frac{2\sqrt{\varepsilon_1 \mu_2}}{\sqrt{\varepsilon_1 \mu_2} + \sqrt{\varepsilon_2 \mu_1}}, \quad t_p = t_s. \tag{5.30}$$

We now write down these equations for the important case of nonmagnetic media when $\kappa_{m1} = \kappa_{m2} = 1$

$$r_s = -r_p = \frac{\sqrt{\varepsilon_1} - \sqrt{\varepsilon_2}}{\sqrt{\varepsilon_1} + \sqrt{\varepsilon_2}} = \frac{n_1 - n_2}{n_1 + n_2}, \quad t_s = t_p = \frac{2\sqrt{\varepsilon_1}}{\sqrt{\varepsilon_1} + \sqrt{\varepsilon_2}} = \frac{2 n_1}{n_1 + n_2}. \tag{5.31}$$

We use Fresnel equations for an arbitrary incident angle in the case of nonmagnetic media where $n_j = \sqrt{\kappa_j}$ and $Z_j = \sqrt{\mu_0 / \varepsilon_0 \kappa_j}$; Snell's law equation (5.25) can be rewritten as

$$\frac{\sin \theta_0}{\sin \theta_2} = \sqrt{\frac{\varepsilon_2}{\varepsilon_1}} = \frac{Z_1}{Z_2}. \tag{5.32}$$

Taking into account this relation, Fresnel's equations (5.28) and (5.29) take the following form:

$$r_p = -\frac{\tan(\theta_2 - \theta_0)}{\tan(\theta_2 + \theta_0)}, \quad t_p = \frac{2 \sin \theta_2 \cos \theta_0}{\sin(\theta_2 + \theta_0) \cos(\theta_2 - \theta_0)},$$

$$r_s = \frac{\sin(\theta_2 - \theta_0)}{\sin(\theta_2 + \theta_0)}, \quad t_s = \frac{2 \sin \theta_2 \cos \theta_0}{\sin(\theta_2 + \theta_0)}. \tag{5.33}$$

Equations 5.33 are valid under the assumption that media "1" and "2" are lossless and their dielectric permittivities ε_1 and ε_2 are real quantities. Thus, the refraction angle, θ_2, and the reflection and transmission coefficients in Equation 5.33 are also real quantities (except for the case of TIR).

Under certain conditions, the reflection coefficient of a wave with *p*-polarization can vanish. The angle of incidence $\theta_0 = \theta_B$, at which $r_p = 0$ and at which all the energy of the incident wave is transmitted into medium "2," is called *Brewster's angle* (see Figure 5.11).

From Equation 5.33 for r_p, it follows that at the angle of incidence θ_0, which is equal to Brewster's angle, the following relation is valid:

$$\tan(\theta_2 + \theta_0) = \infty \quad \text{or} \quad \theta_0 + \theta_2 = \pi/2. \tag{5.34}$$

Substituting the angle $\theta_2 = \pi/2 - \theta_0$ into Equation 5.32, for a Brewster angle $\theta_0 = \theta_B$, we obtain

$$\tan \theta_B = \sqrt{\varepsilon_2 / \varepsilon_1} \quad \text{or} \quad \theta_B = \arctan(n_2 / n_1). \tag{5.35}$$

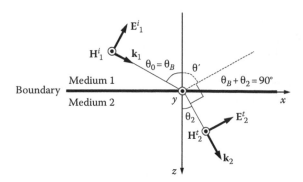

FIGURE 5.11 Wave with p-polarization totally transmitted into medium "2" for incidence angle $\theta_0 = \theta_B$.

For a wave with s-polarization, the reflection coefficient does not vanish for any angle of incidence. Therefore, if a wave of arbitrary elliptical polarization is incident on the media interface at Brewster's angle, the reflected wave contains only the component with the s-polarization, that is, it will appear linearly polarized. For waves with the p- and s-polarizations that are incident from vacuum on the surface of a polystyrene with $\kappa = 2.56$, the angular dependences of the absolute values of the reflection coefficients for the case of air and glass are shown in Figure 5.12.

Exercise 5.5

A light beam is incident at an angle $60°$ on a glass plate with a thickness of $d = 10.0$ cm. Find the shift S of the beam by the plate if the plate is immersed in water. The refractive indices of glass and water are $n_2 = 1.50$ and $n_1 = 1.33$, respectively.

Solution. We calculate the diagonal OC from two right-angle triangles AOC and BOC (see Figure 5.13) and obtain the following relation between d and S:

$$\frac{d}{\cos\theta_2} = \frac{S}{\sin(\theta_0 - \theta_2)} = OC.$$

From this expression, we can find the shift of the beam by the glass plate

$$S = \frac{d\sin(\theta_0 - \theta_2)}{\cos\theta_2} = d\left(\sin\theta_0 - \frac{\cos\theta_0 \cdot \sin\theta_2}{\cos\theta_2}\right).$$

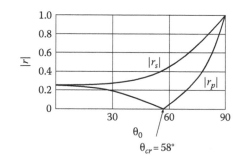

FIGURE 5.12 Angle dependence of the absolute value of the reflection coefficients for waves with p- and s-polarizations that are incident from vacuum on the surface of a polystyrene with $\kappa = 2.56$.

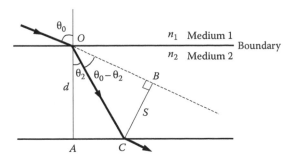

FIGURE 5.13   Parallel displacement of a beam after propagation through a glass plate.

According to Snell's law of refraction,

$$\frac{\sin\theta_0}{\sin\theta_2} = \frac{n_2}{n_1}, \quad \sin\theta_2 = \frac{n_1 \sin\theta_0}{n_2}.$$

Therefore,

$$\cos\theta_2 = \sqrt{1-\sin^2\theta_2} = \frac{1}{n_2}\sqrt{n_2^2 - n_1^2 \sin^2\theta_0}.$$

By substituting the derived relations into the expression for S, we get

$$S = d\left(\sin\theta_0 - \frac{n_1 n_2 \cos\theta_0 \sin\theta_0}{n_2\sqrt{n_2^2 - n_1^2 \sin^2\theta_0}}\right) = d\left(\sin\theta_0 - \frac{n_1 \sin 2\theta_0}{2\sqrt{n_2^2 - n_1^2 \sin^2\theta_0}}\right)$$

$$S = 10\left(\sin 60° - \frac{1.33 \sin 120°}{2\sqrt{1.50^2 - 1.33^2 \sin^2 60°}}\right) \approx 2.67 \text{ cm}.$$

5.6   TOTAL INTERNAL REFLECTION

Consider Snell's law (see Equation 5.25) in the case when medium "2" is optically less dense than medium "1," that is, $n_2 < n_1$. In this case, $\theta_2 > \theta_0$, and therefore, there is a critical value of the angle of incidence $\theta_0 = \theta_{cr}$, at which the refraction angle reaches its limiting value $\theta_2 = \pi/2$, and the transmitted wave propagates along the interface as shown in Figure 5.14:

$$\theta_{cr} = \arcsin\left(\frac{n_2}{n_1}\right) = \arcsin\left(\sqrt{\frac{\varepsilon_2}{\varepsilon_1}}\right). \tag{5.36}$$

At $\theta_0 > \theta_{cr}$, the transmitted wave in the usual sense does not exist, and all the energy of the incident wave is completely returned to the first medium. This phenomenon is called **TIR** and the angle θ_{cr}—angle of TIR.

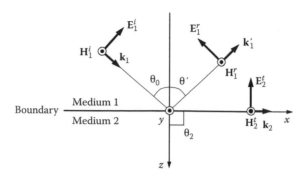

FIGURE 5.14 Total internal reflection for $n_1 > n_2$ and $\theta_0 = \theta_{cr}$.

Let us consider the electric fields in media "1" and "2" at TIR. We introduce a coordinate system in the same way as in Figure 5.14. The electric field in medium "1" is a superposition of the incident and reflected waves

$$\mathbf{E}_1 = \mathbf{E}_1^i + \mathbf{E}_1^r =$$

$$= \mathbf{E}_m^i e^{i(\omega t - k_1 \sin\theta \cdot x - k_1 \cos\theta \cdot z)} + \mathbf{E}_m^r e^{i(\omega t - k_1 \sin\theta \cdot x + k_1 \cos\theta \cdot z)}. \tag{5.37}$$

In the case of total reflection, the amplitudes of the incident and reflected waves are equal in magnitude and differ only in the phase factor $e^{i\psi}$. Thus, Equation 5.37 takes the form

$$\mathbf{E}_1 = |\mathbf{E}_m| \left(e^{-ik_1 \cos\theta_0 \cdot z} + e^{i(k_1 \cos\theta_0 \cdot z + \psi)} \right) e^{i(\omega t - k_1 \sin\theta_0 \cdot x)}$$

$$= |\mathbf{E}_m| \left(e^{-i(k_1 \cos\theta_0 \cdot z + \psi/2)} + e^{i(k_1 \cos\theta_0 \cdot z + \psi/2)} \right) e^{i(\omega t - k_1 \sin\theta_0 \cdot x + \psi/2)}$$

$$= 2|\mathbf{E}_m| \cos\left(k_1 \cos\theta_0 \cdot z + \psi/2 \right) e^{i(\omega t - k_1 \sin\theta_0 \cdot x + \psi/2)}. \tag{5.38}$$

From Equation 5.38, it follows that in medium "1" along the x-direction, we have a traveling wave that propagates with phase velocity

$$v_1 = \frac{\omega}{k_1 \sin\theta_0} = \frac{c}{n_1 \sin\theta_0}. \tag{5.39}$$

The amplitude of this wave has periodic dependence on the coordinate z. This means that the field in the direction of z-axis is a standing wave with a characteristic alternation of maxima (antinodes) and minima (nodes) (refer Section 4.7 about standing waves). Despite the fact that the reflection is total, an electromagnetic field will also be present in medium "2." Let us write down formally the expression for the transmitted wave

$$\mathbf{E}_2 = \mathbf{E}_m^t e^{i(\omega t - k_2 \sin\theta_2 \cdot x - k_2 \cos\theta_2 \cdot z)}. \tag{5.40}$$

At total reflection, $\sin\theta_2 = (n_1/n_2)\sin\theta_0 > 1$. This is impossible, if the refraction angle θ_2 is considered as a real number. However, if we broaden our understanding and treat θ_2 as the complex (or imaginary) number, then the sine of a complex argument can take any values. Cosine of the complex angle θ_2 will thus be an imaginary quantity:

$$\cos\theta_2 = \sqrt{1 - \sin^2\theta_2} = \pm i\sqrt{\left(\frac{n_1}{n_2}\right)^2 \sin^2\theta_0 - 1} = \pm i\alpha. \tag{5.41}$$

Substituting the expressions for $\cos \theta_2$ into Equation 5.40, we obtain

$$\mathbf{E}_2 = \mathbf{E}_m^t e^{\mp \alpha k_2 z} e^{i(\omega t - k_0 \sin \theta_2 \cdot x)}. \tag{5.42}$$

With increasing z, we expect a corresponding decrease in the amplitude of the transmitted wave. Hence, the sign of the first exponent in Equation 5.42 is negative ($\alpha > 0$). The field in medium "2" is a wave traveling along the x-axis with a phase velocity

$$v_2 = \frac{\omega}{k_2 \sin \theta_2} = \frac{c}{n_2 \sin \theta_2}. \tag{5.43}$$

The wave amplitude of Equation 5.42 decreases exponentially along the direction of z-axis. This is an example of a wave for which the directions of propagation of phase and amplitude fronts do not coincide. Such wave is called **an inhomogeneous wave**.

The depth d at which the field amplitude of the transmitted wave decreases by a factor e time is $1/\alpha k_2$, that is, it is of the order of magnitude of the wavelength in medium "2." Thus, in the medium "2," the field at TIR is concentrated in a layer rather close to the interface.

A wave of this type, which propagates parallel to the interface, is referred to as a **surface wave**. This wave is also called the **slow wave** since its phase velocity (see Equation 5.43) in the direction of the x-axis is less than the phase velocity of a homogeneous wave in this medium, which is $v_2 = \omega/k_2$.

A more detailed analysis shows that at TIR, the moduli of amplitude of reflection coefficients for waves with s- and p-polarizations are equal to unity, that is,

$$|r_s| = |r_p| = 1.$$

However, the phase of these coefficients is different and is a function of the angle θ_0:

$$\tan \frac{\varphi_s}{2} = -\frac{(\varepsilon_1 \sin^2 \theta_0 - \varepsilon_2)^{1/2}}{\sqrt{\varepsilon_1} \cos \theta_0}, \quad \tan \frac{\varphi_p}{2} = \frac{\left[\varepsilon_1(\varepsilon_1 \sin^2 \theta_0 - \varepsilon_2)\right]^{1/2}}{\varepsilon_2 \cos \theta_0}.$$

Thus, if the wave, whose polarization plane is inclined to the plane of incidence at some angle, experiences TIR, the reflected wave becomes elliptically polarized.

It can be shown that the average energy flux density $\langle S_z \rangle$ of the wave in the direction of z-axis in the second medium is zero. However, there is an oscillating component of the Poynting vector S_z. Thus, the energy transfer by a surface wave is carried out only in the direction of the x-axis, that is, along the surface of the medium.

Exercise 5.6

When a light beam from medium "1" transmits to medium "2," the angle of refraction is 45°. When the light beam travels from medium "1" to medium "3," the refraction angle is 30° (the angle of incidence is the same for both cases). Find the angle of TIR for a beam that is reflected from medium "3" to medium "2."

Solution. The law of refraction applied to the first and second media is

$$\frac{\sin \theta_0}{\sin \theta_{12}} = \frac{n_2}{n_1}, \quad n_2 = n_1 \frac{\sin \theta_0}{\sin \theta_{12}}.$$

The law of refraction applied to the first and third media is

$$\frac{\sin\theta_0}{\sin\theta_{13}} = \frac{n_3}{n_1}, \quad n_3 = n_1 \frac{\sin\theta_0}{\sin\theta_{13}}.$$

Snell's law applied to the second and third media under the condition of TIR gives

$$\frac{\sin\theta_{cr}}{\sin(\pi/2)} = \frac{n_2}{n_3}, \quad \sin\theta_{cr} = \frac{n_2}{n_3} = \frac{\sin\theta_{13}}{\sin\theta_{12}}.$$

Since $\theta_{12} = 45°$ and $\theta_{13} = 30°$, we get

$$\sin\theta_{cr} = \frac{1}{\sqrt{2}}, \quad \theta_{cr} = 45°.$$

5.7   REFLECTION OF A WAVE FROM A DIELECTRIC PLATE

Consider a dielectric slab (medium "2") of thickness d with refractive index n_2. The refractive indices of the surrounding media "1" and "3" are n_1 and n_3, respectively. We assume that all media are homogeneous and isotropic. In this case, their refractive indices are real constants.

In the following text, we obtain equations for the reflection and transmission coefficients of an electromagnetic wave incident on the slab from the medium "1." We consider the case of normal incidence, for which the coefficients for the p- and s-polarized incident waves are the same. The wave vectors and electric and magnetic field vectors in all three areas are shown in Figure 5.15.

The field vectors in medium "1" are written as superposition of the incident and reflected waves propagating in opposite directions:

$$H_{1y} = H^i_{1y} + H^r_{1y} = Ae^{i(\omega t - k_1 z)} - rAe^{i(\omega t + k_1 z)},$$

$$E_{1x} = E^i_{1x} + E^r_{1x} = Z_1 Ae^{i(\omega t - k_1 z)} + rZ_1 Ae^{i(\omega t + k_1 z)}. \tag{5.44}$$

Here, we have introduced the complex reflection coefficient r. The wave inside the slab (for $0 \leq z \leq d$) is formed as a result of multibeam interference: its field is the sum of an infinite number of partial

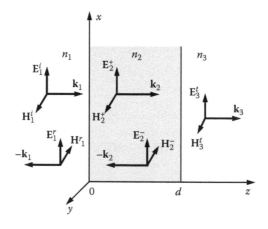

FIGURE 5.15   Orientation of magnetic and electric fields near the interface of a slab of thickness d.

waves appearing at multiple reflections from the layer surfaces. Formally, this field can be written as a superposition of two counterpropagating waves:

$$H_{2y} = H_{2y}^+ + H_{2y}^- = \beta^+ A e^{i(\omega t - k_2 z)} - \beta^- A e^{i(\omega t + k_2 z)},$$

$$E_{2x} = E_{2x}^+ + E_{2x}^- = Z_2 \beta^+ A e^{i(\omega t - k_2 z)} + Z_2 \beta^- A e^{i(\omega t + k_2 z)},$$

(5.45)

where $\beta^\pm$ are complex coefficients that will be determined. Finally, in the region $z \geq d$ (medium "3"), there is only a transmitted wave:

$$H_{3y} = H_{3y}^t = t A e^{i(\omega t - k_3 z)},$$

$$E_{3x} = E_{3x}^t = Z_3 t A e^{i(\omega t - k_3 z)},$$

(5.46)

where t is the complex transmission coefficient.

To find the coefficients r and t, the boundary conditions, which require continuity of the tangential components of the magnetic and electric wave fields, are used. These conditions are written down for each of the interfaces, that is, at $z = 0$ and $z = d$:

$$\begin{cases} H_{1y}^i + H_{1y}^r = H_{2y}^+ + H_{2y}^- \\ E_{1x}^i + E_{1x}^r = E_{2x}^+ + E_{2x}^- \end{cases} \quad \text{at } z = 0,$$

$$\begin{cases} H_{2y}^+ + H_{2y}^- = H_{3y}^t \\ E_{2x}^+ + E_{2x}^- = E_{3x}^t \end{cases} \quad \text{at } z = d.$$

(5.47)

Substituting expressions (5.44) through (5.46) for the field vectors into Equation 5.47, we obtain the following system of equations:

$$\begin{cases} 1 - r = \beta^+ - \beta^-, \\ Z_1(1+r) = Z_2(\beta^+ + \beta^-), \\ \beta^+ e^{-ik_2 d} - \beta^- e^{ik_2 d} = t e^{-ik_3 d}, \\ Z_2\left(\beta^+ e^{-ik_2 d} + \beta^- e^{ik_2 d}\right) = Z_3 t e^{-ik_3 d}. \end{cases}$$

(5.48)

Eliminating the coefficients β^+ and β^- from Equation 5.48, we obtain the following expressions for the reflection and transmission coefficients:

$$r = \frac{Z_2(Z_3 - Z_1)\cos k_2 d + i\left(Z_2^2 - Z_1 Z_3\right)\sin k_2 d}{Z_2(Z_3 + Z_1)\cos k_2 d + i\left(Z_2^2 + Z_1 Z_3\right)\sin k_2 d} = |r| e^{i\psi_r},$$

$$t = \frac{2 Z_2 Z_3 \exp(ik_3 d)}{Z_2(Z_3 + Z_1)\cos k_2 d + i\left(Z_2^2 + Z_1 Z_3\right)\sin k_2 d} = |t| e^{i\psi_t}.$$

(5.49)

From these relations, it follows that the reflection and transmission coefficients for a dielectric non-absorbing slab are periodic functions of the thickness d. Without going into a detailed analysis of this dependence, we note some important special cases.

Case 1. Consider the case when $n_1 < n_2$ and $n_3 < n_2$. In this case, there is an additional phase shift π at the $z = 0$ interface. No such phase shift exists at the $z = d$ interface.

First, we determine the condition when complete transmission of the incident wave energy through the layer takes place, that is, when the reflection coefficient is zero. If $k_2 d = N\pi$, that is, the thickness of layer takes the values

$$d = \frac{N \times \lambda_2}{2} = \frac{N \times \lambda}{2n_2}, \tag{5.50}$$

where

$N = 1, 2, 3, \ldots$

λ_2 and λ are the wavelengths in the slab and in vacuum, respectively

In this case, $\cos k_2 d = \pm 1$, $\sin k_2 d = 0$, and the reflection coefficient is

$$r = \frac{Z_3 - Z_1}{Z_3 + Z_1}. \tag{5.51}$$

Equation 5.51 coincides with the expression for reflection coefficient (5.30) for a single boundary between media with impedances Z_1 and Z_3, that is, the wave behaves in the same way as if slab does not exist between medium "1" and medium "3." If medium "1" and medium "3" on either side of the slab are identical, that is, $Z_1 = Z_3$, we get $r = 0$.

The maxima in the reflected intensity occur when

$$d = \left(N - \frac{1}{2}\right)\frac{\lambda}{2n_2}, \tag{5.52}$$

$$d_{\min} = \frac{\lambda}{4n_2} \quad \text{for } N = 1.$$

Case 2. Consider now the case when $n_1 < n_2$ and $n_3 > n_2$. Under these conditions, there are additional phase shifts π at $z = 0$ and $z = d$ interfaces, and intensity minima in the reflection occur when

$$d = \left(N - \frac{1}{2}\right)\frac{\lambda}{2n_2}. \tag{5.53}$$

Intensity maxima in the reflection occur when

$$d = N\frac{\lambda}{2n_2}. \tag{5.54}$$

For $N = 1, 2, 3, \ldots$ in Equation 5.53 the layer thickness is equal to an odd number of quarter of the wavelength in medium, $d = (2N - 1)\lambda_2/4$, so $\cos k_2 d = 0$ and $\sin k_2 d = \pm 1$. As a result, for the reflection coefficient, we get

$$r = \frac{Z_2^2 - Z_1 Z_3}{Z_2^2 + Z_1 Z_3}. \tag{5.55}$$

From this expression, it follows that the reflection coefficient vanishes under the condition

$$Z_2 = \sqrt{Z_1 Z_3}. \tag{5.56}$$

For nonmagnetic media ($\kappa_{mj} = 1$) $Z = Z_0/n$, where n is the refractive index of medium, the condition (5.56) is written as $n_2 = \sqrt{n_1 n_3}$. The Equations 5.53 and 5.56 are used for the deposition of antireflection coatings on reflective surfaces for the purpose of reducing the reflected light from the surface. The thickness of the antireflection coating should be equal to a quarter of the wavelength in the medium, and its material parameters should satisfy condition (5.56).

Exercise 5.7

A glass plate of thickness 1.20 μm and of refractive index $n = 1.50$ is placed between two media with refractive indices n_1 and n_2. Light of wavelength $\lambda = 0.6$ μm is normally incident on the plate (Figure 5.16). Find the results of interference of light beams 1 and 2 reflected from the upper and lower surfaces of plate in the following cases: (1) $n_1 < n < n_2$, (2) $n_1 > n > n_2$, (3) $n_1 < n > n_2$, and (4) $n_1 > n < n_2$.

Solution. The result of interference depends on how many half-waves, $\lambda/2$, are contained within the difference of two optical path lengths, L_{opt}. If

$$\frac{L_{opt}}{\lambda/2} = 2m$$

(even number), then we have interference maxima. If

$$\frac{L_{opt}}{\lambda/2} = 2m+1$$

(odd number), we get interference minima.

When finding the optical path length, it is necessary to take into account that during the reflection from an optically denser medium, an additional phase shift π appears. This phase shift corresponds to an additional difference $\lambda/2$ in L_{opt}.

Thus, at normal incidence of light on the plate, the optical path length of two beams 1 and 2 will be $L_{opt} = 2dn + (\lambda/2)$, if one of the beams is reflected from the optically denser medium, or $L_{opt} = 2dn$, if both beams are reflected from optically denser medium or both beams are reflected from optically less dense medium. Taking all this into account, for the four cases given earlier we get, respectively,

$$L_{opt} = 2dn, \quad L_{opt} = 2dn, \quad L_{opt} = 2dn + \left(\frac{\lambda}{2}\right), \quad L_{opt} = 2dn + \left(\frac{\lambda}{2}\right).$$

In the first two cases, we get $L_{opt} = 2 \times 1.20 \times 1.50 = 3.60$ μm. Since $\lambda/2 = 0.30$ μm, $L_{opt}/(\lambda/2) = 12$, that is, it is even number, and therefore, the constructive interference of light takes place. In the third and fourth cases, we get $L_{opt} = 2 \times 1.20 \times 1.50 + 0.30 = 3.90$ μm and $L_{opt}/(\lambda/2) = 13$, that is, it is odd number, and therefore, destructive interference is observed.

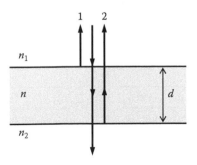

FIGURE 5.16   Interference of two beams reflected from a glass plate.

PROBLEMS

5.1 A point charge Q is surrounded by an uncharged dielectric concentric spherical shell with electrical permittivity ε and internal and external radii R_1 and R_2, respectively. Find the induced bound charge on the surfaces of dielectric layer, sheet density of bound charge, and electric field and electric displacement in the entire space.

$\Bigg($ *Part of the answer*: The induced charge, $q' = -\left(1 - \dfrac{1}{\kappa}\right)Q = -\dfrac{\kappa-1}{\kappa}Q.$

The sheet density, $\sigma'(R_1) = -\dfrac{(\kappa-1)Q}{4\pi\kappa R_1^2}.$

Electric displacement D (everywhere) is equal to $D(r) = \dfrac{Q}{4\pi r^2}.\Bigg)$

5.2 A point charge q is placed at the center of a dielectric sphere of radius a with permittivity κ_1. The sphere is surrounded by an infinite dielectric of permittivity κ_2. Find the surface density of bound charges at the interface of the two dielectrics.

5.3 A metal sphere of radius R is charged with charge Q. The surface of the sphere is covered by an uncharged dielectric shell of thickness h. What is the surface density of bound charges on the outer and inner surfaces of the dielectric shell if the dielectric permittivity of the shell is κ?

5.4 The space between the plates of a parallel-plate capacitor is filled with two dielectric layers with widths d_1 and d_2 and electrical permittivities ε_1 and ε_2. A constant potential difference $\Delta\varphi$ is applied between the capacitor plates. Find the electric field, the electric displacement in each layer, the potential difference across each layer, and the density of bound charges at the surfaces of the dielectric layers. $\Big($ *Answer*: Electric field in each layer is

$$E_1 = \frac{\varepsilon_2\Delta\varphi}{\varepsilon_2 d_1 + \varepsilon_1 d_2}, \quad E_2 = \frac{\varepsilon_1\Delta\varphi}{\varepsilon_2 d_1 + \varepsilon_1 d_2}.$$

Electric displacement, $D_1 = \dfrac{\varepsilon_0\kappa_1\varepsilon_2\Delta\varphi}{\varepsilon_2 d_1 + \varepsilon_1 d_2}, \quad D_2 = \dfrac{\varepsilon_0\kappa_2\varepsilon_2\Delta\varphi}{\varepsilon_2 d_1 + \varepsilon_1 d_2}.$

Potential difference across each layer, $\Delta\varphi_1 = \dfrac{\varepsilon_2 d_1\Delta\varphi}{\varepsilon_2 d_1 + \varepsilon_1 d_2}, \quad \Delta\varphi_2 = \dfrac{\varepsilon_1 d_2\Delta\varphi}{\varepsilon_2 d_1 + \varepsilon_1 d_2}.$

Density of bound charges,

$$\sigma_1' = \frac{\varepsilon_0\kappa_2(\kappa_1-1)\Delta\varphi}{\kappa_2 d_1 + \kappa_1 d_2}, \quad \sigma_2' = \frac{\varepsilon_0\kappa_1(\kappa_2-1)\Delta\varphi}{\kappa_2 d_1 + \kappa_1 d_2}, \quad \sigma_{12}' = \sigma_1' - \sigma_2' = \frac{\varepsilon_0(\kappa_1-\kappa_2)\Delta\varphi}{\kappa_2 d_1 + \kappa_1 d_2}.\Big)$$

5.5 Write expressions for reflection and transmission amplitude coefficients at the boundary between two nonmagnetic media ($\kappa_{m1} = \kappa_{m2} = 1$) with dielectric constants ε_1 and ε_2 for waves with s- and p-polarizations. Determine the conditions under which the reflection coefficients become zero. (*Answer*:

$$r_s = \frac{\cos\theta_0 - \sqrt{n_{21}^2 - \sin^2\theta_0}}{\cos\theta_0 + \sqrt{n_{21}^2 - \sin^2\theta_0}}, \quad r_p = \frac{\sqrt{n_{21}^2 - \sin^2\theta_0} - n_{21}^2\cos\theta_0}{\sqrt{n_{21}^2 - \sin^2\theta_0} + n_{21}^2\cos\theta_0},$$

$$t_s = \frac{2\cos\theta_0}{\cos\theta_0 + \sqrt{n_{21}^2 - \sin^2\theta_0}}, \quad t_p = \frac{\sqrt{n_{21}^2 - \sin^2\theta_0}}{\sqrt{n_{21}^2 - \sin^2\theta_0} + n_{21}^2\cos\theta_0}.$$

The reflection coefficient, r_s, for the perpendicular polarization vanishes only if $n_2 = n_1$. The reflection coefficient of the wave with parallel polarization r_p vanishes not only when $n_2 = n_1$ but also at Brewster's angle (θ_B), $\theta_B = \arctan(n_2/n_1) = \arctan\sqrt{\varepsilon_2/\varepsilon_1}$.)

5.6 A coil with $N = 800$ turns is wound around an iron core that has the form of a torus with an average diameter of $d = 0.10$ m. The core has a gap with the following width: $b = 2.50 \times 10^{-3}$ m. When the current in the winding $I = 5.00$ A, the magnetic field in the gap is $B_0 = 0.9$ T. Find the magnetic permeability of the iron core. (*Answer:* $\kappa_m = 101$.)

5.7 Find the magnetic moment of a diamagnetic helium atom induced by an external magnetic field **B**. *Hint:* The diamagnetism of helium is due to compensation of moments of oppositely directed spins of two electrons in helium. To explain the origin of diamagnetism in helium atom, that is, the absence of the total momentum in the absence of an external magnetic field, classical description assumes that the orbital radii of the two electrons are a and electrons on the orbit have the same velocities but opposite directions. (*Answer:* The magnetic moment induced in helium atom is $\mu_m = (ea/2)(v_2 - v_1)$, where $v_{1,2} = \sqrt{(\omega_L a)^2 + e^2/(2\pi\varepsilon_0 ma)} \mp \omega_L a$ and $\omega_L = eB/2m$.)

5.8 A dielectric layer with refractive index n_2 is in contact with dielectric media "1" and "3" with refractive indices n_1 and n_3, respectively, as shown in Figure 5.16. Using expression (5.49), obtain an expression for the energy reflection and transmission coefficients in the case of normal incidence of the wave on the layer. (*Answer:*

$$R = |r|^2 = \frac{(n_1 - n_3)^2 \cos^2 k_2 d + (n_2 - n_1 n_3/n_2)^2 \sin^2 k_2 d}{(n_1 + n_3)^2 \cos^2 k_2 d + (n_2 + n_1 n_3/n_2)^2 \sin^2 k_2 d},$$

$$T = 1 - R = \frac{4 n_1 n_3}{(n_1 + n_3)^2 \cos^2 k_2 d + (n_2 + n_1 n_3/n_2)^2 \sin^2 k_2 d}.$$)

5.9 An electromagnetic wave of frequency $f = 1.00 \times 10^8$ Hz and electric field amplitude $E_0 = 2.00$ V/m propagate along the y-axis in a dielectric with parameters $\kappa = 5.00$, $\kappa_m = 1.00$, and $\sigma = 0$. Write the equation that describes the electric field of the wave in the medium and determine the amplitude of the magnetic intensity as well as the wavelength, phase velocity, and wave number k of the wave. (*Answer:* $E(t, y) = E_0 \cos[2\pi(ft - (y/\lambda))]$, $H_0 = 337$ A/m, $\lambda = 1.34$ m, $v_{ph} = 1.34 \times 10^8$ m/s, and $k = 4.68$ m^{-1}.)

5.10 Radar detection of a target is produced by pulses whose frequency $f_0 = 200$ MHz. Each next pulse is emitted immediately after receiving the previous pulse. The distance to the target L is 50.0 km. Find the wavelength corresponding to the frequency and the maximum number of pulses N emitted by the radar over a time interval $\Delta t = 1$ s. (*Answer:* $\lambda_0 = 1.50$ m, $N = 3.00 \times 10^3$ pulses.)

5.11 The operating wavelength λ of a radar is 0.30 m. The radar emits 250 pulses per second, and each pulse has a duration of $\tau = 5.00$ μs. How many oscillations are contained in each pulse and what is the maximum range of the radar? (*Answer:* $n = 5.00 \times 10^3$ oscillations, maximum range 600 km.)

5.12 A linearly polarized wave with frequency $f = 10^9$ Hz propagates in a dielectric with parameters $\kappa_1 = 7.00$, $\kappa_{m1} = 1.00$, and $\sigma_1 = 0$ in a direction perpendicular to a flat infinite surface of a second dielectric with the parameters $\kappa_2 = 2.00$, $\kappa_{m2} = 1.00$, and $\sigma_2 = 0$. The amplitude of the electric field of the incident wave is 320 mV/m. Find the distribution of the electric and magnetic intensity in each of the media and plot their dependence on the coordinates. (*Answer:*

$$E_1 = A_1 \exp(-ik_1 z) + A_2 \exp(ik_1 z), \quad H_1 = -\frac{A_1}{Z_1}\exp(-ik_1 z) + \frac{A_2}{Z_1}\exp(ik_1 z),$$

$$E_2 = A_3 \exp(-ik_2 z), \quad H_2 = -\frac{A_3}{Z_2} \exp(-ik_1 z),$$

where $A_1 + A_2 = A_3$ and $A_2 \simeq 97$ mV/m, $A_3 \simeq 415$ mV/m.)

5.13 Find the ratio of the displacement current density to the conduction current density in seawater ($\kappa = 80.0$, $\kappa_m = 1.00$, and $\sigma = 4.00$ Sm/m) for waves with frequencies $f = 10^4$, 10^6, and 10^8 Hz. (*Answer*: $\delta_{disp}/\delta_{cond} = 1.11 \times 10^{-5}$ for $f = 10^4$ Hz, $\delta_{disp}/\delta_{cond} = 1.11 \times 10^{-3}$ for $f = 10^6$ Hz, and $\delta_{disp}/\delta_{cond} = 0.11$ for $f = 10^8$ Hz.)

5.14 A monochromatic light beam of wavelength $\lambda = 698$ nm is incident normally on a glass wedge with refractive index $n = 1.50$ (with respect to the base of the wedge). Determine the angle of the wedge if the distance between two adjacent interference minima in the reflected light is 2.00 mm. Assume that the angle is small. (*Answer*: $\alpha = 24''$.)

5.15 A beam of white light is incident normally on a soap film with refractive index $n = 1.33$. What is the minimum film thickness so that the light reflected by the film will appear green? (Wavelength of green light $\lambda = 550$ nm. *Answer*: $d_{min} \simeq 1.00 \times 10^{-7}$ m.)

5.16 A narrow light beam is incident from air on the horizontal surface of water at an angle α with respect to the normal. A mirror is placed in the water at an angle γ with respect to the horizontal. What is the minimum value of the angle γ in order for the beam not to be able to escape from the water into the air after reflection from the mirror? The refractive index of water is n.

$$\left(Answer: \gamma = \frac{1}{2} \arcsin\left[\frac{1}{n^2}\left(\sqrt{n^2 - \sin^2\alpha} - \sin\alpha\sqrt{n^2 - 1} \right) \right]. \right)$$

6 Electromagnetic Waves in Anisotropic and Optically Active Media

The macroscopic properties of an isotropic medium do not depend on the direction along which they are measured. The direction of wave propagation of an electromagnetic wave in an isotropic medium is *along* a single direction. An anisotropic medium is one in which physical properties vary along different directions. In an anisotropic medium, there are specific directions that are associated with the structure of medium or with an external electric and/or magnetic field. Anisotropy is connected with the presence in space of specific directions (which are known as *axes of symmetry*) that are due to the arrangement of atoms or molecules (in natural crystals) or to structural elements (in artificial media). For anisotropic media, the fundamental equations of electrodynamics—Maxwell's equations—retain their form. However, since in such media the dielectric permittivity and magnetic permeability are not scalar but tensor quantities, the material equations become significantly more complicated. In this case, the relation between the vectors **D** and **E** and between **B** and **H** is given by second-rank tensors. A scalar when multiplied by a vector is transformed to another vector, parallel or antiparallel to the first vector. Multiplication of a second-rank tensor and a vector **A** transforms that vector into another vector **A′**, which is not parallel to vector **A**. Thus, in an anisotropic medium, the vectors of the displacement and the electric field, **D** and **E**, and also the vectors **B** and **H** are not collinear any more. For conductive anisotropic media, electric conductivity σ in Ohm's law can also be a tensor quantity, which means that vectors **E** and **j** are not collinear. Anisotropy of the medium leads to the fact that the magnitude of a wave vector, the group and phase velocity of the wave, and its polarization parameters depend not only on the frequency but also on the direction of the wave propagation with respect to the axis of symmetry.

6.1 STRUCTURE OF A PLANE WAVE IN AN ANISOTROPIC MEDIUM

1. *The description of electrodynamic properties* of an anisotropic medium generally requires knowledge of 3×3 matrices, which form the tensors of the dielectric permittivity and magnetic permeability, $\hat{\kappa}$ and $\hat{\kappa}_m$, respectively. The relations 3.37 for anisotropic media take the forms:

$$\mathbf{D} = \varepsilon_0 \hat{\kappa} \mathbf{E}, \quad \mathbf{B} = \mu_0 \hat{\kappa}_m \mathbf{H}. \tag{6.1}$$

These equations in the Cartesian coordinate system can be written as:

$$D_i = \varepsilon_0 \sum_j \kappa_{ij} E_j, \quad B_i = \mu_0 \sum_j \kappa_{mij} H_j, \tag{6.2}$$

where indices $i, j = x, y, z$. The quantities κ_{ij} and κ_{mij} form 3×3 matrices so that the tensors take the forms

$$\hat{\kappa} = \begin{pmatrix} \kappa_{xx} & \kappa_{xy} & \kappa_{xz} \\ \kappa_{yx} & \kappa_{yy} & \kappa_{yz} \\ \kappa_{zx} & \kappa_{zy} & \kappa_{zz} \end{pmatrix}, \quad \hat{\kappa}_m = \begin{pmatrix} \kappa_{mxx} & \kappa_{mxy} & \kappa_{mxz} \\ \kappa_{myx} & \kappa_{myy} & \kappa_{myz} \\ \kappa_{mzx} & \kappa_{mzy} & \kappa_{mzz} \end{pmatrix}. \tag{6.3}$$

The components of tensors $\hat{\kappa}$ and $\hat{\kappa}_m$ are complex quantities, some of which, under certain conditions, can be zero. For the special case of an isotropic medium, all off-diagonal components of both tensors are equal to zero, and the diagonal elements have identical values, that is, $\kappa_{xx} = \kappa_{yy} = \kappa_{zz} = \kappa$, $\kappa_{mxx} = \kappa_{myy} = \kappa_{mzz} = \kappa_m$. The media that exhibit anisotropy in both the dielectric permittivity and magnetic permeability are rare. Therefore, we will consider separately the two types of anisotropy of electromagnetic properties in a medium.

2. Media with anisotropic dielectric permittivity tensor. In this case, the constitutive relations (6.1) become simpler and have the forms

$$\mathbf{D} = \varepsilon_0 \hat{\kappa} \mathbf{E}, \quad \mathbf{B} = \mu_0 \kappa_m \mathbf{H}, \tag{6.4}$$

where κ_m is a scalar quantity. The components of the tensor $\hat{\kappa}$ in general are complex numbers. It can be shown that tensor κ must be symmetric for transparent (nonabsorbing) media. This means that its off-diagonal components, which are symmetrically positioned about the diagonal, are identical, that is, $\kappa_{ij} = \kappa_{ji}$. Thus, in a transparent medium, the tensor $\hat{\kappa}$ generally has only six independent components.

If we change the coordinate axes, the dielectric tensor components take different values. From one system to another, they transform as tensor components. According to Equation 6.4, the directions of vectors $\mathbf{D}$ and $\mathbf{E}$, generally, do not coincide. Indeed, the relationship between the Cartesian components of these vectors can be represented as

$$D_x = \varepsilon_0 (\kappa_{xx} E_x + \kappa_{xy} E_y + \kappa_{xz} E_z),$$

$$D_y = \varepsilon_0 (\kappa_{yx} E_x + \kappa_{yy} E_y + \kappa_{yz} E_z), \tag{6.5}$$

$$D_z = \varepsilon_0 (\kappa_{zx} E_x + \kappa_{zy} E_y + \kappa_{zz} E_z).$$

It is seen that if there is only one nonzero component of the electric field (e.g., $E_x \neq 0$), the displacement vector has generally all three components

$$D_x = \varepsilon_0 \kappa_{xx} E_x, \quad D_y = \varepsilon_0 \kappa_{yx} E_x, \quad \text{and} \quad D_z = \varepsilon_0 \kappa_{zx} E_x.$$

Any symmetric tensor with the appropriate choice of coordinate system can be reduced to its diagonal form, with all off-diagonal components equal to zero. This means that there is always a unique coordinate system in which the dielectric tensor of an anisotropic transparent medium has the diagonal form

$$\hat{\kappa} = \begin{pmatrix} \kappa_x & 0 & 0 \\ 0 & \kappa_y & 0 \\ 0 & 0 & \kappa_z \end{pmatrix}. \tag{6.6}$$

In this coordinate system, tensor $\hat{\kappa}$ is described by three principal values for which the notation $\kappa_x = \kappa_{xx}$, $\kappa_y = \kappa_{yy}$, and $\kappa_z = \kappa_{zz}$ can be used. It is accepted to choose the principal axes x, y, and z so that the conditions $\kappa_x \leq \kappa_y \leq \kappa_z$ are satisfied. In the principal axes coordinate system, constitutive relations (6.2) and many other equations for anisotropic media take the simplest form.

Consider a plane monochromatic wave in a homogeneous anisotropic medium that is free from charges and currents (i.e., $\rho = 0$, $\mathbf{j} = 0$). Then, the time- and space-dependent field vectors can be represented as

$$\mathbf{D}(t, \mathbf{r}) = \mathbf{D} e^{i(\omega t - \mathbf{k} \cdot \mathbf{r})}, \quad \mathbf{E}(t, \mathbf{r}) = \mathbf{E} e^{i(\omega t - \mathbf{k} \cdot \mathbf{r})},$$

$$\mathbf{B}(t, \mathbf{r}) = \mathbf{B} e^{i(\omega t - \mathbf{k} \cdot \mathbf{r})}, \quad \mathbf{H}(t, \mathbf{r}) = \mathbf{H} e^{i(\omega t - \mathbf{k} \cdot \mathbf{r})}. \tag{6.7}$$

where **D**, **E**, **B**, and **H** are the constant amplitudes. After substituting these expressions in Maxwell's equations (3.45), we obtain the following system of equations:

$$\mathbf{k} \cdot \mathbf{D} = 0, \qquad \mathbf{k} \times \mathbf{H} = -\omega \mathbf{D},$$
$$\mathbf{k} \times \mathbf{E} = \omega \mathbf{B}, \qquad \mathbf{k} \cdot \mathbf{B} = 0. \tag{6.8}$$

Taking into account the constitutive relations (6.4), this can be written as

$$\mathbf{k} \cdot \mathbf{D} = 0, \qquad \mathbf{k} \times \mathbf{H} = -\omega \mathbf{D},$$
$$\mathbf{k} \times \mathbf{E} = \omega \mu_0 \kappa_m \mathbf{H}, \qquad \mathbf{k} \cdot \mathbf{B} = \mathbf{k} \cdot \mathbf{H} = 0. \tag{6.9}$$

From the first and last equations of system (6.9), it follows that $\mathbf{D} \perp \mathbf{k}$ and $\mathbf{H} \perp \mathbf{k}$, that is, the wave is transverse for the electric displacement vector, **D**, and the magnetic intensity vector, **H**, and also the magnetic field vector, **B**, which is collinear with **H**. From the remaining two equations (6.9), it follows that $\mathbf{D} \perp \mathbf{k}$, **H** and $\mathbf{E} \perp \mathbf{H}$. This means that vectors **E**, **D**, and **k** lie in one plane, perpendicular to a vector **H**, but **E** and **D** are not collinear due to the anisotropy of the dielectric permittivity. The wave surface plane is perpendicular to vector **k** and coincides with the plane in which vectors **D** and **H** lie. Since vector **E** does not lie in this plane, the Poynting vector, $\mathbf{S} = \mathbf{E} \times \mathbf{H}$, is not collinear with **k**. Thus, in an anisotropic medium, the direction of the energy density flux does not coincide with the direction of the wave vector. The Poynting vector together with vectors **E**, **D**, and **k** lies in the plane, perpendicular to vectors **H** and **B**.

3. Media with anisotropic magnetic permeability tensors. In this case, the constitutive relations obtain the forms

$$\mathbf{D} = \varepsilon_0 \kappa \mathbf{E}, \qquad \mathbf{B} = \mu_0 \hat{\kappa}_m \mathbf{H}, \tag{6.10}$$

and Maxwell's equations for the constant amplitudes of waves determined by Equations 6.7 will be written as follows (compare with Equations 6.9):

$$\mathbf{k} \cdot \mathbf{E} = 0, \qquad \mathbf{k} \times \mathbf{H} = -\omega \varepsilon_0 \kappa \mathbf{E},$$
$$\mathbf{k} \times \mathbf{E} = \omega \mathbf{B}, \qquad \mathbf{k} \cdot \mathbf{B} = 0. \tag{6.11}$$

From these equations, it follows that $\mathbf{E} \perp \mathbf{k}$, $\mathbf{B} \perp \mathbf{k}$, $\mathbf{E} \perp \mathbf{B}$, and the vectors **E** and **D** are parallel. Thus, vectors **E**, **D**, and **B** lie in the plane of the wave surface perpendicular to the wave vector **k**. Vector **H** is not collinear with **B** and does not lie in this plane. For the same reason, Poynting vector **S** is not collinear with the wave vector **k**. Vectors **B**, **H**, **k**, and **S** lie in one plane, perpendicular to the vectors **E** and **D**.

In Figure 6.1, the relative positions of a wave vector, Poynting vector, and the wave field vectors are presented for waves propagating in a medium with tensor dielectric permittivity (a) and tensor magnetic permeability (b).

Exercise 6.1

Derive the equation that gives the angle between the vectors of the electric field, **E**, and the electric displacement, **D**, of an electromagnetic wave in a medium if the electric displacement vector has components D_x, D_y, and D_z and the dielectric permittivity tensor of the medium is diagonal with components $\kappa_{xx} = \kappa_x$, $\kappa_{yy} = \kappa_y$, and $\kappa_{zz} = \kappa_z$.

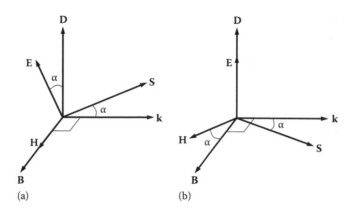

FIGURE 6.1 Relative orientation of six vectors that describe the electromagnetic field in media with anisotropic dielectric permittivity (a) and anisotropic magnetic permeability tensors (b).

Solution. The cosine of an angle between the two vectors is given by the following relationship:

$$\cos\alpha = \frac{E_x D_x + E_y D_y + E_z D_z}{|\mathbf{E}||\mathbf{D}|}.$$

Since the relation between the components of vectors $\mathbf{E}$ and $\mathbf{D}$ has the form $D_i = \varepsilon_0 \kappa_i E_i$ (where $i = x, y, z$), then

$$\cos\alpha = \frac{D_x^2/\kappa_x + D_y^2/\kappa_y + D_z^2/\kappa_z}{\varepsilon_0 \sqrt{E_x^2 + E_y^2 + E_z^2}\ \sqrt{D_x^2 + D_y^2 + D_z^2}}$$

$$= \frac{D_x^2/\kappa_x + D_y^2/\kappa_y + D_z^2/\kappa_z}{\varepsilon_0 \sqrt{(D_x^2/\kappa_x^2 + D_y^2/\kappa_y^2 + D_z^2/\kappa_z^2)/\varepsilon_0^2}\ \sqrt{D_x^2 + D_y^2 + D_z^2}}$$

$$= \frac{D_x^2/\kappa_x + D_y^2/\kappa_y + D_z^2/\kappa_z}{\sqrt{D_x^2/\kappa_x^2 + D_y^2/\kappa_y^2 + D_z^2/\kappa_z^2}\ \sqrt{D_x^2 + D_y^2 + D_z^2}}.$$

For positive values of κ_i, we have $\cos\alpha > 0$, and therefore, $0 < \alpha < \pi/2$. The same angle is formed between the Poynting vector $\mathbf{S}$ and the wave vector $\mathbf{k}$ and also between the vectors of the group and phase velocities. We note that the four vectors $\mathbf{E}$, $\mathbf{D}$, $\mathbf{S}$, and $\mathbf{k}$ lie in the same plane.

6.2 DISPERSION RELATION AND NORMAL WAVES

In this section, we consider the case of a medium with anisotropic dielectric permittivity. We will obtain the wave equation for $\mathbf{E}$ from system (6.9) by expressing vector $\mathbf{H}$ from the third equation ($\mathbf{H} = \mathbf{k} \times \mathbf{E}/\omega\mu_0\kappa_m$) and substituting it in the second equation

$$\mathbf{k} \times (\mathbf{k} \times \mathbf{E}) + k_0^2 \kappa_m \hat{\kappa} \mathbf{E} = 0 \qquad\qquad (6.12)$$

where $k_0 = \omega/c$ is the wave number in vacuum. Using the vector identity in Equation 6.12 $\mathbf{k} \times (\mathbf{k} \times \mathbf{E}) = \mathbf{k}(\mathbf{k} \cdot \mathbf{E}) - k^2\mathbf{E}$, we obtain the equation

$$\mathbf{k}(\mathbf{k} \cdot \mathbf{E}) - k^2\mathbf{E} + k_0^2 \kappa_m \hat{\kappa}\mathbf{E} = 0. \qquad\qquad (6.13)$$

The projections on the three coordinate axes of vector (Equation 6.13) give a system of three linear homogeneous equations that involve the components E_i ($i = x, y, z$). Choosing the principal axes of the tensor $\hat{\kappa}$ as the coordinate axes and introducing the components k_i of the wave vector on these axes, Equation 6.13 can be written as

$$k_i(k_x E_x + k_y E_y + k_z E_z) - (k^2 - k_0^2 \kappa_m \kappa_i)E_i = 0, \quad \text{where } i = x, y, z.$$

The three equations for $i = x, y, z$ are the following:

$$k_x(k_x E_x + k_y E_y + k_z E_z) - (k^2 - k_0^2 \kappa_m \kappa_x)E_x = 0,$$

$$k_y(k_x E_x + k_y E_y + k_z E_z) - (k^2 - k_0^2 \kappa_m \kappa_y)E_y = 0,$$

$$k_z(k_x E_x + k_y E_y + k_z E_z) - (k^2 - k_0^2 \kappa_m \kappa_z)E_z = 0.$$

It is convenient to present the preceding system of equation in the following form:

$$(k_0^2 \kappa_m \kappa_x - k^2 + k_x^2)E_x + k_x k_y E_y + k_x k_z E_z = 0,$$

$$k_y k_x E_x + (k_0^2 \kappa_m \kappa_y - k^2 + k_y^2)E_y + k_y k_z E_z = 0, \qquad (6.14)$$

$$k_z k_x E_x + k_z k_y E_y + (k_0^2 \kappa_m \kappa_z - k^2 + k_z^2)E_z = 0.$$

The system of three homogeneous equations (6.14) has a nontrivial solution if its determinant is equal to zero:

$$\det \begin{pmatrix} k_0^2 \kappa_m \kappa_x - k_y^2 - k_z^2 & k_x k_y & k_x k_z \\ k_y k_x & k_0^2 \kappa_m \kappa_y - k_x^2 - k_z^2 & k_y k_z \\ k_z k_x & k_z k_y & k_0^2 \kappa_m \kappa_z - k_x^2 - k_y^2 \end{pmatrix} = 0. \qquad (6.15)$$

The resulting equation is known as the **dispersion equation** because it describes the dependence of wave vector on frequency (in addition to the wave number k_0, the tensor components κ_x, κ_y, κ_z can also depend on frequency). On the other hand, Equation 6.15 determines a 3D closed surface in the wave vector space with coordinate axes k_x, k_y, and k_z.

We introduce vector $\mathbf{n} = \mathbf{k}/k_0$ with components $n_i = k_i/k_0 = n \cos \gamma_i$, whose magnitude is equal to the refractive index n of the wave propagating in a given direction. Here, γ_i are the angles formed by the wave vector $\mathbf{k}$ with the corresponding coordinate axes (see Figure 6.2).

Expanding the determinant in Equation 6.15, we get the equation that is often referred as the equation for the wave normal lines (here we put $\kappa_m = 1$):

$$n^2(\kappa_x n_x^2 + \kappa_y n_y^2 + \kappa_z n_z^2) - \left[n_x^2 \kappa_x(\kappa_y + \kappa_z) + n_y^2 \kappa_y(\kappa_z + \kappa_x) + n_z^2 \kappa_z(\kappa_x + \kappa_y)\right] + \kappa_x \kappa_y \kappa_z = 0. \quad (6.16)$$

For a wave of fixed frequency, this equation is power four relative to the refractive index. For each direction in space, there are two positive roots of this equation, that is, two values of the wave number k and the refractive index n. For example, for propagation along the principal axis z, it is necessary to put $n = n_z$, $n_x = n_y = 0$ since $k_x = k_y = 0$. Then, Equation 6.16 takes the form

$$n^4 - n^2(\kappa_x + \kappa_y) + \kappa_x \kappa_y = 0, \qquad (6.17)$$

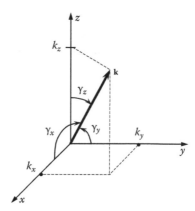

FIGURE 6.2 Orientation of vector **k** in the anisotropic material.

from where we find two values of the refractive index

$$n_1^2 = \kappa_x \quad \text{and} \quad n_2^2 = \kappa_y. \tag{6.18}$$

Thus, in an anisotropic medium in a given direction two independent waves with different wave numbers and different phase velocities can propagate (these two waves are referred to as *normal waves*). It can be shown that the displacement vectors $\mathbf{D}_1$ and $\mathbf{D}_2$ of two linearly polarized normal waves are orthogonal to each other (see Figure 6.3) and have the following dependence on the coordinates and time:

$$\mathbf{D}_j = \mathbf{D}_{j0} \exp\left[i(\omega t - k_j z)\right], \quad j = 1, 2, \tag{6.19}$$

where the wave numbers k_j are determined by the following equation:

$$k_j = k_0 n_j.$$

As it was already noted, for a wave in an anisotropic medium (such as a single crystal), which we will consider further, it is necessary to distinguish the propagation direction of the wave phase (i.e., the direction of the wave vector **k**) and the direction of energy propagation (in *crystal optics* we identify

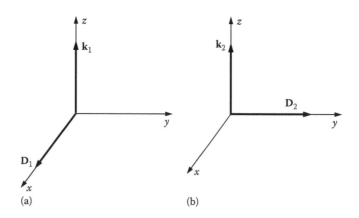

FIGURE 6.3 Two normal waves with four vectors shown for $n_1^2 = \kappa_x$ (a) and $n_2^2 = \kappa_y$ (b).

the latter as the ray direction). In addition to the velocity of the wave front phase, that is, the phase velocity $\mathbf{v}_{ph}$, there is also the velocity of energy propagation, known as ray velocity v_r. It is directed along the Poynting vector $\mathbf{S}$ and is defined as the ratio of wave intensity to its average energy density

$$\mathbf{v}_r = \frac{\langle \mathbf{S} \rangle}{\langle u \rangle}. \tag{6.20}$$

In an anisotropic medium, the ray velocity differs from phase velocity in direction (due mismatch of the orientations of vectors $\mathbf{k}$ and $\mathbf{S}$) as well as in magnitude. It is possible to show that v_r is equal to the group velocity of a wave and it is related to phase velocity by the relationship $v_{ph} = v_r \cos \alpha$, where α is an angle between the vectors $\mathbf{k}$ and $\mathbf{S}$ and is given by

$$\cos \alpha = \frac{E_x D_x + E_y D_y + E_z D_z}{|\mathbf{E}||\mathbf{D}|} = \frac{\varepsilon_0 (\kappa_x E_x^2 + \kappa_y E_y^2 + \kappa_z E_z^2)}{|\mathbf{E}||\mathbf{D}|} > 0. \tag{6.21}$$

Crystals, according to their optical properties, can be divided into the following three groups:

- *Crystals of cubic symmetry*, for which $\kappa_x = \kappa_y = \kappa_z = \kappa$. These behave as an optically isotropic medium.
- *Uniaxial crystals*, for which $\kappa_x = \kappa_y < \kappa_z$ or $\kappa_x < \kappa_y = \kappa_z$. In the first case, we speak about a "positive" crystal, while in the second case, we talk about a "negative" crystal.
- *Biaxial crystals*, for which all three principal values of the tensor of dielectric permittivity are different, that is, $\kappa_x < \kappa_y < \kappa_z$.

Exercise 6.2

Derive from the system of equations (6.8), the equation for the wave normal lines in a nonmagnetic medium ($\kappa_m = 1$) with diagonal dielectric permittivity tensor $\hat{\kappa}$.

Solution. We rewrite the third and the second equations of the system (6.8) in the forms

$$\mathbf{H} = \frac{1}{\omega \mu_0} \mathbf{k} \times \mathbf{E}, \quad \mathbf{D} = -\frac{1}{\omega} \mathbf{k} \times \mathbf{H}.$$

Then, we eliminate the magnetic intensity $\mathbf{H}$ from these two equations

$$\mathbf{D} = -\frac{1}{\mu_0 \omega^2} \mathbf{k} \times (\mathbf{k} \times \mathbf{E}) = -\frac{k^2}{\mu_0 \omega^2} \mathbf{m} \times (\mathbf{m} \times \mathbf{E}) = -\varepsilon_0 n^2 \mathbf{m} \times (\mathbf{m} \times \mathbf{E}).$$

Here, we introduce the unit vector $\mathbf{m} = \mathbf{k}/k$, $k = (\omega/c)n$, and $c = 1/\sqrt{\varepsilon_0 \mu_0}$. Taking into account the relationship

$$\mathbf{A} \times (\mathbf{B} \times \mathbf{C}) = \mathbf{B}(\mathbf{A} \cdot \mathbf{C}) - \mathbf{C}(\mathbf{A} \cdot \mathbf{B}),$$

we get

$$\mathbf{D} = -\varepsilon_0 n^2 \mathbf{m} \times (\mathbf{m} \times \mathbf{E}) = -\varepsilon_0 n^2 \left[\mathbf{m}(\mathbf{m} \cdot \mathbf{E}) - \mathbf{E} \right] = \varepsilon_0 n^2 \left[\mathbf{E} - \mathbf{m}(\mathbf{m} \cdot \mathbf{E}) \right].$$

From this equation, we can find the projections of the electric displacement vector on the chosen coordinate axes:

$$D_x = \varepsilon_0 n^2 \left[\frac{D_x}{\varepsilon_0 \kappa_x} - m_x (\mathbf{E} \cdot \mathbf{m}) \right], \quad D_x \left(\frac{1}{n^2} - \frac{1}{\kappa_x} \right) = -\varepsilon_0 m_x (\mathbf{E} \cdot \mathbf{m}),$$

$$D_x m_x = -\frac{\varepsilon_0 m_x^2 (\mathbf{E} \cdot \mathbf{m})}{\dfrac{1}{n^2} - \dfrac{1}{\kappa_x}}, \quad D_y m_y = -\frac{\varepsilon_0 m_y^2 (\mathbf{E} \cdot \mathbf{m})}{\dfrac{1}{n^2} - \dfrac{1}{\kappa_y}}, \quad D_z m_z = -\frac{\varepsilon_0 m_z^2 (\mathbf{E} \cdot \mathbf{m})}{\dfrac{1}{n^2} - \dfrac{1}{\kappa_z}}.$$

Finally, we add the last three relationships:

$$D_x m_x + D_y m_y + D_z m_z = -\varepsilon_0 (\mathbf{E} \cdot \mathbf{m}) \left(\frac{m_x^2}{\dfrac{1}{n^2} - \dfrac{1}{\kappa_x}} + \frac{m_y^2}{\dfrac{1}{n^2} - \dfrac{1}{\kappa_y}} + \frac{m_z^2}{\dfrac{1}{n^2} - \dfrac{1}{\kappa_z}} \right).$$

Since the vectors $\mathbf{D}$ and $\mathbf{m}$ are perpendicular to each other, then

$$\mathbf{D} \cdot \mathbf{m} = D_x m_x + D_y m_y + D_z m_z = 0.$$

And this gives

$$\frac{m_x^2}{\dfrac{1}{n^2} - \dfrac{1}{\kappa_x}} + \frac{m_y^2}{\dfrac{1}{n^2} - \dfrac{1}{\kappa_y}} + \frac{m_z^2}{\dfrac{1}{n^2} - \dfrac{1}{\kappa_z}} = 0.$$

This equation can be reduced to

$$\frac{\kappa_x m_x^2}{n^2 - \kappa_x} + \frac{\kappa_y m_y^2}{n^2 - \kappa_y} + \frac{\kappa_z m_z^2}{n^2 - \kappa_z} = 0.$$

We make a series of transformations in this equation by adding the term $m_x^2 + m_y^2 + m_z^2 = 1$ and performing the summation

$$\frac{\kappa_x m_x^2}{n^2 - \kappa_x} + m_x^2 + \frac{\kappa_y m_y^2}{n^2 - \kappa_y} + m_y^2 + \frac{\kappa_z m_z^2}{n^2 - \kappa_z} + m_z^2 = 1,$$

$$\frac{n^2 m_x^2}{n^2 - \kappa_x} + \frac{n^2 m_y^2}{n^2 - \kappa_y} + \frac{n^2 m_z^2}{n^2 - \kappa_z} = 1.$$

By the definition $nm_\alpha = n_\alpha$, so we get an equation for the wave normal lines in the form of Equation 6.16 for a nonmagnetic medium, that is, for $\kappa_m = 1$

$$\frac{n_x^2}{n^2 - \kappa_x} + \frac{n_y^2}{n^2 - \kappa_y} + \frac{n_z^2}{n^2 - \kappa_z} = 1.$$

6.3   WAVES IN UNIAXIAL CRYSTALS

Uniaxial crystals are the most important and at the same time the simpler examples of anisotropic media. In uniaxial crystals, there is one preferred direction (*optical axis*), which coincides with one of the principal dielectric axes. The directions of the other two principal axes are arbitrary. If, for example, the optical axis coincides with the z-axis, then $\kappa_x = \kappa_y = \kappa_\perp$ and $\kappa_z = \kappa_\parallel$. The values $\kappa_\parallel$ and $\kappa_\perp$ are called the **longitudinal** and **transverse dielectric constants** of a uniaxial crystal. If the wave does not propagate along the optical axis, the vectors **E** and **D** can be decomposed into components directed along the optical axis, $\mathbf{E}_\parallel$, $\mathbf{D}_\parallel$, and perpendicular to it, $\mathbf{E}_\perp$, $\mathbf{D}_\perp$. For these components, we have

$$\mathbf{D}_\parallel = \varepsilon_0 \kappa_\parallel \mathbf{E}_\parallel, \quad \mathbf{D}_\perp = \varepsilon_0 \kappa_\perp \mathbf{E}_\perp. \tag{6.22}$$

Let the wave vector **k** form an angle θ with the optical axis z. The plane defined by the wave vector **k** and the optical axis z is called the **principal section of the crystal**. For a uniaxial crystal, Equation 6.16 takes the form

$$(n^2 - \kappa_\perp)\left[\kappa_\parallel n_z^2 + \kappa_\perp (n_x^2 + n_y^2) - \kappa_\parallel \kappa_\perp \right] = 0, \tag{6.23}$$

that is, it splits into two equations

$$n_o^2 = \kappa_\perp, \tag{6.24}$$

$$\frac{n_z^2}{\kappa_\perp} + \frac{n_x^2 + n_y^2}{\kappa_\parallel} = 1, \tag{6.25}$$

corresponding to two normal waves of a uniaxial crystal. The wave determined by Equation 6.24 is called **ordinary** as it would propagate in a uniaxial crystal in the same way as it would propagate in an isotropic medium with a refractive index $n = \sqrt{\kappa_\perp}$. The refractive index of the second wave, as follows from Equation 6.25, depends on the angle θ, that is, on the direction of the wave propagation. This wave is called the **extraordinary wave**.

The ordinary wave propagates in a crystal with the same phase velocity $v_o = c/n_o$ in all directions. The phase velocity of the extraordinary wave $v_e = c/n_e$ depends on the propagation angle relative to the optical axis. Thus, if the wave vector lies in the plane of $z0y$, then $k_z = k_0 n \cos \theta$, $k_y = k_0 n \sin \theta$, $k_x = 0$, and for the refractive indices of the ordinary and extraordinary waves, we get

$$n_o = \sqrt{\kappa_\perp}, \quad n_e = \sqrt{\frac{\kappa_\perp \kappa_\parallel}{\kappa_\perp \sin^2 \theta + \kappa_\parallel \cos^2 \theta}}. \tag{6.26}$$

If the wave propagates along the optical axis, then $\theta = 0$ and $n_o = n_e = \sqrt{\kappa_\perp}$, that is, the phase velocities of the ordinary and extraordinary waves are the same ($v_o = v_e$). The greatest difference in refractive indices and phase velocities of the two normal waves is achieved as the wave propagates in a direction perpendicular to the optical axis.

Consider the polarization of the ordinary and extraordinary waves. To each of two normal waves, there corresponds a particular case of orientation of an electric displacement vector **D** (Figure 6.4).

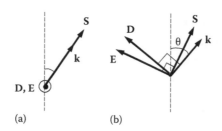

FIGURE 6.4 Orientation of vectors **E**, **D**, **S**, and **k** in (a) ordinary and (b) extraordinary waves.

Case 1. Vector **D** of an ordinary wave is perpendicular to the principal section and the principal optical axis (Figure 6.4a). Then, $\mathbf{D} = \mathbf{D}_\perp = \varepsilon_0 \kappa_\perp \mathbf{E}_\perp$, that is, **D** and **E** are collinear and $\mathbf{E} \perp \mathbf{k}$, as well as in an isotropic medium. Vectors **k** and **S** are also collinear, that is, the directions of a wave vector and Poynting vector coincide.

Case 2. Vectors **D** and **E** of the extraordinary wave lie in the plane of the principal section (Figure 6.4b). If $\theta \neq (0, \pm \pi/2)$, both vectors have both longitudinal and transverse components and are not collinear each other. In this case, $\mathbf{D} \perp \mathbf{k}$ and $\mathbf{E} \perp \mathbf{S}$. The directions of phase and ray velocities coincide only for propagation along the principal optical axis or perpendicular to it.

If an electromagnetic wave enters into a uniaxial crystal from an isotropic medium, it generally generates two linearly polarized waves propagating inside the crystal with different phase and ray velocities. This phenomenon is called a ***birefringence***. In the following text, we consider it in more detail.

Assume that a nonpolarized plane wave is incident at an angle φ on the plane boundary between a uniaxial crystal and vacuum. Assume that the optical axis of the crystal lies in the plane of incidence and is directed at an arbitrary angle to the boundary plane. We use Huygens' principle to determine the ordinary and extraordinary rays in the crystal. Huygens' principle states that any point on a wave front of an electromagnetic wave may be regarded as the source of secondary waves and the surface that is tangent to the secondary waves at a later time t can be used to determine the future position of the wave front at t.

To determine the wave surface of the propagating wave at later times, it is necessary to construct the envelope of the wave surfaces of these secondary waves. For a time during which the right edge of the front AB reaches a point B' on the crystal surface, two ray surfaces—spherical and ellipsoidal—appear around each point of the surface between A and B'. These two surfaces coincide with each other along the optical axis.

Figure 6.5 shows the ray surfaces centered at point A in the case of a positive crystal, when the ellipsoid is inscribed in the sphere. We draw the tangents $A'B'$ and $A''B'$ to the sphere and the

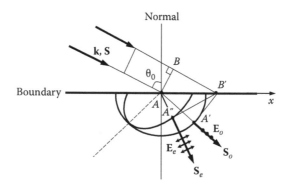

FIGURE 6.5 Refraction on the surface of positive crystal.

ellipsoid, respectively. The lines connecting a point A to points, where the spherical and ellipsoidal surfaces intersect with tangents $A'B'$ and $A''B'$, give the direction of ordinary and extraordinary rays, respectively, with refraction angles that are different in each case. Thus, unlike an ordinary ray, the direction of an extraordinary ray and the direction of a normal to the corresponding wave front do not coincide.

Since the principal section of the crystal coincides with the plane of the page, the electric vector of the ordinary ray $\mathbf{E}_o$ oscillates perpendicular to this plane, while the electric vector of the extraordinary ray $\mathbf{E}_e$ oscillates in the plane of the page.

Exercise 6.3

A linearly polarized wave is incident normally on a uniaxial crystalline plate whose optical axis forms an angle $\alpha = 45°$ with the normal to the surface of the plate. Show that for an arbitrary thickness d of the plate, the originally linearly polarized wave after transmission through the plate will emerge elliptically polarized. Consider different particular cases for the plate thickness.

Solution. Let us write the equations that describe the oscillations of the electric field vector in the coordinate system of optical axis after the wave emerges from the plate (note that refractive indices are n_o and n_e for ordinary and extraordinary waves in the coordinate system of optical axis)

$$E_x = E_0 \cos(\omega t + \varphi_e), \quad E_y = E_0 \cos(\omega t + \varphi_o),$$

where the phase shift for each wave corresponding to the plate thickness is equal to

$$\varphi_e = \frac{2\pi}{\lambda} n_e d, \quad \varphi_o = \frac{2\pi}{\lambda} n_o d.$$

Let us find the projections of these components on the axes of a coordinate system of the plate as the optical axes form angle $\alpha = 45°$ with the normal to the plate and take into account that $\cos\alpha = \sin\alpha = 1/\sqrt{2}$

$$E_{x'} = E_0 \cos(\omega t + \varphi_e)\cos\alpha + E_0 \cos(\omega t + \varphi_o)\sin\alpha$$

$$= \frac{E_0}{\sqrt{2}}\left[\cos(\omega t + \varphi_e) + \cos(\omega t + \varphi_o)\right],$$

$$E_{y'} = -E_0 \cos(\omega t + \varphi_e)\sin\alpha + E_0 \cos(\omega t + \varphi_o)\cos\alpha$$

$$= \frac{E_0}{\sqrt{2}}\left[-\cos(\omega t + \varphi_e) + \cos(\omega t + \varphi_o)\right].$$

Let us perform summation of the trigonometric functions in the square brackets

$$E_{x'} = \sqrt{2}E_0 \cos\left(\omega t + \frac{\varphi_e + \varphi_o}{2}\right)\cos\left(\frac{\varphi_e - \varphi_o}{2}\right),$$

$$E_{y'} = \sqrt{2}E_0 \sin\left(\omega t + \frac{\varphi_e + \varphi_o}{2}\right)\sin\left(\frac{\varphi_e - \varphi_o}{2}\right).$$

And let us then introduce $\Delta\varphi = \varphi_e - \varphi_o$. Then, these relationships can be written as follows:

$$\frac{E_{x'}}{\sqrt{2}E_0 \cos(\Delta\varphi/2)} = \cos\left(\omega t + \frac{\varphi_e + \varphi_o}{2}\right),$$

$$\frac{E_{y'}}{\sqrt{2}E_0 \sin(\Delta\varphi/2)} = \sin\left(\omega t + \frac{\varphi_e + \varphi_o}{2}\right).$$

In order to eliminate time from these equations, let us square them and add them. As a result, we get

$$\frac{E_{x'}^2}{2E_0^2 \cos^2(\Delta\varphi/2)} + \frac{E_{y'}^2}{2E_0^2 \sin^2(\Delta\varphi/2)} = 1.$$

This equation represents an ellipse with axes A and B:

$$A = \sqrt{2}E_0 \cos\left(\frac{\Delta\varphi}{2}\right) = \sqrt{2}E_0 \cos\left[\frac{\pi}{\lambda}(n_e - n_o)d\right],$$

$$B = \sqrt{2}E_0 \sin\left(\frac{\Delta\varphi}{2}\right) = \sqrt{2}E_0 \sin\left[\frac{\pi}{\lambda}(n_e - n_o)d\right].$$

The transmitted light is generally elliptically polarized, and the axes of the ellipse form an angle of 45° with the normal to the plate. In the case when $\pi(n_e - n_o)d/\lambda = \pi/4 + m\pi/2$, where $m = 0, 1, 2...$, the polarization becomes circular since $|A| = |B|$. The thickness of the plate in this case must be equal to $d = (1 + 2m)\lambda/(4(n_e - n_o))$. The rotation direction of the electric field vector can be clockwise or counterclockwise. In the case when $\pi(n_e - n_o)d/\lambda = m\pi/2$ or $d = m\lambda/(2(n_e - n_o))$, the transmitted beam is linearly polarized along either the x- or y-axis.

6.4 REFRACTIVE INDEX ELLIPSOID

The constitutive relations (6.2) for the electric field in an anisotropic medium, written in principal axes coordinate system, have the forms

$$D_x = \varepsilon_0\kappa_x E_x, \quad D_y = \varepsilon_0\kappa_y E_y, \quad D_z = \varepsilon_0\kappa_z E_z. \tag{6.27}$$

Taking into account these equations, the expression for the energy density of the electric field of a wave in an anisotropic medium, $u_e = \mathbf{D}\cdot\mathbf{E}/2$, may be expressed as

$$\frac{D_x^2}{\kappa_x} + \frac{D_y^2}{\kappa_y} + \frac{D_z^2}{\kappa_z} = 2\varepsilon_0 u_e. \tag{6.28}$$

For a transparent medium, energy density u_e is constant. We introduce in Equation 6.28 the following new variables:

$$X = \frac{D_x}{\sqrt{2\varepsilon_0 u_e}}, \quad Y = \frac{D_y}{\sqrt{2\varepsilon_0 u_e}}, \quad Z = \frac{D_z}{\sqrt{2\varepsilon_0 u_e}}. \tag{6.29}$$

Then, we use the relation $\kappa_i = n_i^2$. Hence, Equation 6.28 can be written as

$$\frac{X^2}{n_x^2} + \frac{Y^2}{n_y^2} + \frac{Z^2}{n_z^2} = 1. \tag{6.30}$$

This equation represents the surface of a triaxial ellipsoid, which is called the ***refractive index ellipsoid*** or ***optical indicatrix***. Any ellipsoid described by an equation that has the form of Equation 6.30 has generally two circular sections. The directions, perpendicular to such circular sections, coincide with the optical axes of the crystal. In Figure 6.6, these directions are shown by lines OO' and OO''.

Let us draw from the ellipsoid center a straight line in the direction of the wave front propagation, that is, in the direction of a unit vector $\mathbf{m} = \mathbf{k}/k$. The intersection of the ellipsoid by a plane that is perpendicular to the direction described by the Equation 6.31

$$Xm_x + Ym_y + Zm_z = 0 \tag{6.31}$$

is an ellipse (as shown in Figure 6.7). The lengths of the semiaxes of this ellipse are the refraction indices of waves propagating in the given direction. The directions of these semiaxes are parallel to vectors of the electric displacement $\mathbf{D}_1$ and $\mathbf{D}_2$ of the wave fields of ordinary and extraordinary waves.

Thus, by specifying the wave direction in a crystal and choosing the ellipsoid section perpendicular to this direction, it is possible to find the corresponding indices of refraction from the lengths of the semiaxes of the resulting ellipse. The directions of the semiaxes indicate the oscillation directions of the wave electric field in the crystal and, hence, determine the polarization of the two normal plane waves arising in a crystal.

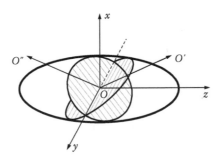

FIGURE 6.6   The refractive index ellipsoid and the directions of the optical axes OO' and OO'' of the crystal.

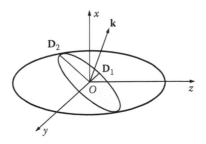

FIGURE 6.7   The intersection of the refractive index ellipsoid by a plane perpendicular to the direction of propagation.

For a better understanding of light propagation in crystals, a number of surfaces that describe the optical properties are introduced. In order to show clearly how the wave velocity depends on the direction of wave propagation, the wave ***normal surface*** is used. The equation for this surface is obtained, if instead of n_x, n_y, n_z in the equation for ellipsoid like Equation 6.30, the components of the phase velocity v_x, v_y, v_z in the corresponding coordinate axes are used as the principal semiaxes of the ellipsoid, that is,

$$\frac{X^2}{v_x^2} + \frac{Y^2}{v_y^2} + \frac{Z^2}{v_z^2} = 1. \tag{6.32}$$

To construct ellipsoid described by this equation, it is necessary to draw lines in the directions x, y, z whose lengths are equal to the values of phase velocities v_x, v_y, v_z in the corresponding direction (Figure 6.8).

For each direction, there are two values of velocities; therefore, there are two surfaces of wave normals. For an ordinary wave, the surface is represented by a sphere; for the extraordinary wave, it has the shape of an ellipsoid of rotation. The ellipsoid and the sphere coincide with each other at the points of intersection with the optical axis. If for all other directions $n_e > n_o$, then $v_e < v_o$, and the crystal is called ***positive***. For ***negative crystals***, $n_e < n_o$ and $v_e > v_o$. For a positive crystal, the ellipsoid of rotation is elongated and inscribed in the sphere (Figure 6.8a), and for the negative, compressed and circumscribed about a sphere (Figure 6.8b).

Exercise 6.4

In an anisotropic nonmagnetic ($\kappa_m = 1$) medium, in a given direction, two linearly polarized normal waves with different wave numbers and phase velocities are propagating. Show that the vectors of electric displacement $\mathbf{D}_1$ and $\mathbf{D}_2$ of these waves are orthogonal to each other.

Solution. We introduce the projection of the phase velocity along the principal axes of the crystal x, y, z and the velocity of the wave along the normal $\mathbf{m}$

$$v_x = \frac{c}{\sqrt{\kappa_x}}, \quad v_y = \frac{c}{\sqrt{\kappa_y}}, \quad v_z = \frac{c}{\sqrt{\kappa_z}}, \quad v = \frac{c}{n}.$$

We then use the relation obtained in Exercise 6.2:

$$D_\alpha m_\alpha = -\frac{\varepsilon_0 m_\alpha^2 (\mathbf{E} \cdot \mathbf{m})}{\dfrac{1}{n^2} - \dfrac{1}{\kappa_\alpha}}, \quad \alpha = x, y, z.$$

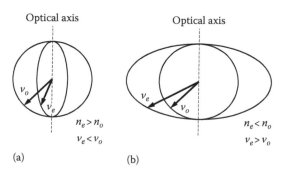

FIGURE 6.8   The wave normal surfaces for (a) positive and (b) negative crystals.

Then,

$$D_\alpha = -\frac{\varepsilon_0 m_\alpha (\mathbf{E} \cdot \mathbf{m})}{\dfrac{1}{n^2} - \dfrac{1}{\kappa_\alpha}} = -\frac{\varepsilon_0 c^2 m_\alpha (\mathbf{E} \cdot \mathbf{m})}{v^2 - v_\alpha^2}.$$

Each of the values of the velocity v or n will correspond to a certain value of the electric displacement $\mathbf{D}_1$ and $\mathbf{D}_2$:

$$\mathbf{D}_1 \cdot \mathbf{D}_2 = D_{1x}D_{2x} + D_{1y}D_{2y} + D_{1z}D_{2z}$$

$$= \varepsilon_0^2 c^4 (\mathbf{E} \cdot \mathbf{m})^2 \left\{ \frac{m_x}{v_1^2 - v_x^2} \cdot \frac{m_x}{v_2^2 - v_x^2} + \frac{m_y}{v_1^2 - v_y^2} \cdot \frac{m_y}{v_2^2 - v_y^2} + \frac{m_z}{v_1^2 - v_z^2} \cdot \frac{m_z}{v_2^2 - v_z^2} \right\}$$

$$= \frac{\varepsilon_0^2 c^4 (\mathbf{E} \cdot \mathbf{m})^2}{v_1^2 - v_2^2} \left\{ \frac{m_x^2}{v_1^2 - v_x^2} - \frac{m_x^2}{v_2^2 - v_x^2} + \frac{m_y^2}{v_1^2 - v_y^2} - \frac{m_y^2}{v_2^2 - v_y^2} + \frac{m_z^2}{v_1^2 - v_z^2} - \frac{m_z^2}{v_2^2 - v_z^2} \right\}$$

$$= \frac{\varepsilon_0^2 c^4 (\mathbf{E} \cdot \mathbf{m})^2}{v_1^2 - v_2^2} \left\{ \frac{m_x^2}{v_1^2 - v_x^2} + \frac{m_y^2}{v_1^2 - v_y^2} + \frac{m_z^2}{v_1^2 - v_z^2} - \left(\frac{m_x^2}{v_2^2 - v_x^2} + \frac{m_y^2}{v_2^2 - v_y^2} + \frac{m_z^2}{v_2^2 - v_z^2} \right) \right\} = 0,$$

where we took into account that for each wave the following equation holds:

$$\frac{m_x^2}{n_\alpha^2 - \kappa_x} + \frac{m_y^2}{n_\alpha^2 - \kappa_y} + \frac{m_z^2}{n_\alpha^2 - \kappa_z} = 0.$$

Thus, $\mathbf{D}_1 \cdot \mathbf{D}_2 = 0$, which means that vectors $\mathbf{D}_1$ and $\mathbf{D}_2$ are orthogonal to each other, that is, $\mathbf{D}_1 \perp \mathbf{D}_2$.

6.5 OPTICALLY ACTIVE MEDIA

The interest in the study of artificial composite media started in the nineteenth century and continues until today. These media possess the *optical activity*: optical activity is the ability of a medium to rotate polarization plane of light that travels through the medium. The rotation of the polarization plane of a light wave in a medium can occur either in the clockwise rotation direction or in the counterclockwise direction. Examples are a *bi-isotropic* medium and its special case—a *chiral* medium.

The definition of chirality was first given by Thompson, better known in science as Lord Kelvin (among other things, he introduced in thermodynamics the concept of the absolute temperature scale). It is based on the fact that a chiral object does not coincide with its mirror image. Numerous experimental data indicate that in nature the symmetry of "right" and "left" is often broken.

There are molecules of the same chemical composition, which differ only by the fact that they are the mirror images of each other (e.g., glucose, tartaric acid). In 1860, the biologist Louis Pasteur explained the nature of optical activity.

In 1888, F. Reynittser discovered liquid crystals (LCs), that form an important class of natural optically active media and find wide practical application in modern optics. Almost all LCs consist of organic compounds.

We distinguish the following three types of LCs, which differ in the degree of order of molecular arrangement.

Nematic LCs are characterized by the existence of an orientational order with the long axes of the molecules directed along a preferred direction. In this case, positional order is absent (unlike real solid crystals, for which there is a long-range order in the arrangement of the centers of the molecules). Molecules themselves slip continuously in the direction of their long axes, revolving around them while keeping the orientation order. The molecules of nematic LCs are not chiral but are identical to their mirror images. There are also nematic phases, in which the molecules have a disklike shape. These phases of LCs are characterized by a structure consisting of oriented disks.

Smectic LCs are characterized by both orientational and positional order. The molecules are arranged so that their axes not only are parallel to each other but also form a layered structure. The layers can slide over each other, and each of the molecules can move in two dimensions: it can slide together with the layer and rotate around its longitudinal axis. We distinguish also ***chiral smectic*** LCs, in which the longitudinal axes of the molecules of one layer are turned by a small angle relative to molecules of the adjacent layer.

Cholesteric LCs have layered structure in which for each layer there is a small change of the molecule axis orientation in relation to the adjacent layer. They are characterized by a distance (period) at which the orientation vector (***director***) rotates by 360°. For many cholesteric LCs, this distance is comparable with the wavelength of visible light and in many respects this fact determines their optical properties.

Heating LCs of this type on a few tenth of degree slightly changes the angle of molecules' rotation in the adjacent layers of molecules. For all LCs, which consist of thousands of molecular layers, this leads to a change in the angle of the reflected light rotation resulting in a change of color.

In some natural solid-state crystals, the two structural—right-handed and left-handed—forms are found. Crystals possess optical activity if they have no center of symmetry. An example of such a crystal is quartz, which in nature can exist in two types—right-handed and left-handed. Let us note that in the optical range, it is possible to create artificial optical activity in a medium. Thus, when an optically inactive substance is placed in an external magnetic field, a rotation of the polarization plane of a wave propagating in the medium is observed.

An artificial chiral medium consists of a set of conducting mirror-asymmetric microelements uniformly distributed in an isotropic nonconducting medium. Interest to the study of the electrodynamic properties of such artificial media is connected with the ability to create composite media structured on microscopic and nanometer scales. Right- and left-handed metal spirals, open rings with straight orthogonal ends, cylinders with conductivity along helical lines, particles in the form of the Greek letter Ω, and others are examples of chiral structural elements. In the scientific literature, these structural elements are often called electromagnetic particles. The man-made media can possess properties that are much broader than the properties of natural chiral media in the optical range. For natural media, the chirality is explained by the geometry of molecules, whereas the elements of artificial media can possess more complex behavior. In particular, they can have resonance properties though their dimensions are small compared to the wavelength. Due to the use of the resonance properties of the media structural elements, the chirality becomes essential and properties of such a medium differ radically from those of nonchiral medium.

For the description of the electromagnetic properties of a chiral medium, the two commonly used material parameters—dielectric permittivity and magnetic permeability—are not enough. It is necessary to introduce an additional dimensionless parameter η that determines the magneto-electric coupling and is called ***chirality parameter***. The most common is the following form of the constitutive equations for a chiral medium:

$$\mathbf{D} = \varepsilon_0 \left(\kappa \mathbf{E} - i\eta Z_0 \mathbf{H} \right),$$

$$\mathbf{B} = \mu_0 \left(\kappa_m \mathbf{H} + i\eta Z_0^{-1} \mathbf{E} \right),$$

        (6.33)

where $\eta > 0$ and $\eta < 0$ for medium on the basis of right-handed and left-handed spirals, respectively, and $Z_0 = \sqrt{\mu_0/\varepsilon_0}$ is an impedance (wave drag) of vacuum. The chiral properties of a medium are connected to its specific structure, and they disappear upon transition to a continuous medium ($l/\lambda \to 0$, where l is the linear dimension of the chiral microcell).

Since usually the distance between adjacent conducting elements of a chiral media is comparable to the wavelength λ of electromagnetic waves (and their linear dimensions l much smaller than the wavelength λ), such a chiral medium has a *spatial dispersion* (i.e., dependence of material parameters on the wave vector).

For the description of the electromagnetic properties of a bi-isotropic medium, it is necessary to add to the constitutive equations one more parameter. These equations take a more general forms than for a chiral medium:

$$\mathbf{D} = \varepsilon_0 \left[\kappa \mathbf{E} + (\chi - i\eta) Z_0 \mathbf{H} \right],$$

$$\mathbf{B} = \mu_0 \left[\kappa_m \mathbf{H} + (\chi + i\eta) Z_0^{-1} \mathbf{E} \right],$$

(6.34)

where χ is a dimensionless parameter called the parameter of *nonreciprocity*. The value of the nonreciprocity parameter in natural environment is small (at optical frequencies $\chi \sim 10^{-4}$). In an artificial composite, the parameter of nonreciprocity can be significantly increased with optimum selection and arrangement of structural elements (electromagnetic particles). Often, a nonreciprocal medium with zero chirality ($\eta = 0$) is called Tellegen's medium and a reciprocal chiral medium ($\chi = 0$) Pasteur's medium. The constitutive equations in the forms of Equations 6.33 and 6.34 mean that the chiral properties of the medium depend on a right-handed spiral. For a medium based of left-handed spirals, it is necessary to change the sign of the chirality parameter.

Exercise 6.5

An electromagnetic wave is incident on a chiral medium. This wave excites in each structural chiral element of the medium an alternating electric current. This current in turn creates electric and magnetic dipole moments in each structural element of the medium. The number of such elements per unit volume is equal to N, and their size is much smaller than the wavelength of the incident electromagnetic wave. Derive the material relationships (6.33) for such a medium.

Solution. Let us assume that several turns of a conducting coil serve as a chiral element of the medium. Let us examine how electric and magnetic dipoles are generated when an electromagnetic wave is incident on such a chiral element. In the coil's vicinity, the wave fields will have the following electric and magnetic longitudinal components: E_z and H_z, directed along coil axis. The electric dipole component p_{ez} depends on the electric field of the electromagnetic wave as well as on the magnetic field that penetrates the turns of the coil and thus creates a circular current, that is,

$$p_{ez} = \varepsilon_0 \alpha_e E_z + \varepsilon_0 \alpha_{em} Z_0 H_z = \varepsilon_0 (\alpha_e E_z + \alpha_{em} Z_0 H_z),$$

where
  α_e is the electric polarizability of the coil
  α_{em} is the parameter of electromagnetic polarizability of the coil
  Z_0 is the impedance of vacuum

By analogy, an expression for the magnetic dipole moment of the open-ended coil p_{mz}, which is created by the circular electric current and is directed along the coil's axis, can be written. At the

same time, a circular current is created by the magnetic field, which passes through the coil, as well
as by the axial electric field. Therefore,

$$p_{mz} = \mu_0 \alpha_m H_z + \mu_0 \alpha_{me} Z_0^{-1} E_z = \mu_0 \left(\alpha_m H_z + \alpha_{me} Z_0^{-1} E_z \right),$$

where
  α_m is the magnetic polarizability of the coil
  α_{me} is the magnetoelectric polarizability parameter of the coil

Using the symmetry of the problem, we can show that these parameters are related as follows:

$$\alpha_{me} = -\alpha_{em} = \pm i\beta,$$

where β is a positive real magnitude. Upper sign in the right-hand side of this expression corresponds to the "right" coils and lower sign to the "left" coils. Taking into account these relationships,
the expressions for dipole moments of the coil take the forms

$$p_{ez} = \varepsilon_0 \left(\alpha_e E_z \mp i\beta Z_0 H_z \right),$$

$$p_{mz} = \mu_0 \left(\alpha_m H_z \pm i\beta Z_0^{-1} E_z \right).$$

Let us write the general relationships for the electric displacement and magnetic field vectors

$$\mathbf{D} = \varepsilon_0 \mathbf{E} + \mathbf{P} = \varepsilon_0 \mathbf{E} + N\mathbf{p}_e, \quad \mathbf{B} = \mu_0 \mathbf{H} + \mathbf{M} = \mu_0 \mathbf{H} + N\mathbf{p}_m.$$

We substitute into these relationships the expressions for the dipole moments for a single coil. As a
result, we get

$$\mathbf{D} = \varepsilon_0 \left(1 + N\alpha_e \right) \mathbf{E} \mp i\varepsilon_0 \beta N Z_0 \mathbf{H},$$

$$\mathbf{B} = \mu_0 \left(1 + N\alpha_m \right) \mathbf{H} \pm i\mu_0 \beta N Z_0^{-1} \mathbf{E}.$$

We then introduce in these expressions the dielectric permittivity of the medium $1 + N\alpha_e = \varepsilon$, the
magnetic permeability of the medium $1 + N\alpha_m = \mu$, and the chirality $\eta = \beta N$. After this, we get the
material relationships of the medium described by Equation 6.33:

$$\mathbf{D} = \varepsilon_0 \left(\varepsilon \mathbf{E} \mp i\eta Z_0 \mathbf{H} \right),$$

$$\mathbf{B} = \mu_0 \left(\mu \mathbf{H} \pm i\eta Z_0^{-1} \mathbf{E} \right).$$

6.6   WAVES IN CHIRAL MEDIA

1. In isotropic nonchiral media, plane waves can have linear, circular, and, in the most general
case, elliptical polarization. Plane waves with linear or circular polarization that have the same
frequency have the same velocity of propagation also. Therefore, any linear combination of these
waves propagates without any distortion. Propagating waves in bi-isotropic and chiral media can
have only circular polarization. Waves with right and left circular polarizations propagate in a given

medium with different velocities, and therefore, their linear combinations (i.e., waves with the linear and elliptic polarizations) cannot be stable.

For a monochromatic wave propagating in a homogeneous chiral medium, Maxwell's equations for electric field $\mathbf{E} \exp(i\omega t)$ and magnetic intensity $\mathbf{H} \exp(i\omega t)$ of the wave take the forms

$$\nabla \times \mathbf{E} = -i\omega\mu_0 \left(\kappa_m \mathbf{H} + i\eta Z_0^{-1} \mathbf{E} \right),$$

$$\nabla \times \mathbf{H} = i\omega\varepsilon_0 \left(\kappa \mathbf{E} - i\eta Z_0 \mathbf{H} \right),$$

$$\nabla \cdot \left(\kappa \mathbf{E} - i\eta Z_0 \mathbf{H} \right) = 0,$$

$$\nabla \cdot \left(\kappa_m \mathbf{H} + i\eta Z_0^{-1} \mathbf{E} \right) = 0.$$

(6.35)

Here, we assume that all wave fields have harmonic time dependence, that is, they are proportional to $\exp(i\omega t)$.

In the following text, we will obtain second-order equations for the wave fields. To do this, we apply to the right- and left-hand side of the first two equations of Equation 6.35 the curl operator:

$$\nabla \times (\nabla \times \mathbf{E}) = -i\omega\mu_0 \left(\kappa_m \nabla \times \mathbf{H} + i\eta Z_0^{-1} \nabla \times \mathbf{E} \right),$$

$$\nabla \times (\nabla \times \mathbf{H}) = i\omega\varepsilon_0 \left(\kappa \nabla \times \mathbf{E} - i\eta Z_0 \nabla \times \mathbf{H} \right).$$

(6.36)

Using the first two equations of Equation 6.35, we replace $\nabla \times \mathbf{H}$ and $\nabla \times \mathbf{E}$ in the right-hand side of Equation 6.36 and get

$$\nabla \times (\nabla \times \mathbf{E}) = \omega^2\varepsilon_0\mu_0 \left[\left(\kappa\kappa_m + \eta^2 \right) \mathbf{E} - 2i\eta\kappa_m Z_0 \mathbf{H} \right],$$

$$\nabla \times (\nabla \times \mathbf{H}) = \omega^2\varepsilon_0\mu_0 \left[\left(\kappa\kappa_m + \eta^2 \right) \mathbf{H} + 2i\eta\kappa Z_0^{-1} \mathbf{E} \right].$$

(6.37)

We now use the relations that have been derived by us earlier

$$\nabla \times (\nabla \times \mathbf{A}) = \nabla(\nabla \cdot \mathbf{A}) - \nabla^2 \mathbf{A},$$

$$\varepsilon_0\mu_0 = \frac{1}{c^2}, \quad k_0 = \frac{\omega}{c}.$$

From the third and fourth equations in Equation 6.35, we have

$$\nabla \cdot (\kappa \mathbf{E}) - iZ_0 \nabla \cdot (\eta \mathbf{H}) = 0,$$

$$\nabla \cdot (\kappa_m \mathbf{H}) + iZ_0^{-1} \nabla \cdot (\eta \mathbf{E}) = 0.$$

(6.38)

As a result, for a homogeneous medium, we obtain that $\nabla \cdot \mathbf{E} = 0$ and $\nabla \cdot \mathbf{H} = 0$. From these relations, we obtain a system of coupled second-order differential equations for vector fields $\mathbf{H}$ and $\mathbf{E}$:

$$\nabla^2 \mathbf{E} + k_0^2 \left(\kappa\kappa_m + \eta^2 \right) \mathbf{E} - 2ik_0^2\eta\kappa_m Z_0 \mathbf{H} = 0,$$

$$\nabla^2 \mathbf{H} + k_0^2 \left(\kappa\kappa_m + \eta^2 \right) \mathbf{H} + 2ik_0^2\eta\kappa Z_0^{-1} \mathbf{E} = 0.$$

(6.39)

For nonchiral media, the chirality parameter $\eta = 0$ and the system (6.39) becomes a system of independent Helmholtz equations for the wave electric and magnetic fields:

$$\nabla^2 \mathbf{E} + k_0^2 \kappa \kappa_m \mathbf{E} = 0,$$
$$\nabla^2 \mathbf{H} + k_0^2 \kappa \kappa_m \mathbf{H} = 0. \tag{6.40}$$

2. From Equations 6.39, it follows that linearly polarized waves are not normal waves of the chiral medium. To find the normal waves, we will represent vectors of an electromagnetic field in the forms

$$\mathbf{E} = \mathbf{E}_R + \mathbf{E}_L, \quad \mathbf{H} = \mathbf{H}_R + \mathbf{H}_L = iZ^{-1}(\mathbf{E}_R - \mathbf{E}_L), \tag{6.41}$$

where

$Z = \sqrt{\kappa_m \mu_0 / \kappa \varepsilon_0}$ is the impedance of the medium
$\mathbf{E}_R$ and $\mathbf{E}_L$ are the wave fields with the right circular and left circular polarizations

We then substitute these expressions into Equation 6.39. As a result, we obtain the system

$$\nabla^2(\mathbf{E}_R + \mathbf{E}_L) + k_0^2(n^2 + \eta^2)(\mathbf{E}_R + \mathbf{E}_L) + 2k_0^2 \eta \kappa_m Z_0 Z^{-1}(\mathbf{E}_R - \mathbf{E}_L) = 0,$$
$$\nabla^2(\mathbf{E}_R - \mathbf{E}_L) + k_0^2(n^2 + \eta^2)(\mathbf{E}_R - \mathbf{E}_L) + 2k_0^2 \eta \kappa(\mathbf{E}_R + \mathbf{E}_L) = 0, \tag{6.42}$$

where $n = \sqrt{\kappa \kappa_m}$. After the simple transformations, we get

$$\nabla^2 \mathbf{E}_R + k_0^2(n^2 + 2n\eta + \eta^2)\mathbf{E}_R + \nabla^2 \mathbf{E}_L + k_0^2(n^2 - 2n\eta + \eta^2)\mathbf{E}_L = 0,$$
$$\nabla^2 \mathbf{E}_R + k_0^2(n^2 + 2n\eta + \eta^2)\mathbf{E}_R - \nabla^2 \mathbf{E}_L - k_0^2(n^2 - 2n\eta + \eta^2)\mathbf{E}_L = 0. \tag{6.43}$$

After adding these two equations and subtracting the second equation from the first one, we get two uncoupled equations for the electric field of the normal waves, which are the wave of the right and left circular polarizations:

$$\nabla^2 \mathbf{E}_R + k_R^2 \mathbf{E}_R = 0, \quad \nabla^2 \mathbf{E}_L + k_L^2 \mathbf{E}_L = 0. \tag{6.44}$$

Similarly, one can obtain a system of equations for the magnetic intensity

$$\nabla^2 \mathbf{H}_R + k_R^2 \mathbf{H}_R = 0, \quad \nabla^2 \mathbf{H}_L + k_L^2 \mathbf{H}_L = 0, \tag{6.45}$$

where we have introduced the propagation constants of the waves described by Equations 6.44 and 6.45:

$$k_R = k_0 n_R = k_0 \left(\sqrt{\kappa \kappa_m} + \eta \right), \quad k_L = k_0 n_L = k_0 \left(\sqrt{\kappa \kappa_m} - \eta \right). \tag{6.46}$$

Here, we introduced the refractive indices of waves with the right and left circular polarizations $n_{R,L} = \sqrt{\kappa \kappa_m} \pm \eta$.

3. As we just obtained, linearly polarized waves are not normal waves of chiral media. Only two linear combinations of such waves that have right or left circular polarization can be normal waves. For the waves propagating in the direction of z-axis, we will present these combinations as follows:

$$\mathbf{E}_R = (\mathbf{i} - i\mathbf{j})E_0 \exp(-ik_R z), \quad \mathbf{H}_R = (i\mathbf{i} + \mathbf{j})Z^{-1}E_0 \exp(-ik_R z),$$

$$\mathbf{E}_L = (\mathbf{i} + i\mathbf{j})E_0 \exp(-ik_L z), \quad \mathbf{H}_L = (-i\mathbf{i} + \mathbf{j})Z^{-1}E_0 \exp(-ik_L z), \tag{6.47}$$

where $\mathbf{i}$ and $\mathbf{j}$ are the unit vectors along the x- and y-coordinate axes. Thus, the two normal waves in chiral media have different propagation constants: k_R and k_L. This means that the phase velocities $v_R = \omega/k_R$ and $v_L = \omega/k_L$ of the waves of the right and left circular polarizations are also different.

We note that the signs in the constitutive equations (6.46) and (6.47) correspond to a chiral medium on the basis of right-handed helices. For media, based on left-handed helices, one has to change the sign in front of the parameter of chirality in these equations. Therefore, for the left- and right-hand media, the propagation constants are

$$k_R = k_0 n_R = k_0\left(\sqrt{\kappa\kappa_m} - \eta\right), \quad k_L = k_0 n_L = k_0\left(\sqrt{\kappa\kappa_m} + \eta\right) \tag{6.48}$$

and the wave phase velocities obey the relation $v_R > v_L$. The relationship between the electric and magnetic fields of waves with right and left circular polarizations has the form

$$\mathbf{E}_R = -iZ\mathbf{H}_R, \quad \mathbf{E}_L = iZ\mathbf{H}_L. \tag{6.49}$$

In the case of bi-isotropic medium, for which the constitutive equations are given by Equations 6.34, the expressions for the propagation constants of the normal waves with right and left circular polarizations are

$$k_{R,L} = k_0 n_{R,L} = k_0\left(\sqrt{\kappa\kappa_m - \chi^2} \pm \eta\right). \tag{6.50}$$

The electromagnetic field in bi-isotropic medium is represented as a superposition of the right and left circular polarized waves:

$$\mathbf{E} = \mathbf{E}_R + \mathbf{E}_L,$$

$$\mathbf{H} = \mathbf{H}_R + \mathbf{H}_L = iZ^{-1}[\mathbf{E}_R \exp(i\theta) - \mathbf{E}_L \exp(-i\theta)], \tag{6.51}$$

where the angle θ is related with a parameter of nonreciprocity χ as $\theta = \arcsin(\chi/n)$.

Exercise 6.6

A linearly polarized electromagnetic wave with the amplitude E_0 and frequency ω is incident on a chiral sample with the thickness l and chirality η. Determine the angle of rotation of the polarization plane of the wave transmitted through the sample.

Solution. Let us assume that the linearly polarized wave is propagating along z-axis and is incident on the sample at point $z = 0$. The electric field vector $\mathbf{E}$ at this point is parallel to the x-axis. The field after transmission through the sample is a superposition of two waves with right and left circular polarizations:

$$E_{Rx} = \frac{1}{2} E_0 \exp(-ik_R l), \quad E_{Ry} = \frac{i}{2} E_0 \exp(-ik_R l),$$

$$E_{Lx} = \frac{1}{2} E_0 \exp(-ik_L l), \quad E_{Ly} = -\frac{i}{2} E_0 \exp(-ik_L l),$$

where propagation constants are equal to $k_{R,L} = k_0 n_{R,L} = k_0 \left(\sqrt{\kappa \kappa_m} \pm \eta \right)$ and $k_0 = \omega/c$.
Let us find the sum of these waves:

$$E_x = E_{Rx} + E_{Lx} = \frac{E_0}{2} \left[\exp(-ik_R l) + \exp(-ik_L l) \right]$$

$$= E_0 \cos \left(k_0 \frac{n_R - n_L}{2} l \right) \exp \left(-ik_0 \frac{n_R + n_L}{2} l \right) = E_0 \cos(k_0 \eta l) \exp\left(-ik_0 \sqrt{\varepsilon \mu} l\right),$$

$$E_y = E_{Ry} + E_{Ly} = i \frac{E_0}{2} \left[-\exp(-ik_R l) + \exp(-ik_L l) \right]$$

$$= E_0 \sin \left(k_0 \frac{n_R - n_L}{2} l \right) \exp \left(-ik_0 \frac{n_R + n_L}{2} l \right) = E_0 \sin(k_0 \eta l) \exp\left(-ik_0 \sqrt{\varepsilon \mu} l\right).$$

Since the phases of the fields E_x and E_y are the same, that is, $\varphi_x = \varphi_y = k_0 \sqrt{\kappa \kappa_m} l$, then after transmission through the sample, the wave field is linearly polarized. The angle of rotation of polarization plane with respect to y-axis, that is, of vector $\mathbf{E}$ with $E = \sqrt{E_x^2 + E_y^2}$, is defined by the relationship

$$\tan \theta = \frac{E_y}{E_x} = \frac{\sin(k_0 \eta l)}{\cos(k_0 \eta l)} = \tan(k_0 \eta l), \quad \theta = k_0 \eta l.$$

PROBLEMS

6.1 Write the equation of motion for a valence electron of an elongated molecule that is part of an anisotropic crystal in the presence of an electric field of an electromagnetic wave. Assume that the valence electrons of the elongated molecules can move only in one direction. Derive an expression for the dielectric permittivity tensor for the case in which the symmetry axes of a crystal do not coincide with the chosen coordinate system. *Answer:* $\hat{\kappa} = \begin{pmatrix} 1 + \chi_{xx} & \chi_{xy} & \chi_{xz} \\ \chi_{yx} & 1 + \chi_{yy} & \chi_{yz} \\ \chi_{zx} & \chi_{zy} & 1 + \chi_{zz} \end{pmatrix}$

where $\hat{\chi} = \begin{pmatrix} \chi_{xx} & \chi_{xy} & \chi_{xz} \\ \chi_{yx} & \chi_{yy} & \chi_{yz} \\ \chi_{zx} & \chi_{zy} & \chi_{zz} \end{pmatrix}$ is the dielectric susceptibility tensor.

6.2 Determine the decrease in the intensity of natural light that passes through two Nicol prisms (crystal polarizers); the plane of polarization of the two crystal polarizers forms an angle α. Each Nicol prism absorbs a fraction η of the energy of the incident light.

6.3 A linearly polarized wave is incident along the normal to the surface of a plate made of a uniaxial crystal. The optical axis of the crystal is parallel to the surface of the plate.

The plate thickness is equal to d. Determine the polarization state of the transmitted wave. (*Answer*: For an arbitrary thickness of the crystal, the polarization of the transmitted wave is elliptic.)

6.4 Write down the dispersion equation in the form of the determinant (6.15) for a wave propagating along one of the principal axes in a uniaxial crystal. Solve the dispersion equation and find the velocity of the normal waves. $\left(\text{*Answer*: Velocity of normal waves, } v_1 = \dfrac{c}{\sqrt{\kappa_m \kappa_x}}, \quad v_2 = \dfrac{c}{\sqrt{\kappa_m \kappa_y}}. \right)$

6.5 A plane monochromatic wave propagates in an anisotropic medium that is characterized by the permittivity tensor $\hat{\kappa}$ and scalar magnetic permeability κ_m (see Equation 6.4). Determine the structure of the wave, the angle between the vectors $\mathbf{E}$ and $\mathbf{D}$, as well as phase and ray velocities. $\left(\text{*Answer*: } \mathbf{H} = \sqrt{\dfrac{\varepsilon_0}{\mu_0}} \dfrac{n}{\kappa_m} (\mathbf{e} \times \mathbf{E}), \quad \mathbf{D} = -\dfrac{n}{c}(\mathbf{e} \times \mathbf{H}), \text{ where } \mathbf{D} = \varepsilon_0 \hat{\kappa} \mathbf{E} \text{ and } \mathbf{H} = \mathbf{B}/\mu_0 \kappa_m, \right.$

$\left. \theta = -\arctan(\kappa_{xz}/\kappa_{xy}), \text{ and } v_{ph} = v_r \cos \theta. \right)$

6.6 A linearly polarized wave is incident normally on a plate of thickness d made of a positive uniaxial crystal. The optical axis of the crystal lies in the plane of the plate. The electric field vector $\mathbf{E}$ of the incident wave forms an angle α with the optical axis. Show that the transmitted wave is elliptically polarized. $\left(\text{*Answer*: } \dfrac{E_x^2}{E_o^2} + \dfrac{E_z^2}{E_e^2} - \dfrac{2E_x E_z}{E_o E_e} \cos(\Delta\varphi) = \sin^2(\Delta\varphi), \text{ where} \right.$

$\left. E_e = E \cos \alpha, \, E_o = E \sin \alpha, \text{ and } \Delta\varphi = (2\pi/\lambda)(n_e - n_o)d. \right)$

6.7 Take the solution obtained for Problem 6.6 and find the thicknesses of the plate that would result in linear and circular polarizations of the wave passing through the uniaxial plate. $\left(\text{*Answer*: If } L_e - L_o = (n_e - n_o)d = m\dfrac{\lambda}{2} + \dfrac{\lambda}{4}, \text{ then } \dfrac{E_x^2}{E_o^2} + \dfrac{E_z^2}{E_e^2} = 1. \text{ For } \alpha = 45° E_o = E_e, \text{ therefore} \right.$

$E_x^2 + E_z^2 = E_o^2$, which corresponds to circular polarization. If $L_e - L_o = (n_e - n_o)d = m\lambda/2$, then

$E_z = \pm\dfrac{E_e}{E_o} E_x$, which corresponds to linear polarization. $\Big)$

6.8 A beam of partially polarized light that consists of a linear polarized component with intensity I_{pol} and natural light of intensity I_{nat} is transmitted through a polaroid sheet. Initially, the polaroid is oriented so that its transmission plane is parallel to the oscillation plane of the linearly polarized component of the beam. By turning the polarizer by an angle $\varphi = 60°$, the intensity of the transmitted beam decreases by a factor of 2. Determine the ratio of the intensities I_{nat}/I_{pol} of the natural and linearly polarized light that make up the partially polarized light as well as the degree of polarization P of the beam. (*Answer*: $I_{nat}/I_{pol} = 1$ and $P = 1/2$.)

6.9 A quartz plate has a thickness $d_1 = 1.00$ mm and has the optical axis perpendicular to the surface of the plate. The plate rotates the plane of polarization of monochromatic linearly polarized light by an angle $\varphi_1 = 20°$.

(a) What should be the thickness of the quartz plate d so that when it is placed between two polarizers with parallel axes the transmitted light is completely extinguished?

(b) Calculate the length l of a tube filled with a sugar solution of concentration of $C = 0.4$ g/cm^3 to be placed between the two polarizers to obtain the same effect (the specific rotation of the sugar solution is $\alpha = 0.665°/\text{m·kg·m}^{-3}$). (*Answer*: (a) $d = 4.50$ mm and (b) $l = 0.34$ m.)

7 Electromagnetic Waves in Conducting Media

An electromagnetic wave propagating through a medium loses some of its energy as a result of its interaction with the medium. As a result, the energy lost by the wave is transferred predominantly into the thermal energy of that material. A material, whose electrical conductivity is very small (it is zero for all practical purposes), is called a "dielectric" or an "insulator." In dielectrics, the loss of electromagnetic wave energy is associated with the phenomenon called the polarization of the dielectric. In media with nonzero conductivity thermal losses also take place, but the mechanism is fundamentally different from the loss mechanism in dielectrics. The most common conducting media are metals. Metals have strong absorption over a wide wavelength range. However, they become sufficiently transparent for waves in the far ultraviolet region of the spectrum. The interaction of electromagnetic waves with free electrons plays a major role in the absorption of light in metals. Free electrons moving in conductors under the influence of the electric field of the wave exert a force on the positive ions located at the crystal lattice sites of the material. In this case, an effect similar to friction arises, which is accompanied by the deformation of the crystal lattice and release of a certain amount of heat. Due to the large amplitude of oscillations of free electrons, a large portion of the wave energy is spent on the excitation of the electrons. Thus, the absorption coefficient of metals is very large compared to that of insulators.

7.1 DIELECTRIC PERMITTIVITY AND IMPEDANCE OF A METAL

1. The absorption of electromagnetic waves by a medium must be taken into account if we assume that the medium has a conductivity σ. An electromagnetic wave propagating in a conducting medium excites alternating currents, which leads to a partial conversion of the wave energy into Joule heat. We assume that the conducting medium is isotropic, that is, we assume that the material parameters κ, κ_m, and σ are scalars. The current density j is related with the electric field $\mathbf{E}$ through Ohm's law $\mathbf{j} = \sigma\mathbf{E}$, $\mathbf{B} = \kappa_m\mu_0\mathbf{H}$, and $\mathbf{D} = \kappa\varepsilon_0\mathbf{E}$. As a result, the two Maxwell's equations (3.34) take the following form:

$$\nabla \times \mathbf{E} = -\kappa_m\mu_0 \frac{\partial \mathbf{H}}{\partial t},$$

$$\nabla \times \mathbf{H} = \kappa\varepsilon_0 \frac{\partial \mathbf{E}}{\partial t} + \sigma\mathbf{E}. \tag{7.1}$$

We will assume that the solutions of these equations have the form of monochromatic harmonic waves proportional to $\exp(i\omega t)$. As a result, Equations 7.1 for complex amplitudes $\mathbf{E}(\mathbf{r})$ and $\mathbf{H}(\mathbf{r})$ become

$$\nabla \times \mathbf{E} = -i\omega\mu_0\kappa_m\mathbf{H},$$

$$\nabla \times \mathbf{H} = i\omega\varepsilon_0\kappa\mathbf{E} + \sigma\mathbf{E}. \tag{7.2}$$

This system can be reduced to the form

$$\nabla \times \mathbf{E} = -i\omega\mu_0\kappa_m\mathbf{H},$$

$$\nabla \times \mathbf{H} = i\omega\varepsilon_0\tilde{\kappa}\mathbf{E}, \tag{7.3}$$

where the new parameter $\tilde{\kappa}$, the complex permittivity of the conducting medium, was introduced and it is defined as

$$\tilde{\kappa} = \kappa - i\frac{\sigma}{\varepsilon_0 \omega} = \kappa' - i\kappa'', \quad \tan\delta_e = \frac{\kappa''}{\kappa'} = \frac{\sigma}{\varepsilon_0 \kappa \omega} = \frac{\sigma}{\varepsilon \omega}. \tag{7.4}$$

In this equation, we also introduced the electric loss tangent that is determined by the ratio of real part of the complex permittivity to its imaginary part. For various media in different frequency ranges, the value $\tan\delta_e = \sigma/\varepsilon_0\kappa\omega$ can vary several orders of magnitude, and it is considered as one of the important parameters of a material.

The solutions of the wave equations (7.1) can be expressed in the form of plane monochromatic waves, which have the following forms:

$$\mathbf{E}(t,r) = \mathbf{E}_0 \exp\left[i(\omega t - \mathbf{k}\cdot\mathbf{r})\right],$$
$$\mathbf{H}(t,r) = \mathbf{H}_0 \exp\left[i(\omega t - \mathbf{k}\cdot\mathbf{r})\right], \tag{7.5}$$

where $\mathbf{E}_0$ and $\mathbf{H}_0$ are the vector amplitudes of the electric and magnetic fields. In these solutions of Maxwell's equations, the fields $\mathbf{E}$ and $\mathbf{H}$ are complex functions. The measurable values of both fields are described by the magnitudes of the fields in Equations 7.5.

Substituting Equations 7.5 into Equation 7.1 and taking into account Equation 7.4, we obtain the dispersion equation that relates the wave vector with the angular frequency and medium parameters:

$$k^2 = k_0^2 \kappa_m \left(\kappa - \frac{i\sigma}{\omega\varepsilon_0}\right), \tag{7.6}$$

where $k_0 = \omega/c$ is the wave number for vacuum. From this expression, it follows that the wave number for a wave propagating in a conducting medium has to be complex, that is, $k = k' - ik''$. Separating the real and imaginary parts of Equation 7.6, we get

$$\left(k'\right)^2 - \left(k''\right)^2 = k_0^2 \kappa\kappa_m, \quad 2k'k'' = \frac{k_0^2 \kappa_m \sigma}{\omega\varepsilon_0}. \tag{7.7}$$

From the solution of these equations, we obtain the following expressions for the real and imaginary parts of the complex wave number:

$$k' = k_0 \left(\frac{\kappa\kappa_m}{2}\left[\sqrt{1+\left(\frac{\sigma}{\varepsilon_0\kappa\omega}\right)^2} + 1\right]\right)^{1/2},$$

$$k'' = k_0 \left(\frac{\kappa\kappa_m}{2}\left[\sqrt{1+\left(\frac{\sigma}{\varepsilon_0\kappa\omega}\right)^2} - 1\right]\right)^{1/2}. \tag{7.8}$$

These relations are useful for very wide frequency range ($0 < \omega < 10^{14}$ s^{-1}) for both weakly and strongly absorbing media.

The complex impedance of a conducting medium can be presented in the form

$$Z = \sqrt{\frac{\tilde{\kappa}_m \mu_0}{\tilde{\kappa}\varepsilon_0}} = Z' + iZ'' = |Z|e^{i\zeta}, \tag{7.9}$$

where $\tilde{\kappa}$ and $\tilde{\kappa}_m$ generally are complex numbers.

In what follows, we assume that a plane wave propagates in a conducting medium along the x-axis. In this case, taking into account the complex character of the wave number in an absorbing medium, the solutions of wave Equations 7.5 have the forms

$$\mathbf{E}(t,x) = \mathbf{E}_0 e^{-k''x} e^{i(\omega t - k'x)},$$
$$\mathbf{H}(t,x) = \mathbf{H}_0 e^{-k''x} e^{i(\omega t - k'x)}, \tag{7.10}$$

where $\mathbf{E}_0$ and $\mathbf{H}_0$ are the amplitudes of wave fields at $x = 0$. From this equation, it follows that in the direction of wave propagation, the amplitudes of the wave fields decrease exponentially:

$$\mathbf{E}(x) = \mathbf{E}_0 e^{-k''x} = \mathbf{E}_0 e^{-x/\delta},$$
$$\mathbf{H}(x) = \mathbf{H}_0 e^{-k''x} = \mathbf{H}_0 e^{-x/\delta}. \tag{7.11}$$

The imaginary part of the wave number k'' describes a rate of decay of the wave amplitude in the direction of its propagation, and the reciprocal value $\delta = 1/k''$ determines the depth of the wave penetration inside the medium. Note that δ is often called **attenuation coefficient**.

Exercise 7.1

Find the frequency dependence of the conductivity in a metal. What is the dependence of the conductivity in the region of high and low frequencies?

Solution. Let us write the equation of motion for a single electron in the electric field of an electromagnetic wave:

$$m\frac{d\mathbf{v}(t)}{dt} = e\mathbf{E}(t) - b\mathbf{v}(t).$$

The right-hand side of this equation includes the electric force and a friction-like force that is responsible for the electrical resistance.

Note that there are no friction forces acting on a single electron. A single electron is constantly colliding with the defects of the crystalline lattice and also with other electrons, and therefore, it is moving chaotically when there is no applied electric field. In an external field, the electron motion becomes more directed with some average velocity that we can find from the equation given earlier.

The electric field of an electromagnetic wave oscillates with an angular frequency ω, and therefore,

$$m\frac{d\mathbf{v}(t)}{dt} + b\mathbf{v}(t) = e\mathbf{E}\exp(i\omega t).$$

We will seek a solution of this equation in the form $\mathbf{v}(t) = \mathbf{v}\exp(i\omega t)$. After its substitution into the equation of motion, we get the equation

$$m(i\omega + 2\gamma)\mathbf{v} = e\mathbf{E},$$

where we introduced damping constant $2\gamma = b/m$, which has the physical meaning of collision frequency of the electron with the lattice irregularities. From this expression, we find

$$\mathbf{v}(\omega) = \frac{e\mathbf{E}}{m} \times \frac{1}{i\omega + 2\gamma} = \frac{e\mathbf{E}}{m} \times \frac{2\gamma - i\omega}{\omega^2 + 4\gamma^2}.$$

The current density is given by $\mathbf{j}(\omega) = Ne\,\mathbf{v}(\omega) = \tilde{\sigma}(\omega)\,\mathbf{E}$, where N is the electron concentration per unit volume and $\tilde{\sigma} = \sigma' - i\sigma''$ is the complex conductivity. Thus, we get

$$\tilde{\sigma} = \frac{Ne^2}{m} \cdot \frac{2\gamma - i\omega}{\omega^2 + 4\gamma^2} = \frac{Ne^2}{2m\gamma} \cdot \frac{1 - i\omega/2\gamma}{(\omega/2\gamma)^2 + 1} = \sigma_0 \frac{1 - i\omega\tau}{1 + (\omega\tau)^2},$$

where

$\sigma_0 = Ne^2\tau/m$ is the static conductivity

$\tau = 1/2\gamma = m/b$ is the average time between collisions

Let us introduce the plasma frequency $\omega_p = \sqrt{Ne^2/\varepsilon_0 m} = \sqrt{\sigma_0/\varepsilon_0 \tau}$ and separate real and imaginary parts of complex conductivity:

$$\sigma' = \frac{Ne^2}{m} \cdot \frac{2\gamma}{\omega^2 + 4\gamma^2} = \frac{2\varepsilon_0\gamma\omega_p^2}{\omega^2 + 4\gamma^2} = \frac{\sigma_0}{1 + (\omega\tau)^2},$$

$$\sigma'' = \frac{Ne^2}{m} \cdot \frac{\omega}{\omega^2 + 4\gamma^2} = \frac{\varepsilon_0\omega\omega_p^2}{\omega^2 + 4\gamma^2} = \frac{\sigma_0\omega\tau}{1 + (\omega\tau)^2}.$$

The conductivity σ, which is related with the real currents of the charge carriers in the conducting medium, is given by the real part of the complex conductivity, that is,

$$\sigma(\omega) = \sigma'(\omega) = \frac{2\varepsilon_0\gamma\omega_p^2}{\omega^2 + 4\gamma^2} = \frac{\sigma_0}{1 + (\omega\tau)^2}.$$

For sufficiently low frequencies ($\omega\tau \ll 1$), the conductivity depends on the frequency:

$$\sigma(\omega) = \sigma_0\left(1 - (\omega\tau)^2\right).$$

We see that when the frequency tends to zero, the conductivity of the metal tends to its static limit σ_0.

At higher frequencies ($\omega\tau \gg 1$), the real part of conductivity of the metal tends to zero:

$$\sigma' = \frac{\sigma_0}{(\omega\tau)^2}.$$

At higher frequencies ($\omega\tau \gg 1$), the imaginary part of the conductivity of metals is substantially larger than its real part and the total conductivity tends to zero:

$$\sigma'' = \frac{\sigma_0}{\omega\tau}.$$

Because of such frequency dependence, the metal becomes transparent for high-frequency radiation (e.g., for X-rays).

Previously, we used extensively the notion of a complex electric permittivity of a medium. The real part of this parameter has described refraction of light and the imaginary part—its absorbance. However, for a medium with finite conductivity σ, the use of electric permittivity for the region of lower frequencies becomes inconvenient. At $\omega \to 0$, the imaginary part of the electric permittivity $\kappa'' = \sigma'/\omega\varepsilon_0$ tends to infinity. This is why for physics of metals it is more convenient to work with complex conductivity whose real part is equal to $\sigma' = \sigma = \varepsilon_0\omega\kappa''$. The use of complex electric permittivity and conductivity usually gives the same results for a wide range of frequencies. In one of these approaches, the polarization of a medium is used (which is convenient for dielectrics); otherwise, the electric current is used (which is convenient for conductors).

7.2 SKIN EFFECT

1. From expressions (7.11), it follows that at a depth equal to δ inside the conducting medium, the wave amplitude decreases by a factor of e, that is, the value

$$\delta = \frac{c}{\omega}\left(\frac{\kappa\kappa_m}{2}\left[\sqrt{1+\left(\sigma/\varepsilon_0\kappa\omega\right)^2}-1\right]\right)^{-1/2} \tag{7.12}$$

determines the penetration depth of a wave field in a conducting medium, for which absorption is given by the imaginary part of the dielectric permittivity. We assume that the magnetic permeability of the medium is a scalar, is real, and is close to unity (this is true for the majority of metals, which are not ferromagnetic).

Since the intensity of a wave is proportional to the square of its amplitude ($I \sim |E|^2$), the energy of the wave penetrating in the medium is restricted in a layer with thickness of a few δ. For media with high conductivity, the thickness of this layer can be rather small. For example, for copper ($\sigma = 5.70 \times 10^7$ S/m) a wave with frequency $f = \omega/2\pi = 100$ MHz has a $\delta \approx 6.60$ μm. For this reason, the area near the surface of the medium, where most of the wave energy is located, is called the *skin layer*. The parameter δ is defined as the *thickness of the skin layer*, and the effect of localization of the electromagnetic wave energy in a layer near the medium surface is called the *skin effect*.

Let us consider two limiting cases of Equation 7.12: the cases of low and high conductivity. In the case of low conductivity or sufficiently high frequency ($\sigma \ll \varepsilon_0\kappa\omega$), we obtain

$$\delta = \frac{2\varepsilon_0 c}{\sigma}\sqrt{\frac{\kappa}{\kappa_m}} = \frac{2}{\sigma}\sqrt{\frac{\varepsilon_0\kappa}{\mu_0\kappa_m}} = \frac{2}{\sigma Z}. \tag{7.13}$$

In this limit, the depth of field penetration in a medium does not depend on the wave frequency; however, the lower is the conductivity σ, the deeper is the penetration of wave into the medium. In the case $\sigma \to 0$, the medium becomes nonconducting and the depth of an electromagnetic field penetration $\delta \to \infty$.

In the case of high conductivity or sufficiently low frequencies ($\sigma \gg \varepsilon_0\kappa\omega$), the condition $\kappa'' \gg \kappa'$ is satisfied. In this case, it is possible to neglect the real part of the dielectric permittivity of the medium and consider it to be purely imaginary and equal to $\tilde{\kappa} = -i\sigma/\omega\varepsilon_0$. The condition $\kappa'' \gg \kappa'$ is satisfied for metals in the microwave range of frequencies. The expression for the skin thickness takes the following form:

$$\delta = c\sqrt{\frac{2\varepsilon_0}{\kappa_m\sigma\omega}} = \sqrt{\frac{2}{\mu_0\kappa_m\sigma\omega}}, \tag{7.14}$$

from which it follows that an increase in the medium conductivity and/or the wave frequency results in a decrease in the penetration depth. When $\omega \to 0$, the penetration depth $\delta \to \infty$. For sufficiently high frequency, the thickness of the skin layer can be very small. As an example, we will give the values of the skin layer thickness of copper for several frequencies.

Frequency	δ
60 Hz	8.57 mm
10 kHz	0.66 mm
100 kHz	0.21 mm
1 MHz	66 μm
10 MHz	21 μm

The complex refractive index $\tilde{n}$ of a highly conducting medium is given by

$$\tilde{n} = \sqrt{\tilde{\kappa}\kappa_m} = \sqrt{-i\frac{\sigma\kappa_m}{\varepsilon_0\omega}} = (1-i)\sqrt{\frac{\sigma\kappa_m}{2\varepsilon_0\omega}}, \tag{7.15}$$

where we have used the equality $\sqrt{-i} = (1-i)/\sqrt{2}$. The refractive index modulus is

$$|\tilde{n}| = \sqrt{\frac{\sigma\kappa_m}{\omega\varepsilon_0}} \gg 1. \tag{7.16}$$

This means that the angle of refraction at the interface of a dielectric and a highly conducting medium is close to zero. Therefore, we conclude that, irrespective of the angle of incidence θ_0, the refracted wave goes into the conducting medium perpendicularly to its surface (this will be discussed in more details in Section 7.3 and Figure 7.1).

The complex impedance of a highly conducting medium has the following form:

$$Z = \sqrt{\frac{\kappa_m\mu_0}{\tilde{\kappa}\varepsilon_0}} = \sqrt{-\frac{\omega\kappa_m\mu_0}{i\sigma}} = (1+i)\sqrt{\frac{\omega\kappa_m\mu_0}{2\sigma}}, \tag{7.17}$$

from where it is seen that in this case $Z' = Z''$ and $\tan\zeta = Z''/Z' = 1$, that is, the oscillations of the wave field vectors $\mathbf{E}$ and $\mathbf{H}$ are shifted in phase by $\zeta = \pi/4$.

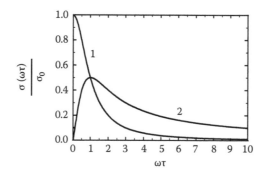

FIGURE 7.1 Dependence of real (curve 1) and imaginary (curve 2) parts of complex conductivity of Equation 6.64 on frequency ω; $\sigma_0 = \sigma(\omega = 0)$.

2. The theory discussed earlier is valid only under the condition that the skin layer thickness δ is much larger than the mean free path l of the electrons in the medium. It is assumed that due to collisions with the crystal lattice ions, electrons continuously lose energy to overcome the ohmic resistance of the conductor. This energy is transformed into the Joule heat. The relation δ ≫ l holds for a rather wide range of materials and wavelengths. However, at very low temperatures the situation changes: the mean free path length of the electrons increases considerably, and hence, the conductivity increases significantly. As a result, there is a sharp decrease in the skin layer thickness and condition δ ≫ l fails. When the mean free path l of the electrons in a conductor becomes greater than a skin layer thickness δ, at rather high frequencies, the skin effect acquires a number of features, thanks to which it received the name ***anomalous***. In the following, we discuss the origin of some of these features. Since the electromagnetic wave field along the mean free path of the electrons is nonuniform, the resulting current at each point depends not only on the electric field at this point but also on the electric field in its vicinity, which has dimensions of the order of l. Therefore, for the solution of Maxwell's equations instead of a local Ohm's law $\mathbf{j}(\mathbf{r}) = \sigma(\mathbf{r})\mathbf{E}(\mathbf{r})$, it is necessary to take into account the nonlocal nature of this relation as current density in point $\mathbf{r}$ depends on electric fields in the vicinity of point $\mathbf{r}$.

Under these conditions, electrons moving in various directions become nonequivalent from the point of view of their contribution to the electric current. At $l \gg$ δ, the main contribution to the current comes from those electrons that move in a skin layer parallel to the surface of a metal or under very small angles with respect to the surface and thus spend more time in the strong field area. This results in an attenuation of the electromagnetic wave in the layer near to the surface. The quantitative result in this case differs from the case of a normal skin effect. In particular, for the anomalous skin effect, the decrease in the wave field in the near-surface layer does not have anymore an exponential dependence on distance as in Equation 7.11.

In the infrared range of frequencies, the electron may not be able to travel a distance l during one period. In this case, the wave field on the electron path during one period can be considered as uniform. This leads again to Ohm's law and the skin effect becomes normal. Thus, at low and very high frequencies, the skin effect is always normal. In the radio frequency range, depending on the relationships between l and δ, we may have either the normal or the anomalous skin effect.

In practice, the skin effect is often undesirable. For example, alternating current in wires with a strong skin effect flows mainly on the surface layer. Thus, the central section of the wire is not used. As a result, the effective resistance of the wire and thus the power losses increase.

Exercise 7.2

Derive an expression for the plasma frequency of electrons in a metal using the continuity equation. Determine the penetration depth of an electromagnetic wave incident on a nonmagnetic metal with conductivity σ = 5.70 × 10⁷ S/m at a frequency equal to the metal plasma frequency (the nonmagnetic metal used here is copper with electron concentration equals to $N = 8.00 \times 10^{28}$ m⁻³).

Solution. The charge density of electrons in a metal is equal to ρ = Ne, where N is the electron concentration. The continuity equation (3.40) relates the change of charge density with current density as

$$\nabla \cdot \mathbf{j} + \frac{\partial \rho}{\partial t} = 0,$$

where
    $\mathbf{j} = Ne\,\mathbf{v}$ is the current density
    $\mathbf{v}$ is the average velocity of electrons

Let us differentiate this equation with respect to time:

$$\frac{\partial}{\partial t}(\nabla \cdot \mathbf{j}) + \frac{\partial^2 \rho}{\partial t^2} = 0 \quad \text{or} \quad \nabla \cdot \frac{\partial \mathbf{j}}{\partial t} + \frac{\partial^2 \rho}{\partial t^2} = 0 \tag{7.18}$$

Let us write the equation of motion for an electron in the absence of collisions (i.e., without losses):

$$m\frac{d\mathbf{v}}{dt} = e\mathbf{E}.$$

Let us rewrite this equation taking into account that $\mathbf{v} = \mathbf{j}/Ne$:

$$\frac{d\mathbf{j}}{dt} = \frac{Ne^2\mathbf{E}}{m}.$$

After applying to this equation the operator ∇ and using the first of Maxwell's Equations 3.34, $\nabla \cdot \mathbf{E} = \rho/\varepsilon_0$, we get

$$\nabla \cdot \frac{d\mathbf{j}}{dt} = \frac{Ne^2}{m}\nabla \cdot \mathbf{E} = \frac{Ne^2}{m} \times \frac{\rho}{\varepsilon_0}.$$

Substituting $\nabla(d\mathbf{j}/dt)$ into the second equation of the set (7.18), we get

$$\frac{\partial^2 \rho}{\partial t^2} + \omega_p^2 \rho = 0,$$

where we introduced the plasma frequency $\omega_p = \sqrt{Ne^2/\varepsilon_0 m}$. From this equation, it follows that free electrons oscillate with this frequency with respect to the lattice of immobile positive ions.

The depth of penetration of the wave field (the width of skin layer) into the metal with high conductivity for $\sigma \gg \varepsilon_0\kappa\omega$ is defined by the expression

$$\delta = c\sqrt{\frac{2\varepsilon_0}{\kappa_m\sigma\omega}} = c\sqrt{\frac{2\varepsilon_0}{\sigma\omega_p}},$$

where we took into account that the magnetic permeability of the metal is equal to unity and $\omega = \omega_p$. Since for copper $\omega_p = 1.60 \times 10^{16}$ s^{-1}, then we get for the skin layer thickness $\delta \approx 1.32 \times 10^{-6}$ cm.

7.3 WAVE INCIDENCE ON A METAL SURFACE

1. Consider a plane monochromatic wave incident on the plane boundary between a nonabsorbing dielectric medium 1 and a conducting medium 2 (Figure 7.2). Here, we consider the propagation characteristics of the wave passing through the interface. Regardless of the polarization of the incident wave, the coordinate dependence of the field vectors of the transmitted wave is described by the expression

$$\mathbf{E}_2^t(x,z), \mathbf{H}_2^t(x,z) \sim e^{-i(k_{2x}x + k_{2z}z)} = e^{-ik_2(\sin\theta_2 \cdot x + \cos\theta_2 \cdot z)}, \tag{7.19}$$

where θ_2 is the angle of refraction.

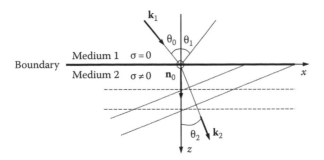

FIGURE 7.2 Orientation of the planes of equal phases (dotted lines) and equal amplitudes (dashed lines) for a nonuniform electromagnetic wave propagating in a medium.

The boundary conditions for the tangential components of the field vectors are

$$E_{1x,y} = E_{2x,y}, \quad H_{1x,y} = H_{2x,y}. \tag{7.20}$$

In order to satisfy these conditions, it is necessary to satisfy the relation

$$k_{1x} = k_{2x} \quad \text{or} \quad k_1 \sin \theta_0 = k_2 \sin \theta_2. \tag{7.21}$$

The wave vector $\mathbf{k}_2$ in the absorbing medium 2 will be complex; however, its projection k_{2x} is real (the wave vector $\mathbf{k}_1$ and the angle of incidence θ_0 are also real). This means that $\sin \theta_2$ and the refraction angle θ_2 have to be complex numbers. The projection $k_{2z} = k_2 \cos \theta_2$ will also be complex: $k_{2z} = k'_{2z} - ik''_{2z}$. Thus, expressions (7.19) can be written as

$$\mathbf{E}^t_2, \mathbf{H}^t_2 \sim e^{-k''_{2z}z} \cdot e^{-i(k_{2x} \cdot x + k'_{2z} \cdot z)}. \tag{7.22}$$

This expression shows that irrespective of the angle of incidence, the refracted wave attenuates strictly in the direction of the normal to the interface. Here, the planes, parallel to the interface, are surfaces of equal amplitudes. Surfaces of equal phases are described by the equation

$$k_{2x}x + k'_{2z}z = \text{const.} \tag{7.23}$$

The normal to the surfaces of equal phase is directed along the z-axis at an angle θ_2, which has the meaning of a real refraction angle ($\tan \theta_2 = k_{2x}/k'_{2z}$). Thus, the planes of equal amplitude and of equal phase do not coincide, that is, the refracted wave in an absorbing medium is **nonuniform**.

2. In a highly conducting medium $k'_{2z} \gg k_{2x}$ and the angle $\theta_2 \cong 0$, that is, the refracted wave propagation direction is practically perpendicular to the interface. Surfaces of equal phase become parallel to the interface, and the refracted wave is approximately uniform. Thus, in the case of $\mathbf{k}_2 \uparrow \uparrow \mathbf{n}_0$, where $\mathbf{n}_0$ is a unit vector normal to the interface, directed inside of medium 2 as it is shown in Figure 7.2. It means that the field vectors $\mathbf{E}_2$ and $\mathbf{H}_2$, which are perpendicular to $\mathbf{k}_2$, are parallel to the interface, that is, they are equal to the tangential components of fields, $\mathbf{E}_{2\tau}$ and $\mathbf{H}_{2\tau}$.

These two fields are connected by a vector ratio

$$\mathbf{E}_{2\tau} = Z_2(\mathbf{H}_{2\tau} \times \mathbf{n}_0), \tag{7.24}$$

where Z_2 is the complex impedance of medium 2 defined by Equation 7.17 and, in this case, it is called the **surface impedance**.

Since at the interface the tangential components of the wave field obey the condition of continuity

$$E_{1\tau} = E_{2\tau}, \quad H_{1\tau} = H_{2\tau},$$

then these components, $\mathbf{E}_{1\tau}$ and $\mathbf{H}_{1\tau}$, of fields in the dielectric at the interface are also related by the following form (7.24):

$$\mathbf{E}_{1\tau} = Z_2(\mathbf{H}_{1\tau} \times \mathbf{n}_0). \tag{7.25}$$

This condition is called in some books as the approximate **Leontovich boundary condition**. It is often used in the calculation of electromagnetic fields in the vicinity of conducting surfaces.

In conclusion, we will discuss the limiting case of a conducting medium, when its conductivity is considered to be infinitely large (*ideal conductor*). For $\sigma \to \infty$, the module impedance of medium 2, according to Equation 7.17, tends to zero, that is,

$$|Z_2| = \sqrt{\frac{\omega \kappa_m \mu_0}{\sigma}} \to 0, \tag{7.26}$$

meaning the vanishing of the electric field (and therefore of the magnetic field) of the wave inside the conductor. Thus, the field of the incident wave does not penetrate in a perfect conductor, the thickness of the skin layer δ is zero, and the Joule heat losses inside the conductor can be neglected.

Let us discuss the boundary conditions at the surface of a perfect conductor. For simplicity, we consider normal incidence of a wave on the surface of a perfect conductor (Figure 7.3).

Since for the conductor $E_{2\tau} = H_{2\tau} = 0$, the condition $E_{1\tau} = H_{1\tau} = 0$ has to be satisfied. The value $E_{1\tau}$ is the sum of electric field of the incident and reflected waves:

$$E_{1\tau} = E_{1\tau}^i + E_{1\tau}^t = 0. \tag{7.27}$$

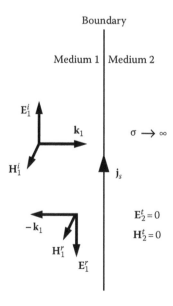

FIGURE 7.3 Orientation of the vectors of electric and magnetic fields for incident and reflected waves in the case of normal incidence on an ideal conductor.

From the last equation it follows that $E_{1\tau}^t = -E_{1\tau}^i$. Thus, the incident wave undergoes a total reflection, and a standing wave is formed in medium 1 in front of the interface between media 1 and 2 (consult Section 4.7 about the standing waves). Vector **E** reverses direction without changing its magnitude, but the direction of the vector **H**, according to the right-hand screw rule, remains unchanged and

$$H_{1\tau} = H_{1\tau}^i + H_{1\tau}^t = 2H_{1\tau}^i \neq 0. \tag{7.28}$$

This creates the appearance of a violation of the boundary conditions—continuity of the tangential component of the magnetic field. However, in this case, when we apply the boundary condition for the magnetic field, it is necessary to take into account the density of surface current j_s. This current flows in a thin near-surface layer, which has a thickness that tends to zero when $\sigma \to \infty$. Thus, for an interface between a dielectric and an ideal conductor, the boundary condition for the magnetic field is $2H_{1\tau}^i = j_s$.

Exercise 7.3

A plane electromagnetic wave is incident obliquely on the surface of a nonmagnetic metal ($\kappa_m = 1$) with conductivity σ at an incidence angle θ_0. Find the power losses of the waves with p- and s-polarizations on a unit area of metal surface (power density losses).

Solution. The energy of the wave that penetrates into the metal will be ultimately turned into heat. Therefore, we will determine the energy flux that penetrates into the metal. According to the boundary conditions, the tangential components of the electric and magnetic fields must be continuous at the interface of the two media, that is,

$$E_{1\tau} = E_{2\tau} = E_\tau, \quad H_{1\tau} = H_{2\tau} = H_\tau.$$

If at the metal surface the wave field is known, then to find the power losses density, it is necessary to calculate the average value of the Poynting vector that is directed into the metal:

$$\mathbf{S} = \frac{1}{2}\mathrm{Re}(\mathbf{E}_\tau \times \mathbf{H}_\tau^*).$$

Here, the tangential components $\mathbf{E}_\tau$ and $\mathbf{H}_\tau$ at the surface of the conductive medium are related by the Leontovich boundary condition:

$$\mathbf{E}_\tau = Z_m(\mathbf{H}_\tau \times \mathbf{n}_0),$$

where $\mathbf{n}_0$ is the unit vector of the normal to the metal surface that is directed inside of the metal (see Figure 7.2) and

$$Z_m = \sqrt{\frac{\kappa_m \mu_0}{\tilde{\kappa}\varepsilon_0}} = \sqrt{-\frac{\omega \kappa_m \mu_0}{i\sigma}} = (1+i)\sqrt{\frac{\omega \kappa_m \mu_0}{2\sigma}}$$

is the metal impedance where we took into account that the complex conductivity of the metal is given by $\tilde{\kappa} = -i\sigma/\varepsilon_0\omega$, $\kappa_m = 1$ (see Equation 7.17), and

$$\sqrt{-\frac{1}{i}} = \sqrt{i} = \left(e^{i\pi/2}\right)^{1/2} = e^{i\pi/4} = \cos\frac{\pi}{4} + i\cdot\sin\frac{\pi}{4} = \frac{1+i}{\sqrt{2}}.$$

Let us substitute these relationships into the expression for the Poynting vector:

$$\mathbf{S} = \frac{1}{2}\mathrm{Re}\left[Z_m \left(\mathbf{H}_\tau \times \mathbf{n}_0 \right) \times \mathbf{H}_\tau^* \right] = \frac{1}{2}\left| \mathbf{H}_\tau \right|^2 \mathrm{Re}(Z_m)\mathbf{n}_0,$$

where the real part of impedance is equal to $\mathrm{Re}(Z_m) = \sqrt{\omega \kappa_m \mu_0 / 2\sigma}$. Thus, for the power loss density, we get

$$P_A = \frac{1}{2}\left| \mathbf{H}_\tau \right|^2 \mathrm{Re}(Z_m) = \frac{1}{2}\left| \mathbf{H}_\tau \right|^2 \sqrt{\frac{\omega \kappa_m \mu_0}{2\sigma}} = \left| \mathbf{H}_\tau \right|^2 \sqrt{\frac{\omega \kappa_m \mu_0}{8\sigma}}.$$

Since at the interface the tangential components of the field vectors are continuous, then

$$P_A = \left| \mathbf{H}_0 \right|^2 \sqrt{\frac{\omega \kappa_m \mu_0}{8\sigma}}$$

in the case of an incident wave with p-polarization (vector $\mathbf{H}_0$ is perpendicular to incidence plane) and

$$P_A = \left| \mathbf{H}_0 \right|^2 \sqrt{\frac{\omega \kappa_m \mu_0}{8\sigma}} \cos^2 \theta_0$$

in the case of an incident wave with s-polarization (vector $\mathbf{H}_0$ lies in the plane of incidence).

7.4 SURFACE WAVES AT THE INTERFACE BETWEEN A DIELECTRIC AND A CONDUCTOR

1. **Surface waves** propagate along the interface between two different media. Their fields are localized near the interface and decrease in the direction perpendicular to the interface. A surface wave is a particular solution of Maxwell's equations, which was first obtained by Sommerfeld early in the twentieth century. We assume that the plane $z = 0$ separates two semi-infinite media with dielectric permittivities κ_1 (half-space 1, $z < 0$) and κ_2 (half-space 2, $z > 0$). We assume that these media are nonmagnetic ($\kappa_{m1} = \kappa_{m2} = 1$). For each medium, one can write Maxwell's equations for the wave field vectors:

$$\nabla \times \mathbf{E}_j = -i\omega\mu_0\mathbf{H}_j,$$
$$\nabla \times \mathbf{H}_j = i\omega\kappa_j\varepsilon_0\mathbf{E}_j, \tag{7.29}$$

where $j = 1, 2$ is the index of the corresponding half-space. As we already demonstrated, by applying $\nabla \times$ operation to Equations 7.29, these equations are reduced to the following Helmholtz equations:

$$\nabla^2\mathbf{E}_j + k_0^2\kappa_j\mathbf{E}_j = 0,$$
$$\nabla^2\mathbf{H}_j + k_0^2\kappa_j\mathbf{H}_j = 0, \tag{7.30}$$

where $k_0 = \omega/c$ is the wave number in a vacuum.

We search for solutions of Equations 7.30 in the form of a surface waves traveling along the x-axis. Let us write this solution for the tangential components of the magnetic field

$$H_{jy}(x,z,t) = H_{jy}(z) \cdot e^{i(\omega t - kx)}, \tag{7.31}$$

where k is the wave number, the same in both regions 1 and 2. Substituting Equation 7.31 into the second Equation 7.30, we obtain

$$\frac{\partial^2 H_{jy}}{\partial z^2} - g_j^2 H_{jy} = 0, \tag{7.32}$$

where we have introduced the transverse wave vector components (along the normal to the interface) in each of the two media:

$$g_j^2 = k^2 - \kappa_j k_0^2. \tag{7.33}$$

The solutions of Equation 7.32 have the form

$$\begin{cases} H_{1y} = U_1 e^{g_1 z} \cdot e^{i(\omega t - kx)}, & z < 0, \\ H_{2y} = U_2 e^{-g_2 z} \cdot e^{i(\omega t - kx)}, & z > 0, \end{cases} \tag{7.34}$$

where $U_{1,2}$ are the constants. As energy of the wave is limited, the wave fields in both nonamplifying media have to decrease exponentially when we move away from the interface. Therefore, in Equations 7.34, the values $g_{1,2}$, which have the meaning of damping constants, have to be positive.

Writing down the equations of system (7.29), it is easy to show that the system has two independent solutions, which correspond to two waves with different orthogonal polarizations.

The first solution determines the relation of ***transverse magnetic (TM) wave*** with the components of a vector field E_x, H_y, E_z, and the second solution determines the ***transverse electric (TE) wave*** with the components H_x, E_y, H_z.

First, we consider the solution in the form of a TM wave. At the interface ($z = 0$), the following boundary conditions have to be satisfied: (1) the continuity of the tangential components of the electric and magnetic fields and (2) the continuity of the normal components of the dielectric displacements:

$$H_{1y} = H_{2y}, \quad E_{1x} = E_{2x}, \quad \kappa_1 E_{1z} = \kappa_2 E_{2z}.$$

It follows from these relations that the magnetic field vector at the interface is the same in both media. To express values E_x, E_z through a magnetic field H_y, we will write down the second equation of system (7.29) in projections on the x- and z-axes taking into account that the fields do not depend on y-coordinate:

$$-\frac{\partial H_{jy}}{\partial z} = i\omega \kappa_j \varepsilon_0 E_{jx}, \quad \frac{\partial H_{jy}}{\partial x} = i\omega \kappa_j \varepsilon_0 E_{jz}. \tag{7.35}$$

Using Equation 7.34 and calculating the derivatives of H_{jy}, we obtain

$$E_{jx} = \pm \frac{ig_j}{\omega \kappa_j \varepsilon_0} H_{jy}, \quad E_{jz} = -\frac{k}{\omega \kappa_j \varepsilon_0} H_{jy}, \tag{7.36}$$

where the signs "+" and "−" in the expression for E_{jx} correspond to media 1 and 2, respectively. The field vectors of the surface wave are shown in Figure 7.4.

Taking into account Equations 7.36 and the relation $H_{1y} = H_{2y}$, we see that the boundary condition $E_{1x} = E_{2x}$ is satisfied only when

$$\frac{g_1}{g_2} = -\frac{\kappa_1}{\kappa_2}, \tag{7.37}$$

from where, taking into account that $g_1 > 0$ and $g_2 > 0$, it follows that *the surface wave can propagate only along the interface between media with dielectric permittivities of different signs.* If $\kappa_1 > 0$, then $\kappa_2 = -|\kappa_2| < 0$ (in the case of a complex permittivity, this condition concerns to its real part). The medium with negative dielectric permittivity is called *surface active.*

Substituting expressions (7.33) for the damping constants into Equation 7.37, we obtain the following expression for the wave number:

$$k = k_0 \sqrt{\frac{\kappa_1 |\kappa_2|}{|\kappa_2| - \kappa_1}} = k_0 \sqrt{\frac{\kappa_1 \kappa_2}{\kappa_2 + \kappa_1}}, \tag{7.38}$$

which is a dispersion relation for waves of this type, since it determines the dependence of the wave number k on the angular frequency ω (on the right side of Equation 7.38, not only k_0 but generally κ_1 and κ_2 are functions of frequency).

From Equation 7.38, it follows that for the wave number k to be real, the expression under the square root should be positive, that is, the surface wave will propagate only if condition $|\kappa_2| > \kappa_1$ is satisfied.

For the surface TE wave (with components of fields E_y, H_x, and H_z), the dispersion relation has the following form, which is analogous to Equation 7.37:

$$\frac{g_1}{g_2} = -\frac{\kappa_{m1}}{\kappa_{m2}}. \tag{7.39}$$

With positive values of g_1 and g_2, this condition can be satisfied only in the case when the magnetic permeabilities of the adjacent media, κ_{m1} and κ_{m2}, have different signs. Thus, at the boundary between two nonmagnetic dielectrics, the surface wave can only be of TM type.

2. Surface waves can be excited at the interface of two dielectrics at frequencies that for one of the two media correspond to a range of anomalous dispersion (near a line of resonance absorption). Also, the surface waves can be excited at the boundary of media that have high conductivity— for example, metals and semiconductors. In this case, the negative sign of the value $\kappa' = \mathrm{Re}\kappa$ is due to the dominant contribution to the material polarizability of the electron gas. The free electrons move in such a way that they generate a field opposite to the external electric field.

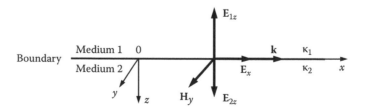

FIGURE 7.4 Orientation of the electric and magnetic field vectors for a TM surface wave at the interface of two media.

Therefore, electrons partially screen the external field, and the latter penetrates into the conductor to a depth that is considerably smaller than the wavelength. However, at sufficiently high frequencies, the electrons cannot respond to the rapidly changing external electric field, and the metal becomes transparent. The characteristic frequency at which this occurs is called the ***plasma frequency***.

The simplest equation that describes the dispersion properties of the metal, that is, the function $\kappa(\omega)$, is known as the ***Drude formula***:

$$\kappa(\omega) = 1 - \frac{\omega_p^2}{\omega(\omega + i\gamma)}, \tag{7.40}$$

where
 ω_p is the plasma frequency
 γ is the frequency of electron collisions

For metals, in the optical frequency range ($\omega \sim 10^{15}$ s^{-1}), $\omega_p \sim 10^{16}$ s^{-1}, $\gamma \sim 10^{13} - 10^{14}$ s^{-1}, that is, $\gamma < \omega < \omega_p$. In particular for an *air–metal* interface ($\kappa_1 = 1$, $\kappa_2' \simeq 1 - \omega_p^2/\omega^2$, and $\gamma = 0$) from Equation 7.38, we obtain

$$k \simeq k_0 \sqrt{\frac{1 - \omega^2/\omega_p^2}{1 - 2\omega^2/\omega_p^2}},$$

and taking into account that $\omega \ll \omega_p$, this equation is simplified to

$$k \simeq k_0 \left(1 + \frac{\omega^2}{2\omega_p^2} \right). \tag{7.41}$$

Figure 7.5 shows the dispersion curve $\omega(k)$ for surface waves described by Equations 7.38 and 7.41 (curve 1). For comparison, the dependence $\omega = ck$ for a vacuum is also shown (curve 2).

It is seen that the deviation of curve 2 from curve 1 increases with frequency, indicating a decrease in both the phase and group velocity of the surface wave compared with these parameters of a wave in a vacuum. Curve 3 represents the solution of dispersion equation $\omega^2 = \omega_p^2 + c^2 k^2$ for volume waves that can propagate in the metal in the transparency range $\omega > \omega_p$, and for large k as ck becomes substantially larger than ω_p, the $\omega(k)$ line tends to ck (line 2).

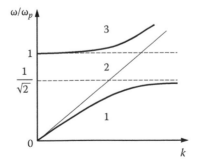

FIGURE 7.5 Dependences ω versus k for the surface (curve 1) and volume (curve 3) waves; curve 2 shows $\omega(k)$ for the case of the vacuum.

Because of the difference between the phase velocities in vacuum and other media, the surface wave cannot be excited by a wave incident from a vacuum. There are two effective methods of the excitation of surface waves: (1) using a **prism** and (2) using a **grating**. The prism method, also known as the **method of broken total internal reflection**, can be realized in two geometries: Otto's geometry and Kretschmann's geometry (Figure 7.6a and b).

In Otto's geometry, a light wave before entering medium 1 with a dielectric permittivity κ_1 falls on the prism that has a dielectric permittivity $\kappa_p > \kappa_1$. For an angle of incidence θ, greater than the critical angle for total internal reflection, a *slow* wave propagates along the interface between media 3 and 1. Its phase velocity is less than that of the waves in an infinite medium. For sufficiently small thickness of the gap d (of the order of the wavelength), between the prism and the surface-active medium 2 with a dielectric constant κ_2, the interface between media 1 and 2 also participates in the wave process. A surface wave travels along interfaces 1 and 2 and its phase velocity is matching the phase velocity of the *slow* wave at the interface between media 1 and 3.

Since Snell's law and the dispersion relation (7.38) are symmetric relative to κ_1 and κ_2, it is possible to change the arrangement and get Kretschmann's geometry (Figure 7.6b). In this case, the thin film 2 is placed on the bottom face of a prism, and the film is serving as the active medium. Because of the strong absorption in the active medium, the film thickness has to be rather small so that most of the wave energy reaches interfaces 1–2.

Excitation of surface electromagnetic waves can also be carried out by means of the diffraction grating, which is placed on the surface-active medium and which diffracts radiation under particular angles.

Exercise 7.4

Find the relationship between the material parameters of the transverse components of the wave number of a surface wave at the interface between a transparent dielectric and a metal. Find the frequency range for which a surface wave exists. The electric permittivity κ_1 of the dielectric is positive, and the electric permittivity $\kappa_2(\omega)$ of the metal is described by the Drude model without taking into account losses.

Solution. The transverse components of the wave vector **k** of the surface wave are given by the following expressions:

$$g_1^2 = k^2 - k_0^2 \kappa_1, \quad g_2^2 = k^2 - k_0^2 \kappa_2.$$

The relationship of the wave number with the material parameters can be written as

$$k^2 = k_0^2 \frac{\kappa_1 \kappa_2}{\kappa_2 + \kappa_1},$$

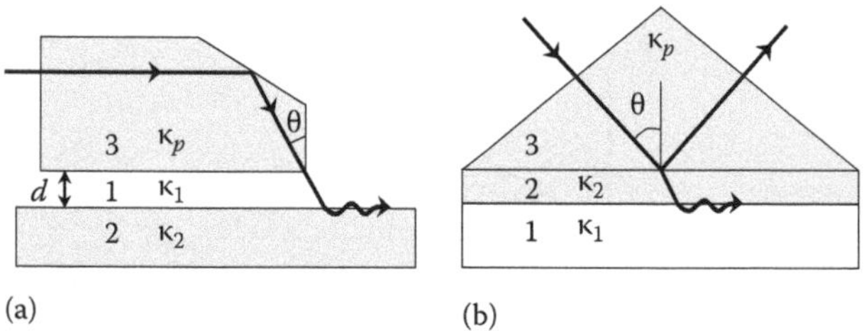

(a) (b)

FIGURE 7.6 Schematic of the excitation of surface waves in geometries by Otto (a) and Kretschmann (b).

where $k_0 = \omega/c$. For a metal $\kappa_2(\omega) = 1 - (\omega_p/\omega)^2$, where ω_p is the plasma frequency. In order for a surface wave to exist, the condition $\kappa_2(\omega) < 0$ must be satisfied, which is valid only in the frequency range $\omega < \omega_p$. Let us find out if a surface wave exists in the entire region or only in a part of it. To answer this question, let us substitute the expression for k^2 into the expressions for $g_{1,2}^2$:

$$g_1^2(\omega) = -k_0^2 \frac{\kappa_1^2}{\kappa_2(\omega) + \kappa_1}, \quad g_2^2(\omega) = -k_0^2 \frac{\kappa_2^2(\omega)}{\kappa_2(\omega) + \kappa_1}.$$

According to the equation for surface wave (7.34), in order for this wave to exist, both g_1 and g_2 must be real. Therefore, the following inequalities must take place: $g_1^2 > 0$ and $g_2^2 > 0$. Thus, $\kappa_2 + \kappa_1 < 0$ or $\kappa_2 < -\kappa_1$. Taking into account the dependence $\kappa_2(\omega)$, we get

$$1 - \frac{\omega_p^2}{\omega^2} < -\kappa_1, \quad \frac{\omega_p^2}{\omega^2} > 1 + \kappa_1, \quad \omega < \frac{\omega_p}{\sqrt{1 + \kappa_1}},$$

that is, the frequency range where a surface wave exists narrows down with the increase of the κ_1. If the first medium is a vacuum, then the frequency range is given by the inequality $\omega < \omega_p/\sqrt{2}$.

7.5 SUPERCONDUCTIVITY

In 1911, the Dutch scientist Kamerlingh Onnes discovered that at a temperature of 4.15 K the electrical resistance of mercury abruptly decreases to zero. Further studies have shown that many other metals and alloys behave in a similar way. This phenomenon was called superconductivity, and the materials that exhibit it are called superconductors. The temperature T_c, at which there is an abrupt transition to the superconducting state, is called the **critical temperature**. The state of a superconductor above the critical temperature is called **normal** and below **superconducting**.

1. A theory for superconductivity that is known as the BCS theory was proposed in 1957 by Bardeen, Cooper, and Schrieffer. This theory of superconductivity is complicated because superconductivity is a **macroscopic quantum effect**. We will limit our discussion to a simplified presentation of the basic concepts of the BCS theory.

The *free* electrons in a metal, in addition to the Coulomb repulsion among them, also experience a special type of attraction, which in the superconducting state prevails over the electron–electron Coulomb repulsion. As a result, the conduction electrons form what is known as the **Cooper pairs**. The electrons in each pair have oppositely directed **spins**. Therefore, the spin of the pair is equal to zero, and therefore, the pair is a **boson**—a particle with an integer spin. According to quantum statistics, bosons tend to accumulate in their **ground energy state**, from which it is relatively difficult to make a transition to an excited state. In other words, below the critical temperature T_c, **Bose condensation** of Cooper pairs of electrons takes place. Cooper pairs of a Bose condensation in the ground state can exist indefinitely long. Cooper pairs move coherently and form the superconducting current.

In the following, we discuss the Cooper pair formation in a little bit more detail. An electron moving in a metal polarizes the crystal lattice that consists of positive ions. As a result of this polarization, the electron appears to be surrounded by a "cloud" of positive charge moving through the lattice together with the electron. The electron together with the surrounding cloud forms a positively charged system, which will attract another electron. Thus, the crystal lattice plays the role of the intermediate medium, the presence of which causes the attraction between the two electrons.

Using the language of quantum mechanics, the attraction between electrons is explained as a result of the exchange between electrons by quanta of the lattice vibrations known as phonons. An electron moving in a metal excites phonons. This excitation energy is transferred to another electron that absorbs the phonon. As a result of such exchange of phonons, there is an additional interaction between electrons, which has the character of attraction. At low temperatures, this attraction for materials that are superconductors exceeds the Coulomb repulsion. This interaction is pronounced most strongly for electrons with opposite momenta and spins. As a result, two such electrons are coupled and form a Cooper pair. This pair should not be imagined as two electrons joined together. On the contrary, the distance between the electrons of a pair is very large; it is about 10^{-4} cm, that is, it exceeds the interatomic distances in a crystal by four orders of magnitude.

Not all conduction electrons form Cooper pairs. At a temperature T_c, other than zero, there is a probability that a pair will be destroyed. Therefore, at temperature below T_c, along with the pairs, there are always the *normal* electrons moving in the crystal in the usual way. The closer T is to T_c, the higher is the percentage of normal electrons, and at $T = T_c$ all electrons become normal. Consequently, at a temperature above T_c, the superconducting state does not exist. The formation of Cooper pairs leads to a restructuring of the electron energy spectrum in the metal. The excitation of the electrons in a superconductor requires the distraction of at least one pair. So, to destroy a Cooper pair, the energy equal to the Cooper pair binding energy W_0 is needed. This energy represents the minimum amount of energy that the system of electrons of a superconductor can absorb. If the current through the superconductor is low, the electron velocities are small and the energy that a Cooper pairs are getting from the current is below W_0. The electronic system will not be excited, and this means electron moves without electrical resistance. If current increases above the so-called critical current I_c or the temperature increases above T_c, the material returns to its normal (non superconducting, but rather metallic) state.

2. The magnetic field does not penetrate into an interior of the superconductor (this phenomenon is called the **Meissner effect**). If a superconducting sample is cooled in a magnetic field, at the time of the transition to the superconducting state, the field is expelled from the sample, that is, the magnetic field B in the material vanishes. It is formally possible to consider that the superconductor possesses zero magnetic permeability ($\kappa_m = 0$, and $B = \kappa_m H = 0$).

Since in the superconductor there is no magnetic field inside the superconductor, electric currents cannot flow in its volume, that is, $\mathbf{j} = 0$ inside the bulk of the superconductor. From Ampere's law ($\nabla \times \mathbf{H} = \mathbf{j}_s$), it follows that all currents in the superconductor have to flow on its surface. These surface currents generate a magnetic field that cancels the externally applied field inside the superconductor. The expulsion of the magnetic field from a superconductor is referred to as the Meissner effect.

A sufficiently strong external magnetic field destroys the superconducting state. The value of magnetic field at which this occurs is called the **critical field** and is denoted as B_c. Its value depends on a sample temperature. The condition $B_c = 0$ corresponds to the critical temperature T_c. When the temperature decreases, the value of B_c increases and approaches the value of the critical field B_{c0} at zero temperature. A sketch of the dependence of B_c on T is shown in Figure 7.7. If we increase the

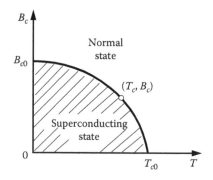

FIGURE 7.7 Phase diagram of *critical field–critical temperature* that shows the regions of normal and superconducting states.

current flowing through a superconductor to a value above I_c, the superconducting state is destroyed. This value is called the **critical current**. I_c depends on temperature. The dependence of I_c on T is similar to the dependence of B_c on T.

3. In 1935, brothers F. London and H. London proposed two equations to describe the electrodynamics of superconductors:

$$\mathbf{E} = \frac{d}{dt}(\Lambda\mathbf{j}), \quad \mu_0\mathbf{H} = -\nabla \times (\Lambda\mathbf{j}), \qquad (7.42)$$

The first equation of system (7.42) is easily obtained from Newton's second law after multiplying it by en_s:

$$m\frac{d\mathbf{v}_s}{dt} = e\mathbf{E}, \quad m\frac{d}{dt}(en_s\mathbf{v}_s) = e^2 n_s\mathbf{E}, \quad \frac{d}{dt}(\Lambda\mathbf{j}) = \mathbf{E}, \qquad (7.43)$$

where
  m is the electron mass
  n_s and $\mathbf{v}_s$ are the density and velocity of *superconducting* electrons
  $\Lambda = m/e^2 n_s$ is the parameter considered

Here, it is assumed that the density of *superconducting* electrons n_s is small at $T = T_c$ and $n_s = n$ at $T \ll T_c$.

The second equation of system (7.42) can be derived from the first equation, by taking the "curl" from both of its parts and using one of Maxwell's equations, $\nabla \times \mathbf{E} = -(\partial \mathbf{B}/\partial t)$:

$$\nabla \times \mathbf{E} = \nabla \times \frac{\partial}{\partial t}(\Lambda\mathbf{j}), \quad -\frac{\partial \mathbf{B}}{\partial t} = \frac{\partial}{\partial t}\nabla \times (\Lambda\mathbf{j}),$$

$$\frac{\partial}{\partial t}\left(\mu_0\mathbf{H} + \nabla \times (\Lambda\mathbf{j})\right) = 0, \quad \mu_0\mathbf{H} + \nabla \times (\Lambda\mathbf{j}) = 0. \qquad (7.44)$$

If we take curl from both sides of Ampere's law $\nabla \times \mathbf{H} = \mathbf{j}$, written for a static field, and take into account the curl of curl vector identity $\nabla \times (\nabla \times \mathbf{H}) = \nabla(\nabla \cdot \mathbf{H}) - \nabla^2\mathbf{H}$, then after substitution of $\nabla \times \mathbf{j}$ from the second London equation, we get

$$\nabla \times (\nabla \times \mathbf{H}) = \nabla \times \mathbf{j}, \quad -\nabla^2\mathbf{H} = \nabla \times \mathbf{j}$$

$$-\nabla^2\mathbf{H} = -\frac{\mu_0}{\Lambda}\mathbf{H}, \quad \nabla^2\mathbf{H} - \left(\frac{1}{\delta^2}\right)\mathbf{H} = 0, \quad \delta^2 = \frac{\Lambda}{\mu_0} = \frac{m}{\mu_0 e^2 n_s}, \qquad (7.45)$$

$$\mathbf{H}(z) = \mathbf{H}(0)\exp\left(-\frac{z}{\delta}\right).$$

From the last equation, it follows that the static magnetic field penetrates inside the superconductor to a depth $\delta = \sqrt{m/\mu_0 e^2 n_s}$. Thus, the London equations (7.42) are compatible with the Meissner effect.

Exercise 7.5

Find the complex conductivity of a superconductor in the field of a monochromatic electromagnetic wave.

Solution. The equation of motion for *superconducting* electrons is determined by the first of the London equations (7.42):

$$\mathbf{E} = \Lambda \frac{d\mathbf{j}_s}{dt},$$

where
$\Lambda = m/e^2 n_s$ is the parameter considered
n_s is the concentration of *superconducting* electrons

In the electric field $\mathbf{E} \sim \exp(i\omega t)$ of the electromagnetic wave, the current $\mathbf{j}_s$ is also monochromatic, $\mathbf{j}_s \sim \exp(i\omega t)$, and from the equation given earlier, we get

$$\mathbf{j}_s = -\frac{i}{\Lambda \omega} \mathbf{E}.$$

For *normal* electrons with concentration n_n, the equation of motion after taking into account the electric field force and the average friction-like force will have the following form:

$$m \frac{d\mathbf{v}_n}{dt} = e\mathbf{E} - m \frac{\mathbf{v}_n}{\tau}, \quad \text{or} \quad \frac{m}{n_n e} \frac{d\mathbf{j}_n}{dt} = e\mathbf{E} - \frac{m}{n_n e} \frac{\mathbf{j}_n}{\tau},$$

where we took into account that $\mathbf{j}_n = n_n e \mathbf{v}_n$. Let us introduce into the last equation the parameter Λ and rewrite as

$$\mathbf{E} = \frac{\Lambda n_s}{n_n} \frac{d\mathbf{j}_n}{dt} + \frac{\Lambda n_s}{n_n} \frac{\mathbf{j}_n}{\tau}.$$

In the field of a monochromatic wave, the current of *normal* electrons is equal to $\mathbf{j}_n \sim \exp(i\omega t)$. Therefore,

$$\mathbf{j}_s = -\frac{i}{\Lambda \omega} \mathbf{E}, \quad \mathbf{j}_n = \frac{n_n}{n_s} \frac{\tau}{\Lambda} \frac{1 - i\omega\tau}{1 + (\omega\tau)^2} \mathbf{E}.$$

The total current density is equal to $\mathbf{j} = \mathbf{j}_s + \mathbf{j}_n$, and therefore,

$$\mathbf{j} = \sigma \mathbf{E}, \quad \sigma = \sigma' - i\sigma'',$$

where the real and imaginary parts of the complex conductivity of high-frequency superconductor are given by the relationships

$$\sigma' = \frac{n_n}{n_s} \frac{\tau}{\Lambda} \frac{1}{1 + (\omega\tau)^2}, \quad \sigma'' = \frac{1}{\Lambda \omega} \left(1 + \frac{n_n}{n_s} \frac{(\omega\tau)^2}{1 + (\omega\tau)^2} \right).$$

7.6   QUANTUM EFFECTS IN SUPERCONDUCTIVITY

1. Here, we will give a very short introduction of certain phenomena that are characteristic for super-conductors, but we will not go into detailed explanations as these phenomena are quantum mechanical in nature. Consider a superconducting ring that carries a circulating superconducting current. Assume that the electrons move in a circle of radius r with velocity v (see Figure 7.8). The current energy is represented by the expression $W = I\Phi/2$, where I is the current intensity and Φ the magnetic flux through the circle, generated by this current. If N is the total number of electrons in the ring and T is the period of revolution, then $I = Ne/T = Nev/2\pi r$. Thus, $W = Nev\Phi/4\pi r$. On the other hand, the same energy is equal to $W = Nmv^2/2$. Equating both expressions, we will obtain $\Phi = 2\pi rmv/e$. If electrons are moving as Cooper pairs, the momentum of each pair is $p = 2mv$, so $\Phi = \pi rp/e$.

The momentum of Cooper pairs can only take quantized values according to the relationship $pr = n\hbar$, where n is an integral number and $\hbar = h/2\pi$ is a reduced Planck constant and h is a Planck constant. Consequently,

$$\Phi = \Phi_0 n, \quad n = 0, 1, 2, \ldots. \tag{7.46}$$

This formula expresses the ***quantization of the magnetic flux*** in superconductors, and the ***quantum of the magnetic flux*** is given by the expression

$$\Phi_0 = \frac{\pi\hbar}{e} = 2.07\times 10^{-15} \ \mathrm{J\,s/C}. \tag{7.47}$$

The formula of this type was obtained by F. London before BCS theory of superconductivity. However, F. London obtained a value for Φ_0, which is twice as large as the value given by Equation 7.48. This originates from the fact that at that time the phenomenon of electron pairing was not yet known. Therefore, for the electron momentum, London used the expression $\mathbf{p} = mv$ instead of the expression $\mathbf{p} = 2mv$, as it should be for Cooper pairs. Experiment has shown the correctness of Equations 7.47 and 7.48 and by that confirmed the existence of the electron Cooper pairs.

It is known that a persistent electric current can be excited in a superconductive ring. For example, one such experiment lasted for 2.5 years, yet no current decay was observed. At first glance, there is nothing surprising because no Joule heat is produced in a superconductor, and therefore, there is no current decay. The real situation is more complicated as the electrons in a superconducting ring are moving with a centripetal acceleration and therefore according to Maxwell's equations must radiate, which should lead to energy loss and thus current decay. Experiment shows, however,

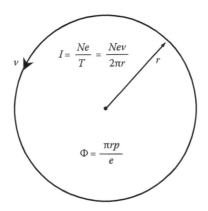

FIGURE 7.8   Quantization of the magnetic flux of circular superconducting current.

that no current decay is observed. This apparent contradiction is eliminated in the same way as in the elimination of contradiction for the radiation in the Bohr model of a hydrogen atom. Because there was no radiation, Bohr introduced the quantum postulate of stationary states of an atom and de Broglie explained that by the formation of a circular standing wave for each atomic orbit (de Broglie wave). In a similar fashion, in a superconducting ring that carries a current, radiation does not appear because of the quantization of the electric current. This quantization is observed already at the macroscopic scale, as a de Broglie standing wave is formed in a current-carrying ring of macroscopic size.

2. On the basis of the theory of superconductivity, B. Josephson in 1962 predicted two surprising effects that were later found subsequently when a superconducting current tunnels through a thin dielectric layer separating two superconductors. In a so-called *Josephson junction*, conduction electrons pass through the insulator (e.g., metal oxide film with a thickness of about 1 nm) due to the tunneling effect (see Figure 7.9). If the current through a Josephson junction does not exceed a certain critical value, there is no voltage drop across the junction. This first phenomenon is the so-called *stationary Josephson effect*. On the other hand, if the current through a Josephson junction exceeds a critical value, a voltage drop U appears across the junction and the junction radiates electromagnetic waves. This second phenomenon is the so-called *nonstationary Josephson effect*.

The radiation angular frequency ω is related to the voltage drop U at the junction by the relation $\omega = 2eU/\hbar$ (here e is the electron charge). The emission of radiation is explained by the fact that the Cooper pairs, which participate in the superconducting current, while passing through the junction obtain an excess energy relatively to the ground state of the superconductor. When they return to the ground state, they emit a quantum of electromagnetic energy $\hbar\omega = 2eU$. The Josephson effect is used for precise measurements of very weak magnetic fields (as low as 10^{-15} T), currents (as low as 10^{-10} A), and voltages (as low as 10^{-15} V). Josephson junctions are used as functional elements for rapid logic devices and amplifiers.

For longer periods, the superconducting state of various metals and compounds could be obtained only at very low temperatures achievable by using liquid helium (with boiling point at 4.2 K) as a coolant. Since the beginning of 1986, the observed maximum value of critical temperature in superconductors was 23 K. In 1986–1987, a number of high-temperature superconductors (HTSs) with critical temperature about 100 K were found. This temperature is achieved using liquid nitrogen as a coolant. Unlike helium liquid that is expensive, liquid nitrogen is produced inexpensively on an industrial scale. This fact considerably expands the area of potential practical applications of superconductors. Intensive research in the field of superconductivity began with the discovery of HTS in systems such as La–Su–O (with $T_c = 35$ K) and La–Va–Su–O (with $T_c = 92$ K). Critical temperatures 110 K in the system Bi–Sr–Sa–Su–O and 125 K in Tl–Sa–Va–Su–O were achieved. HTSs have rather high upper critical fields ($B_{c2} \approx 100 - 200$ T), which corresponds to their complete transition to the normal state. However, the lower critical fields corresponding to the beginning of the transition from the superconducting state to the normal state are small ($B_{c1} \approx 10^{-2}$ T).

In the present, low values of the lower critical field B_{c1} and low critical currents of superconductors restrict their practical application (e.g., in electric power transmission). Currently, research is

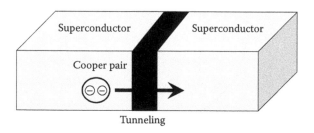

FIGURE 7.9   Josephson junction.

carried out to find new materials, which simultaneously have high values of all critical parameters— T_c, B_c, and j_c. In addition, active search is carried out on HTSs with ambient critical temperature (about 300 K). Any such discovery will produce a genuine technological revolution (e.g., the use of superconducting power lines will completely eliminate the power losses in transmission wires).

Exercise 7.6

In a bulk superconductor, there exists a hole of radius $r = 100$ μm and $n = 10$ is the number of magnetic flux quanta that are captured in it. Determine the strength of the magnetic field in the hole.

Solution. Since the magnetic flux that is trapped by the hole consists of n magnetic flux quanta, then

$$\Phi = \Phi_0\, n = B\pi r^2 = \mu_0 H \pi r^2,$$

where the magnetic flux quantum $\Phi_0 = \pi\hbar/e = 2.07 \times 10^{-15}$ J s/C. Thus,

$$H = \frac{\Phi_0 n}{\mu_0 \pi r^2} \simeq 0.5 \text{ A/m.}$$

PROBLEMS

7.1 Derive an expression for the refractive index of an isotropic plasma (in the absence of an external magnetic field), taking into account the motion of free electrons and positively charged ions. Take into account the energy loss of the particles (due to inelastic collisions) by introducing into the equations of motion of the particles the friction-like damping forces $\mathbf{F}_e = -\delta_e \mathbf{v}_e$ and $\mathbf{F}_i = -\delta_i \mathbf{v}_i$ for the electrons and ions, respectively. $\left(Answer\!: n = 1 - \dfrac{1}{2}\left(\dfrac{\omega_{pe}^2}{\omega^2 - i\gamma_e\omega} + \dfrac{\omega_{pi}^2}{\omega^2 - i\gamma_i\omega} \right),\right.$ $\left. \gamma_e = \delta/m,\ \gamma_i = \delta_i/M. \right)$

7.2 Determine the ratio of the density of the displacement current to the conduction current density and loss tangent, $\tan \delta_e$, in seawater ($\varepsilon = \kappa\varepsilon_0$, $\kappa = 80.0$, $\varepsilon_0 = 8.85 \times 10^{-12}$ F/m, $\sigma = 4.00$ S/m) for waves with frequencies in the range of $f = 10^5$ and 10^{11} Hz. (*Answer*: At $f = 10^5$ Hz, $j_{disp}/j_{cond} = 1.11 \times 10^{-4}$ and $\tan \delta_e = 9 \times 10^3$, and at $f = 10^{11}$ Hz, $j_{disp}/j_{cond} = 111$ and $\tan \delta_e = 9 \times 10^{-3}$.)

7.3 A plane wave with a frequency of 2 GHz propagates in a medium with parameters $\kappa = 2.40$, $\kappa_m = 1.00$, $\tan \delta_e = 10^{-1}$. Find the wavelength in the medium, the phase velocity, and the attenuation coefficient of the wave, $\delta = 1/k''$. (*Answer*: $\lambda \simeq 9.70 \times 10^{-2}$ m, $v_{ph} \simeq 1.94 \times 10^8$ m/s, $\delta \simeq 0.31$ m.)

7.4 Calculate the thickness of the skin layer (the penetration depth of the wave inside the medium) and the absolute value of the wave resistance (impedance) for a metal with a conductivity $\sigma = 5.00 \times 10^7$ S/m and magnetic permeability $\kappa_m = 1$ at a frequency $f = 10$ GHz. (*Answer*: $\delta \simeq 7 \times 10^{-7}$ m, $Z = 0.03(1 + i)\Omega$.)

7.5 Determine the thickness of a copper shield ($\sigma = 5.90 \times 10^7$ S/m, $\kappa_m = 1$) that provides amplitude attenuation of 10^4 for an electromagnetic field at each of two frequencies: $f = 50$ Hz and $f = 50$ MHz. (*Answer*: At $f = 50$ Hz we get $l \simeq 8.60$ cm and at $f = 50$ MHz we get $l \simeq 86.0$ μm.)

7.6 An electromagnetic wave with angular frequency $\omega = 10^5$ s^{-1} is propagating along a copper bar. The conductivity of the copper is $\sigma = 5.60 \times 10^7$ S/m, and its magnetic permeability is $\mu = \mu_0\kappa_m$, $\kappa_m = 1$. Determine the wave resistance (impedance) of the bar, the depth of penetration of the field into the bar, and the phase velocity and wavelength of the wave in the bar. $\left(Answer\colon Z = \dfrac{1+i}{\sqrt{2}} \times 4.73 \times 10^{-4} \ \Omega, \text{ the penetration depth } \delta = 0.53 \text{ mm}, \ v_{ph} = 53.3 \text{ m/s},\right.$

$\left.\lambda = 3.30 \text{ mm.}\vphantom{\dfrac{1}{2}}\right)$

7.7 Determine the fraction of the wave energy converted into heat for a wave incident normally from vacuum on the surface of a metal with conductivity $\sigma = 5.70 \times 10^7$ S/m. The wave frequency $f = 10$ GHz. $\left(Answer\colon \dfrac{\Delta I}{I_0} \approx 2.98 \times 10^{-4}.\right)$

7.8 A TM-type surface electromagnetic wave propagates along the flat interface between a metal and a vacuum. Determine the ratio of the penetration depth of the surface wave into the metal and into the vacuum. The angular frequency $\omega = \omega_p/2$, and the permittivity of the metal is described by the Drude model with $\omega_p = 10^{16}$ s^{-1} and $\gamma = 0$.

7.9 Using the London equation (7.45), determine the distribution of magnetic intensity and the distribution of current density in a superconducting plate that is placed in a homogeneous magnetic field parallel to the surface of the plate. The magnetic intensity is H_0 and the thickness of the plate is d. $\left(Part \ of \ the \ answer\colon H(z) = H_0 \dfrac{\cosh(z/\delta)}{\cosh(d/2\delta)}.\right)$

Section III

Electromagnetic Waves in Periodic
and Waveguiding Structures

Section III

8 Waves in Periodic Structures

Diffraction by periodic structures, with period comparable with the wavelength of the incident electromagnetic radiation, has been studied extensively. Diffraction gratings, interference filters, and multilayer dielectric mirrors are some examples of such structures. Important properties of crystals are determined by the periodicity of their crystal structure. For example, the presence of periodicity in crystals leads to the appearance of "allowed" and "forbidden" bands in their energy spectrum. In this case, the period of the crystal lattice, d, is comparable to the de Broglie wavelength of electrons λ_e: $d \sim \lambda_e \sim 10^{-10}$ m.

The propagation of electromagnetic waves in media with periodically changing properties results in the emergence of new phenomena. These are most noticeable when the wavelength of the wave, which propagates in the medium, becomes comparable to the medium's spatial period. For example, in diffraction experiment, in which the wave is incident normally on a flat periodic lattice, a diffraction pattern is formed by the transmitted wave. The pattern exhibits a characteristic alternation of maxima and minima in the intensity of the diffracted wave. In an experiment with a photonic crystal with the wave propagating in the periodic medium along a periodicity axis, the characteristic allowed and forbidden photonic bands in the reflection and transmission spectra are formed. In this chapter, we will consider a number of common diffraction principles in periodic structures.

8.1 DIFFRACTION PHENOMENA

Diffraction is a phenomenon that is connected with waves of any nature such as electromagnetic waves or material waves. This phenomenon consists in the deviation of wave propagation near obstacles. Experiments show that light (electromagnetic waves in the visible range) under certain conditions can penetrate into the area of the geometric shadow as determined by geometrical optics. For example, if we place an opaque plane with a circular aperture in the path of a light beam, we observe a bright circular spot on a screen behind the aperture with a diameter that is related to the aperture diameter. If we reduce the diameter of the aperture, the diameter of the diffraction pattern on the screen will decrease accordingly up to some limit. If the diameter of the aperture approaches the light wavelength, the diffraction pattern diameter on the screen behind the aperture will begin to increase as the diameter of the hole is reduced. The most distinct diffraction patterns are observed in cases where the sizes of obstacles are comparable to the wavelength of light. The parameter λ/b (where b is a characteristic size of an obstacle) plays an important role in diffraction phenomena. If $\lambda/b \rightarrow 0$, diffraction phenomena are negligibly small and can be ignored; in the case $\lambda/b \rightarrow 1$, the diffraction phenomena become dominant and must be taken into account.

For the precise calculation of a diffraction pattern, it is necessary to solve Maxwell's equations with boundary conditions that are defined by the boundaries and properties of the obstacles. The simplest description of diffraction is based on the Huygens–Fresnel principle, according to which each point of a wave front becomes a source of secondary spherical wave as shown in Figure 8.1. These secondary waves are coherent with respect to each other and therefore interfere and form a new wave front of the propagating wave (Figure 8.1). Resulting pattern at any point in space is the interference of the secondary waves emitted by all wave surfaces.

In accordance with the Huygens–Fresnel principle, every element dA of the electromagnetic wave front A becomes a source of secondary spherical waves. The amplitude of those waves is proportional to the area dA and it decreases with distance from that element as $1/r$. As a result, the

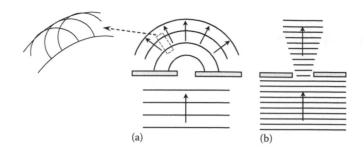

(a)   (b)

FIGURE 8.1   Wave front of the plane wave before and after a screen with a hole: (a) the diameter of the hole is small compared to the wavelength and (b) the diameter of the hole is substantially larger than the wavelength of the incident light. The inset shows the formation of a wave front based on the Huygens–Fresnel principle.

contribution from the element dA located at the point P of the surface A to the intensity at point r can be determined by the following equation:

$$dE(P_1) = E_0(P)\frac{\exp(-i\mathbf{k}\cdot\mathbf{r})}{r}K(\varphi)dA. \tag{8.1}$$

Here, $E_0(P)$ is the amplitude of the field of elementary secondary source with area dA located at an arbitrary point P of the wave front of the incident wave; the second term describes the coordinate dependence of the spherical wave emitted by an elementary source. Factor $K(\varphi)$ accounts for the dependence of dE on the angle φ between the normal to the unit area and the direction of the radius—vector $\mathbf{r}$ pointed in the direction of observation point (see Figure 8.2):

$$K(\varphi) \approx \frac{i}{2\lambda}(1+\cos 2\varphi). \tag{8.2}$$

The summation of the secondary waves at the observation point P_1 is performed using Huygens–Fresnel integral, which has the form

$$E(P_1) = \int_A E_0(P)\frac{\exp(-i\mathbf{k}\cdot\mathbf{r})}{r}K(\varphi)dA. \tag{8.3}$$

Let us write now Equation (8.3) in the coordinate representation, assuming that the diffraction angles are small $\varphi \ll 1$, that is, $K(\varphi) \approx i/\lambda$. Therefore, the amplitude of the field in the integral of Equation 8.3 does not depend on the position of a light front and is as follows:

$$E(x_1, y_1, z) = \frac{i}{\lambda}\int_{-\infty}^{\infty}\int_{-\infty}^{\infty} E_0(x, y)\frac{\exp(-i\mathbf{k}\cdot\mathbf{r})}{r}dxdy, \tag{8.4}$$

where $r = \left[(x_1 - x)^2 + (y_1 - y)^2 + z^2\right]^{1/2}$.

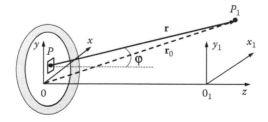

FIGURE 8.2   Construction of the diffraction pattern.

In the Fresnel approximation, the diffraction pattern is observed in the region that is close to the source, that is, close to the region of $E_0(P)$ and close to axis z. Very often, this region is named the near-field region. As the diffraction angles are small for this region, the diffraction broadening of the image is small too and the Equation (8.4) can be simplified. To this end, in the near-field region, we have

$$\frac{(x_1 - x)}{z} \ll 1, \quad \frac{(y_1 - y)}{z} \ll 1.$$

It can be assumed that $\mathbf{k} \cdot \mathbf{r} = kr$, where $k = 2\pi/\lambda$ and modulus of the radius vector r in the exponent can be represented in the approximate form that follows from the Tailor series for r:

$$r \approx z + \frac{(x_1 - x)^2 + (y_1 - y)^2}{2z}. \tag{8.5}$$

As we use Equation 8.5 to replace r in Equation 8.4, we will keep the second term of Equation 8.5 only in the exponent of Equation 8.4, but for a slowly changing denominator, we assume that $r \approx z = const$. As a result, we obtain

$$E(x_1, y_1, z) = \frac{i}{\lambda z} e^{-ikz} \int_{-\infty}^{\infty} \int_{-\infty}^{\infty} E_0(x, y) \exp\left(-ik\frac{(x_1 - x)^2 + (y_1 - y)^2}{2z}\right) dx\, dy. \tag{8.6}$$

In the **Fraunhofer approximation**, the diffraction pattern is observed at large distances from the object (the so-called *far-field*). Therefore, unlike in the **near-field diffraction** (i.e., **Fresnel diffraction**), the image size, even at low diffraction angles, substantially exceeds the size of the object, that is, x_1 and y_1 are not small compared to z, but x and y are small compared to x_1, y_1, and z. The expansion of the expression for r is now carried out as follows:

$$r = \left[z^2 + (x_1 - x)^2 + (y_1 - y)^2\right]^{1/2} = \left[z^2 + x_1^2 + y_1^2 + (x^2 - 2xx_1) + (y^2 - 2yy_1)\right]^{1/2}$$

$$= \left[r_0^2 + (x^2 - 2xx_1) + (y^2 - 2yy_1)\right]^{1/2}$$

$$\approx r_0 \left(1 - \frac{xx_1}{r_0^2} - \frac{yy_1}{r_0^2}\right),$$

where $r_0 = (z^2 + x_1^2 + y_1^2)^{1/2}$. In this expansion, we have neglected the small terms in the expansion, which are proportional to $(x/r_0)^2 + (y/r_0)^2$. Now, unlike the Fresnel approximation, we restricted ourselves to linear terms in the expansion of r in the small parameter (the ratio of the object's size to the distance to it). Equation 8.4 in this case takes the form

$$E(x_1, y_1) = \frac{i}{\lambda r_0} e^{-ikr_0} \int_{-\infty}^{\infty} \int_{-\infty}^{\infty} E_0(x, y) \exp\left[\frac{ik}{r_0}(xx_1 + yy_1)\right] dx\, dy. \tag{8.7}$$

Expressions (8.6) and (8.7) allow us to determine the diffraction pattern of the light front for various simple objects—apertures, slits, obstacles, and lattices.

For a qualitative description of the diffraction phenomena, the method of Fresnel zones is often used. To understand it, let us consider the following example. Let the light wave from a distant source be incident normally on an opaque screen, which has a small circular hole of radius R. Let the observation point P be located on the axis of symmetry and at a distance L from the screen. The wave surfaces of the incident light are parallel to the screen plane and one of them coincides with the screen.

In accordance with the Huygens–Fresnel principle, each point of the wave surface becomes a source of secondary spherical waves. All secondary waves interfere at the observation point P, and the result of their interference determines the intensity of the resultant diffracted wave. In order to determine the diffracted wave, it is necessary to split the wave surface within the aperture into rings known as *Fresnel zones*. The partition is carried out as follows: the distance from the central point O to the observation point P is equal to L. The distance from the boundary of the first zone to the observation point P is chosen to be equal to $L + \lambda/2$. The distance from the boundary between the first zone and the second zone to the observation point P is chosen to be equal to $L + 2(\lambda/2)$ and so on (Figure 8.3). Thus, the difference in distance from adjacent Fresnel zone boundaries to the observation point P differs by $\lambda/2$, and thus, the vibrations from adjacent zones come to point P with opposite phases.

The entire area of the hole is divided into concentric rings, each of which is a Fresnel zone (the central zone is a circle) (see Figure 8.4). From geometrical considerations, it follows that the radii of the Fresnel zones are equal to

$$\rho_m = \sqrt{m\lambda L + \frac{m^2\lambda^2}{4}} \approx \sqrt{m\lambda L},$$

where we took into account that $\lambda \ll L$.

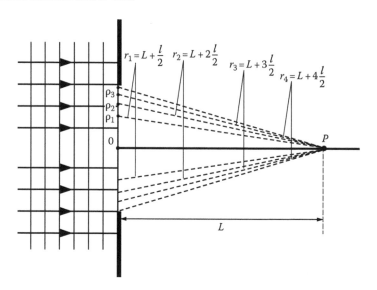

FIGURE 8.3   Partitioning of the wave surface of the Fresnel zone.

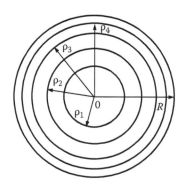

FIGURE 8.4   The radii of the Fresnel zones for a circular hole.

The number of zones that fit inside a hole of radius R is equal to $m = R^2/\lambda L$. The areas of all Fresnel zones are identical and equal to the area of the central zone, which has the form of a circle. The result of the interference of secondary waves at point P depends on the number of open zones. Since the areas of all zones are identical, they give the same contribution to the resultant intensity at point P. However, each successive zone is a little further from the point of the previous, and hence, its contribution will be slightly smaller than that of the previous zone. If A_m is the amplitude of the oscillations created at the point P by the mth zone, then $A_{m+1} < A_m$. If we take into account that the vibrations from neighboring zones come to a point P with opposite phases, then the total amplitude of oscillations is determined by the expression

$$A = A_1 - A_2 + A_3 - A_4 + \cdots + A_m = A_1 - (A_2 - A_3) - (A_4 - A_5) - \cdots - (A_{m-1} - A_m).$$

Since all the expressions in brackets are positive, the total amplitude is always smaller than the amplitude, which is generated by the first Fresnel zone. It can be shown that if there is no screen, that is, there is infinite number of Fresnel zones, then the total amplitude of oscillations is two times smaller than the amplitude due to the first zone. Diffraction effects are observed when the hole fits only a small number of Fresnel zones.

The diffraction pattern consists of alternating maxima and minima in the intensity of the diffracted light behind a barrier that is a consequence of an interference of light waves diffracting on a barrier. Often, the diffraction pattern is concentrated in a very narrow area of space at the boundary between the light and shade of a barrier. In this case, it is rather difficult to observe the diffraction pattern. That is why in many cases, the propagation of light is described without taking into account the wave properties of light, by applying the laws of *geometrical optics*. However, the laws of ray optics cannot explain a wide range of phenomena. Thus, we must ask the following question: under which conditions it is possible to apply the laws of geometrical optics and when is it necessary to use the wave theory of diffraction?

Optical phenomena, in which the wavelength of light is small compared with the size of the obstacle, can be explained by geometrical optics. The condition $\lambda \to 0$ is basic for the transition from wave optics to geometrical optics. In geometrical optics, phenomena are characterized by a rectilinear propagation of rays.

Exercise 8.1

A parallel light beam of wavelength $\lambda = 0.50 \ \mu m$ is incident normally on an opaque diaphragm with a circular aperture of radius $r = 1.00$ mm. A screen is placed in the path of the rays passing through the aperture. Determine the maximum distance L_{max} from the center of the aperture to the screen when at the center of the diffraction pattern a dark spot is still observed.

Solution. The distance at which we will see a dark spot is determined by the number of Fresnel zones that fit the hole. If the number of zones is even, a dark spot is in the center of the diffraction pattern. The number of Fresnel zones that fit the hole decreases with the distance from the screen. The smallest number of zones is equal to two. Therefore, the maximum distance at which there still will be a dark spot in the center of the screen is determined by the condition according which two Fresnel zones should fit the hole.

From Figure 8.5, it follows that the distance from the observation point O on the screen to the edge of the hole for more than $2(\lambda/2)$ is greater than the distance $R_0 = L_{max}$. According to the Pythagorean theorem,

$$r^2 = \left(L_{max} + \frac{2\lambda}{2}\right)^2 - L_{max}^2 = 2\lambda L_{max} + \lambda^2.$$

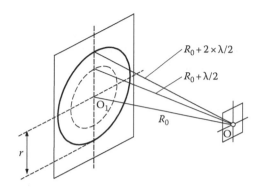

FIGURE 8.5 Diffraction from a circular hole.

Taking into account that $\lambda \ll L_{max}$ and that the term containing λ^2 can be neglected, the last equation can be rewritten as

$$r^2 = 2\lambda L_{max}.$$

From this equation, we obtain $L_{max} = r^2/2\lambda = 1.00$ m.

8.2 DIFFRACTION BY A SLIT

Let us first consider the simplest case of diffraction of light—by a *single slit* in a flat opaque plane. Assume that a plane monochromatic wave with angular frequency ω (wavelength λ) is incident normally on a long narrow slit of width b. It is useful to introduce the dimensionless parameter $b^2/l\lambda$, where l is the distance from the slit to the point of observation. It is possible to distinguish the following three situations:

$$\frac{b^2}{l\lambda} \ll 1 \quad \left(\text{or } l \gg \frac{b^2}{\lambda} \right) \quad \text{Fraunhofer diffraction,}$$

$$\frac{b^2}{l\lambda} \approx 1 \quad \left(\text{or } l \approx \frac{b^2}{\lambda} \right) \quad \text{Fresnel diffraction,} \qquad (8.8)$$

$$\frac{b^2}{l\lambda} \gg 1 \quad \left(\text{or } \lambda \ll \frac{b^2}{l} \right) \quad \text{Geometrical optics.}$$

The Fraunhofer diffraction can be observed by placing two lenses between the light source S and the observation point P so that the points S and P are at the focal plane of the corresponding lenses (Figure 8.6). Now, we will discuss this type of diffraction in more details.

We arrange the x-axis x in the plane of the slit perpendicular to its edges and the origin of coordinates at the slit center. When the wave front reaches the slit, all its points become sources of secondary coherent spherical waves according to the **Huygens–Fresnel principle**. Consider an elemental area with position coordinate x and width dx. We compare at the observation point P the phase difference between the wave from the element, at the center of slit, and the element at the x as it is shown on the inset of Figure 8.6. We assume that the phase of the secondary wave, which

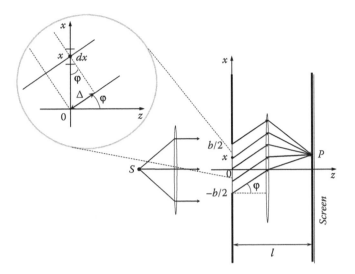

FIGURE 8.6 Schematics of setup for the Fraunhofer diffraction.

arrives at point P from the element at $x = 0$, is equal to zero. As a result, the phase of the wavelet at P from the element at x is

$$\psi(x) = -\frac{2\pi}{\lambda}\Delta = -\frac{2\pi}{\lambda}x\sin\varphi = -k_0 x\sin\varphi, \qquad (8.9)$$

where the angle φ determines the direction of the diffracted wave emitted by an element at x. The amplitude of the wave at P generated by the element with coordinate x in the direction φ is given by

$$dE(\varphi) = \frac{E_0}{b}\exp\left[i(\omega t - k_0 x\sin\varphi)\right]dx. \qquad (8.10)$$

We use the relation $\gamma = (\pi/\lambda)\sin\varphi = (k_0/2)\sin\varphi$ and integrate expression (8.10) across the slit width. For the resultant electric field at the observation point, defined by the angle φ, we obtain

$$E(\varphi) = \frac{E_0}{b}\int_{-b/2}^{b/2}\exp\left[i(\omega t - k_0 x\sin\varphi)\right]dx$$

$$= \frac{E_0}{b}\exp(i\omega t)\int_{-b/2}^{b/2}\exp(-2i\gamma x)dx$$

$$= E_0\exp(i\omega t)\frac{\exp(i\gamma b) - \exp(-i\gamma b)}{2i\gamma b}. \qquad (8.11)$$

Applying the equation

$$\sin u = \frac{1}{2i}\left[\exp(iu) - \exp(-iu)\right],$$

we arrive at the following expression for resultant amplitude:

$$E(\varphi) = E_0 \exp(i\omega t)\frac{\sin(\gamma b)}{\gamma b} = E_0 \exp(i\omega t)\sin\left(\frac{\pi b}{\lambda}\sin\varphi\right)\Big/\frac{\pi b}{\lambda}\sin\varphi. \qquad (8.12)$$

The time factor does not affect the intensity distribution in space, and for the intensity $I(\varphi) = E(\varphi)E^*(\varphi)$, we obtain

$$I(\varphi) = I_0 J_1(\varphi), \qquad (8.13)$$

where

$$I_0 = |E_0|^2, \quad J_1(\varphi) = \left(\frac{\sin\alpha}{\alpha}\right)^2, \quad \alpha = \frac{\pi b}{\lambda}\sin\varphi$$

I_0 is the intensity of the incident wave function
$J_1(\varphi)$ is the diffraction pattern of a single slit

The intensities of the diffraction maxima decrease fast with the order of diffraction. The ratio of the intensity maxima for the first four orders of diffraction is

$$I_0 : I_1 : I_3 : I_4 = 1 : \left(\frac{2}{3\pi}\right)^2 : \left(\frac{2}{5\pi}\right)^2 : \left(\frac{2}{7\pi}\right)^2 \approx 1 : 0.047 : 0.017 : 0.008.$$

Therefore, the main fraction of the diffracted intensity is concentrated inside the first maximum that is located between the first two minima with $n = \pm 1$, that is, in the interval determined by the angles $-\varphi_1 < \varphi < \varphi_1$ where $\sin\varphi_1 = \lambda/b$. The width of the central maximum decreases with the increase in the slit width, and for $b \gg \lambda$, we get $\varphi_1 = \lambda/b$. The angular dependence of the intensity $J_1(\varphi) = I(\varphi)/I_0$ is shown in Figure 8.7.

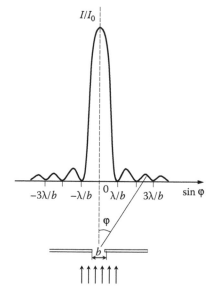

FIGURE 8.7   Angular dependence of the intensity $J_1(\varphi) = I(\varphi)/I_0$ of the Fraunhofer diffraction pattern from a single slit.

In this diffraction pattern, we have a main maximum centered around the direction $\varphi = 0$ and a number of secondary maxima whose directions are determined from the condition

$$b\sin\varphi = \pm\frac{2n+1}{2}\lambda, \qquad (8.14)$$

where $n = 1, 2, 3, 4, \ldots$ is the index of secondary maximum.

According to Equation 8.13, the condition for the observation of diffraction minima (which are located between the maxima) is given by

$$b\sin\varphi = \pm n\lambda. \qquad (8.15)$$

From this, it follows that a decrease in the slit width b results in a broadening of the diffraction pattern.

For $b = \lambda$, $\sin\varphi_{min} = 1$ for $n = 1$ and only one minimum can be observed at $\varphi_{min} = \pi/2$, which means that the characteristic diffraction pattern with alternating maxima and minima disappears if $b \leq \lambda$. The increase in the slit width leads to a narrowing of the diffraction pattern. The limiting value of the slit width b_{max} is determined by the resolving power of the observer's eye. Assuming that the angular position of the first minimum is equal to the smallest angle resolved by the observer's eye (i.e., $\lambda/b_{max} \approx 10^{-3}$), we find that $b_{max} \approx 10^3\lambda$. Thus, for observation of the diffraction of a slit, its width has to be in the interval $\lambda < b < 10^3\lambda$. For visible light, λ is about 0.5 μm, so the slit width should be in the range 0.5 μm $< b <$ 500 μm.

Exercise 8.2

A parallel light beam of wavelength λ is incident normally on a slit with width b. Find the angular width $\Delta\varphi_0$ and the linear width Δx_0, widths of the central diffraction maximum observed on a screen, which is placed at a distance $l \gg b$ from the slit. Assume that the slit width is much larger than the wavelength.

Solution. The central intensity maximum of the diffraction pattern lies between the two first minima. Intensity minima after diffraction from the slit are observed at angles given by Equation 8.15: $b\sin\varphi = \pm n\lambda$, where n is the diffraction order, and for $n = \pm 1$, we have

$$\sin\varphi_{\pm 1} = \frac{\pm\lambda}{b}.$$

As the slit *width* is much larger than the wavelength, therefore the angles $\varphi_{\pm 1}$ are small and are approximately $\varphi_{\pm 1} = \pm\lambda/b$. Thus, the separation angle between these minima is

$$\Delta\varphi_{\pm 1} = \varphi_{-1} - \varphi_{+1} = \frac{2\lambda}{b}.$$

The linear width of the central maximum is given by the expression

$$\Delta x_{\pm 1} = l\Delta\varphi_{\pm 1} = \frac{2l\lambda}{b}.$$

8.3   DIFFRACTION BY A 1D LATTICE

A diffraction grating is an optical device used for the separation in space of electromagnetic waves with different wavelengths in a light beam. If the properties of a structure change periodically only in one direction, the array is 1D (linear). If the array is periodic along two or three directions, the array is called 2D or 3D, respectively. It is possible to distinguish the following two (idealized) array types:

1. An ***amplitude array*** that imposes periodic changes in the amplitude of the transmitted wave without influencing its phase
2. A ***phase array*** that imposes periodic changes in the phase of the transmitted wave but does not influence its amplitude

In practice, ***amplitude–phase*** diffraction gratings, which change both the amplitude and the phase of the transmitted wave, are often used.

The simplest 1D amplitude diffraction grating is composed of N identical equidistant parallel slits in an opaque screen. The widths of the slit and opaque parts of the screen between two adjacent slits are b and a, respectively. This grating is a 1D array with period $d = a + b$. We assume that a plane monochromatic wave is incident normally on the grating. Since the light waves traveling from each slit are coherent, they will interfere among themselves. The diffraction pattern in the transmitted wave is the result of interference between these waves. The diffraction pattern consists of a number of narrow bright fringes separated by wide dark bands.

In the following text, we consider the features of the diffraction pattern that is shown in Figure 8.8. The path difference between adjacent rays from each slit is $d\sin\varphi$. For the rays 1 and 3, the path difference is $2d\sin\varphi$, and for rays 1 and N, it is $(N-1)d\sin\varphi$. Angle φ is defined (see also Figure 8.6) as the angle between the direction of the diffracted light and the normal to the plane of the grating. In accordance with Equations 8.4 and 8.5, the electric field of the diffracted wave from the first slit has the form

$$E_1 = E_0 \frac{\sin\alpha}{\alpha}\exp(i\omega t), \tag{8.16}$$

with α defined in Equation 8.13. The phases of the other diffracted waves differ from the phase of the first diffracted beam by the angle $\Delta\varphi_n = (n-1)k_0 d\sin\varphi$, where $n = 1, 2, 3, \dots, N$, so

$$E_2 = E_0 \frac{\sin\alpha}{\alpha}\exp\left[i(\omega t - k_0 d\sin\varphi)\right],$$

$$E_3 = E_0 \frac{\sin\alpha}{\alpha}\exp\left[i(\omega t - 2k_0 d\sin\varphi)\right], \tag{8.17}$$

$$\dots$$

$$E_N = E_0 \frac{\sin\alpha}{\alpha}\exp\left[i(\omega t - (N-1)k_0 d\sin\varphi)\right].$$

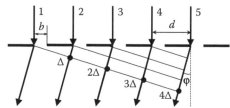

FIGURE 8.8   The geometric path difference of rays for Fraunhofer diffraction on a 1D periodic array, $\Delta = d\sin\varphi$.

The interference of waves 1, 2, ..., N is observed on a screen, which is placed at a sufficient large distance from the grating. Since the secondary waves have the same amplitude, in order to find the total field from all the slits, we need to calculate the following sum:

$$E = E_0 \frac{\sin \alpha}{\alpha} \sum_{n=1}^{N} \exp\left[i(\omega t - (n-1)k_0 d \sin \varphi)\right]$$

$$= E_0 \frac{\sin \alpha}{\alpha} \exp(i\omega t) \sum_{n=1}^{N} \exp\left[-2i(n-1)\beta\right], \quad (8.18)$$

where $k_0 d \sin \varphi = 2\beta$. Calculating the sum on the right, which is a geometrical progression, we obtain

$$E = E_0 \frac{\sin \alpha}{\alpha} \cdot \frac{1 - \exp(-2iN\beta)}{1 - \exp(-2i\beta)} \exp(i\omega t). \quad (8.19)$$

Knowing the expression for the electric field, we can calculate the intensity of the diffraction pattern:

$$I = (EE^*) = E_0^2 \left(\frac{\sin \alpha}{\alpha}\right)^2 \frac{1 - e^{-i2N\beta}}{1 - e^{-i2\beta}} \cdot \frac{1 - e^{i2N\beta}}{1 - e^{i2\beta}}$$

$$= I_0 \left(\frac{\sin \alpha}{\alpha}\right)^2 \frac{2 - \left(e^{i2N\beta} + e^{-i2N\beta}\right)}{2 - \left(e^{i2\beta} + e^{-i2\beta}\right)} = I_0 \left(\frac{\sin \alpha}{\alpha}\right)^2 \cdot \frac{2 - 2\cos 2N\beta}{2 - 2\cos 2\beta}$$

$$= I_0 \left(\frac{\sin \alpha}{\alpha}\right)^2 \cdot \left(\frac{\sin N\beta}{\sin \beta}\right)^2 = I_0 J_1(\varphi) J_2(\varphi), \quad (8.20)$$

where I_0, $J_1(\varphi)$ are determined by Equation 8.5 and

$$J_2(\varphi) = \left(\frac{\sin N\beta}{\sin \beta}\right)^2, \quad \beta = \frac{\pi d}{\lambda} \sin \varphi.$$

Thus, the distribution of light intensity in the diffraction pattern from a grating (1D lattice) is given by

$$I(\varphi) = I_0 J_1(\varphi) J_2(\varphi) = I_0 \frac{\sin^2\left(\frac{\pi b}{\lambda}\sin\varphi\right)}{\left(\frac{\pi b}{\lambda}\sin\varphi\right)^2} \cdot \frac{\sin^2\left(N\frac{\pi d}{\lambda}\sin\varphi\right)}{\sin^2\left(\frac{\pi d}{\lambda}\sin\varphi\right)}. \quad (8.21)$$

The angular distribution of the relative intensity of the diffracted light, $I(\varphi)/I_0$, for a grating with parameters $N = 5$ and $d/b = 3$, is given in Figure 8.9. This distribution is determined by angular dependences of two factors: $J_1(\varphi)$ and $J_2(\varphi)$. The function $J_1(\varphi)$ is the diffraction pattern of a single slit that was considered earlier (see Figure 8.5), while function $J_2(\varphi)$ determines the interference contribution of all slits.

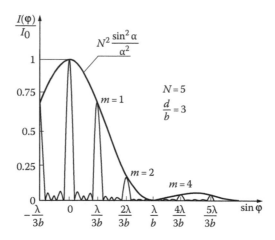

FIGURE 8.9   The angular dependence of the intensity $I(\varphi)/I_0 = J_1(\varphi)J_2(\varphi)$ of the electromagnetic field for the diffraction on 1D lattice with the parameters $N = 5$ and $d/b = 3$.

The function plotted in Figure 8.7 $J_1(\varphi)$ has the following properties:

- At $\alpha = 0$ (i.e., at $\varphi = 0$), it has the main maximum, in which $J_{1max} = 1$.
- At $\alpha = \pm n\pi$ (i.e., at $b\sin\varphi = \pm n\lambda$), it has a number of equidistant minima, in which $J_{1min} = 0$.
- At $\alpha = \pm(n + 1/2)\pi$ (i.e., at $b\sin\varphi = (n + 1/2)\lambda$), it has a number of the secondary maxima, in which $J_{1max} = 1/(n+1/2)^2 \pi^2$.

Consider the function $J_2(\varphi)$. The function $J_2(\varphi) = J_2(\beta)$ has its main maxima at values close to $\beta = m\pi$ ($m = 0, \pm1, \pm2, \pm3, \ldots$) as both numerator, $\sin N\beta$, and denominator, $\sin\beta$, tend to zero and their ratio tends to N, that is, at $\beta = m\pi$, when $\sin\varphi = m\lambda/d$ (or $\varphi = \arcsin(m\lambda/d)$), the value of all main maxima of the function $J_2(\varphi)$ is

$$J_{2max} = \left(\frac{\sin N\beta}{\sin\beta}\right)^2_{\beta=m\pi} = N^2. \tag{8.22}$$

In spite of the fact that J_{2max} does not depend on m, the value of $I(\varphi)/I_0 = J_1(\varphi)J_2(\varphi)$ depends on m as it corresponds to different values of angle $\varphi = \arcsin(m\lambda/d)$ and hence to the different values of $J_1(\varphi)$. This is why the values of $I(\varphi)/I_0$ at the main maxima of $J_2(\varphi)$ in Figure 8.9 are different for the different values of m.

The minima of the function $J_2(\varphi)$ ($J_2 = 0$) occur at $\sin(N\beta) = 0$ but $\sin\beta \neq 0$, that is, at $N\beta = p\pi$ or $\beta = p\pi/N$, where $p = \pm1, \pm2, \pm3, \ldots$ except $p = mN$, with m introduced earlier. This means that between two main maxima, there are $N-1$ minima with $J_2 = 0$, as it is shown in Figure 8.9. The minima, which are next to the main maxima, lie at the points $\beta_1 = \pi(m \pm 1)/N$. As the number of slits in the grating increases, the distance between the minima and the main maxima decreases proportionally to $1/N$, that is, the value of $I(\varphi)/I_0$ increases as given by Equation 8.22, while the width of the diffraction maxima decreases.

Since the single-slit diffraction intensity is given by $J_1(\varphi)$ and the grating intensity distribution is practically determined by the J_{2max}, then

$$I = I_0(J_1 \cdot J_2) = N^2 I_0 J_1. \tag{8.23}$$

Only those main maxima of the function $J_2(\varphi)$, which are inside the central maximum of function J_1, will be strong. Since the width of a slit b is usually very small, the central maximum is quite wide.

The angular width of this maximum is equal to $2\lambda/b$; therefore, within this maximum of J_1, we have several main diffraction peaks of the grating.

White light is a superposition of simple harmonic waves with various wavelengths, which diffract independently from a grating. Therefore, for each component, the corresponding diffraction conditions will be satisfied at different angles $\varphi = \arcsin(m\lambda/d)$ that are determined by the wavelength of the component. This means that the monochromatic components of white light incident on the grating will be spatially separated as it is shown in Figure 8.10. The set of the main diffraction maxima of mth order ($m \neq 0$) for all monochromatic components of the incident light forms the diffraction spectrum of mth order. After the grating, the wave with longer wavelength, λ_1, (solid lines) has its maxima at larger angles, while the wave with shorter wavelength, λ_2, has its maxima at smaller angles (dashed line); both waves have the center maximum at $m = 0$.

The position of the main diffraction peak of zero order does not depend on the wavelength (for the central maximum, $\varphi = 0$). For white light, this maximum will look like a strip of white color. The diffraction spectrum of mth order ($m \neq 0$), on the other hand, contains all the colors of a rainbow.

One of the main characteristics of the grating is its **resolving ability** $R = \lambda/\delta\lambda$. Here, λ is the wavelength of a spectral line corresponding to the spectral component and $\delta\lambda$ is the smallest difference between closely spaced wavelengths that the diffraction grating can resolve. Spectral lines with wavelengths λ and $\lambda' = \lambda + \delta\lambda$ are considered resolved if the central maximum for one wavelength coincides with the first diffraction minimum for the other wavelength. Therefore, in order to find the resolution of the diffraction grating, we determine the position of the center of the mth maximum for the wavelength $\lambda + \delta\lambda$. It is given by the condition

$$d \sin \varphi_{max} = m(\lambda + \delta\lambda). \tag{8.24}$$

The minima situated closest to the central maxima $\beta = m\pi$ are given by $\beta = (m \pm 1/N)\pi$, where $\beta = (\pi d/\lambda)\sin\varphi$. The edges of mth maximum for wavelength λ are given by

$$d \sin \varphi_{min} = \left(\frac{m \pm 1}{N}\right)\lambda. \tag{8.25}$$

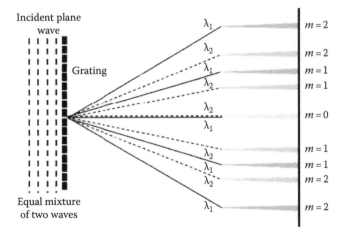

FIGURE 8.10 Example of diffraction pattern of light that consists of two waves with wavelengths λ_1 and λ_2 ($\lambda_1 > \lambda_2$).

The center of the maximum for wavelength $\lambda + \delta\lambda$ coincides with the edge of the maximum for wavelength λ if

$$m(\lambda + \delta\lambda) = \left(\frac{m+1}{N}\right)\lambda, \quad m\delta\lambda = \frac{\lambda}{N}.$$

From this expression, we get

$$R = \frac{\lambda}{\delta\lambda} = mN. \tag{8.26}$$

Exercise 8.3

White light is incident normally on a diffraction grating with slit spacing d. For different wavelengths, the diffraction angles are different (the longer is the wavelength, the larger is the angle of diffraction). Determine (1) the angular dispersion and (2) the resolution of the diffraction grating.

Solution. All diffraction maxima except the central one (zero order) will correspond to the separate spectral components of the white light. For large orders of diffraction, overlapping of spectra of neighboring orders is possible. *Angular dispersion* is a measure of the *angular separation* of two spectral lines, which differ by the unit of wavelength. It is defined by the expression

$$D = \frac{\delta\varphi}{\delta\lambda}.$$

This equation gives the angular separation $\delta\varphi$ of two spectral lines, which differ in wavelength by $\delta\lambda$.

In order to find the angular dispersion of the diffraction grating, we must differentiate expression (8.14):

$$d \cdot \sin\varphi = m\lambda,$$

$$d \cdot \cos\varphi \, d\varphi = md\lambda, \quad D = \frac{d\varphi}{d\lambda} = \frac{m}{d \cdot \cos\varphi}.$$

For small angles of diffraction, $\cos\varphi \approx 1$, and therefore, we can use the approximation $D \approx m/d$. We see that the angular dispersion is proportional to the order of diffraction and inversely proportional to the grating period.

The resolution of a diffraction grating is determined by Equation 8.26:

$$R = \frac{\lambda}{\delta\lambda} = mN.$$

The resolution is proportional to the order of diffraction and to the number of slits in the grating.

8.4 DIFFRACTION BY A 3D LATTICE

We discussed diffraction for the case when the secondary waves are transmitted through the grating slits. It is also possible to construct a 1D periodic lattice where the transmitting slits will be replaced by reflecting stripes so that the diffraction pattern will be formed by rays reflected from different stripes. In the case of an opaque screen with periodically located holes, the 2D lattice would produce a 2D diffraction pattern. In a similar fashion, the holes can be replaced by light reflecting circles to observe the interference of the reflected light. Finally, diffraction can be observed in volume

(3D) structures, that is, a structure in space, which is periodic in three directions. A typical example of this structure is the periodic lattice of a crystal. In this case, atoms of the crystalline structures will serve as scatterers of light. Moreover, the size of atoms is substantially smaller than the characteristic distance between the atoms (this distance is known as the lattice constant d), so crystalline materials are suitable 3D lattices to observe diffraction.

The lattice constants d of different materials are in the range of $0.2 \div 0.5$ nm $((2 \div 5) \times 10^{-10}$ m) and are too small compared to the wavelength of visible light $\lambda \approx 500$ nm (5×10^{-7} m). Therefore, for visible light, crystalline materials act as optically homogeneous media, but they are suitable for the observation of diffraction of shorter wavelengths, for example, X-ray radiation whose wavelength is comparable to the period of crystal lattices. The first X-ray diffraction from a crystal was observed in 1913 by Max von Laue.

In the description of diffraction by 3D periodic structures, we use the angles between the coordinate axes and the directions of propagation of the incident and diffracted light, respectively (we remind again that in the case of crystalline materials the diffracted light is the light scattered by the atoms of the crystal). If a, b, and c are the periods of the crystal lattice along the x-, y-, z-axes, then the main diffraction maxima satisfy the following conditions:

$$a(\cos\alpha - \cos\alpha_0) = m\lambda,$$

$$b(\cos\beta - \cos\beta_0) = n\lambda, \tag{8.27}$$

$$c(\cos\gamma - \cos\gamma_0) = p\lambda.$$

This system of equations is called the Laue equations. In these equations, α_0, β_0, γ_0 and α, β, γ are the angles between the directions of propagation of the incident and diffracted light, respectively, and the coordinate axes x, y, z, which we assume coincide with the symmetry axes of the crystal. In a Cartesian coordinate system, the cosines of these angles satisfy the following important relations:

$$\cos^2\alpha_0 + \cos^2\beta_0 + \cos^2\gamma_0 = 1, \tag{8.28}$$

$$\cos^2\alpha + \cos^2\beta + \cos^2\gamma = 1. \tag{8.29}$$

The set of integral numbers m, n, p determines the corresponding diffraction peak. Each set of numbers (m, n, p) in the far-field diffraction pattern corresponds to one bright spot (diffraction intensity maximum) as it is shown in Figure 8.11. Thus, the result of diffraction from a crystal observed on a screen is a set of bright points.

It is important to remember that we can control the direction of the incident light, so we can always know angles α_0, β_0, γ_0. To observe the diffraction peak of order (m, n, p) at the given values of angles α_0, β_0, γ_0, it is necessary that the wavelength of an incident light has a certain value in order to satisfy Equation 8.27. In order to find that wavelength, one must first find $\cos\alpha$, $\cos\beta$, and $\cos\gamma$ from Equation 8.27:

$$\cos\alpha = \frac{m\lambda}{a} + \cos\alpha_0,$$

$$\cos\beta = \frac{n\lambda}{b} + \cos\beta_0,$$

$$\cos\gamma = \frac{p\lambda}{c} + \cos\gamma_0.$$

FIGURE 8.11 Example of a far-field diffraction pattern of a crystal.

Taking power 2 of the left- and the right-hand parts of each equation, one gets

$$\cos^2\alpha = \left(\frac{m\lambda}{a} + \cos\alpha_0\right)^2 = \left(\frac{m\lambda}{a}\right)^2 + 2\left(\frac{m\lambda}{a}\right)\cos\alpha_0 + \cos^2\alpha_0,$$

$$\cos^2\beta = \left(\frac{n\lambda}{b} + \cos\beta_0\right)^2 = \left(\frac{n\lambda}{b}\right)^2 + 2\left(\frac{n\lambda}{b}\right)\cos\beta_0 + \cos^2\beta_0,$$

$$\cos^2\gamma = \left(\frac{p\lambda}{c} + \cos\gamma_0\right)^2 = \left(\frac{p\lambda}{c}\right)^2 + 2\left(\frac{p\lambda}{c}\right)\cos\gamma_0 + \cos^2\gamma_0.$$

Adding these three equations, taking into account Equations 8.28 and 8.29 and dividing by λ, one gets the equation

$$\left[\left(\frac{m}{a}\right)^2 + \left(\frac{n}{b}\right)^2 + \left(\frac{p}{c}\right)^2\right]\lambda + 2\left[\left(\frac{m}{a}\right)\cos\alpha_0 + \left(\frac{n}{b}\right)\cos\beta_0 + \left(\frac{p}{c}\right)\cos\gamma_0\right] = 0,$$

which determines the wavelength λ as the function of the incident angles α_0, β_0, γ_0:

$$\lambda = -2\frac{(m/a)\cos\alpha_0 + (n/b)\cos\beta_0 + (p/c)\cos\gamma_0}{(m/a)^2 + (n/b)^2 + (p/c)^2}. \tag{8.30}$$

For monochromatic radiation with a given wavelength, λ, the conditions that a diffraction peak of order (m, n, p) occurs are satisfied only for specific angles α_0, β_0, γ_0 of light incidence on the crystal.

At $\lambda \geq 2d_0$ (here, d_0 is the largest of the three spatial periods a, b, c), all diffraction maxima, except for zeroth order, must be absent. Light with such wavelength propagates in an optically nonuniform medium without diffraction. Therefore, the condition $\lambda \geq 2d_0$ is called the condition of optical homogeneity of the medium. For $\lambda < 2d_0$, the diffraction pattern as determined by Equation 8.30 is well pronounced, and it is used to determine the lattice structure parameters: a, b, and c.

The calculation of angular positions of maxima in the diffraction pattern from a crystal lattice can also be carried out by the following simplified method. The spatially periodic arrangement of atoms in the crystal lattice is replaced by a periodic arrangement of the parallel atomic planes. The distance between adjacent lattice planes is called *interplanar distance*; it defines the lattice period in the crystallographic direction. Consider the reflection of x-rays from two adjacent atomic planes (Figure 8.12).

Rays 1 and 2 are coherent and will interfere with each other. Diffraction from the crystal lattice results from the interference of rays 1 and 2, which are reflected in a mirrorlike fashion from the atomic planes. Figure 8.12 shows that the path difference of these rays ($\Delta_{12} = AB + BC$)

$$\Delta_{12} = 2d_0 \sin \theta, \tag{8.31}$$

and therefore, the condition for the observation of diffraction maxima is *Bragg's equation*

$$2d_0 \sin \theta = m\lambda, \tag{8.32}$$

where the angle θ is called *Bragg's angle*.

X-ray diffraction by a crystal lattice is the basis of *structure analysis of crystals*—one of the most powerful methods for studying the crystal structure of matter.

Exercise 8.4

In one of the methods of X-ray structural analysis (Debye–Scherrer method), polycrystalline samples and monochromatic radiation are used to find the distance between planes. Find an expression that gives the crystal plane separation as a function of the X-ray wavelength.

Solution. The majority of solids are polycrystalline, that is, they consist of large number of randomly distributed small crystallites. When illuminating such a polycrystalline material by monochromatic X-ray radiation of wavelength λ, there will be always a great number of crystallites whose parallel atomic planes with interplane distance d_j satisfy Bragg's equation: $\theta_j = \arcsin(m\lambda/2d_j)$.

The directions of the diffracted waves lie on the surface of a cone with an apex angle equal to $4\theta_j$. The rays will be reflected from the given system of parallel planes forming an angle $2\theta_j$ with respect to the incident ray. Therefore, all reflected rays constitute a cone with an apex angle equal to $4\theta_j$. If we place a long strip of photographic film in the form of a cylinder with radius R in the path of the reflected waves, we get a set of double symmetric lines each of which corresponds to a diffraction maximum due to a system of atomic planes (Figure 8.13). If the distance between two lines (the length of arc that rests upon an angle $4\theta_j$) is equal to $2L_j$, then it can be related with the radius of a circle R: $2L_j = 4\theta_j R$, which gives us

$$\theta_j = \frac{L_j}{2R} \quad (\text{in radians}).$$

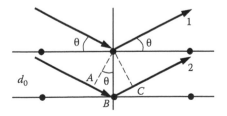

FIGURE 8.12 Schematic diagram of Bragg's diffraction (Equation 8.31) of x-rays from a crystalline lattice.

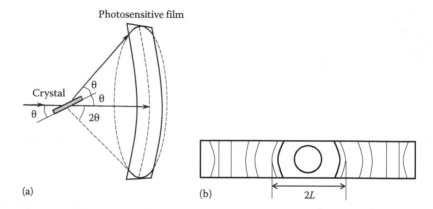

FIGURE 8.13 Diffraction from a polycrystalline material: crystallites whose parallel atomic planes satisfy the Bragg's equation (a) and their diffraction maxima on photographic film in the form of a cylinder (b).

By measuring the distance $2L_j$ between two symmetric lines on the X-ray film, we can determine θ_j and $\sin \theta_j$. Therefore, the distance between the corresponding adjacent atomic planes is given by the expression

$$d_j = \frac{m\lambda}{2\sin\theta_j} = \frac{m\lambda}{2\sin(L_j/2R)}.$$

The central spot on the film corresponds to the X-ray beam that was not diffracted.

8.5 WAVES IN CONTINUOUS PERIODIC MEDIA

Consider a transparent nonmagnetic ($\kappa_m = 1$) medium in which the dielectric permittivity for $x < 0$ is equal to ε_0 (vacuum), while for $x \geq 0$, it is a continuous periodic function of x:

$$\varepsilon(x) = \varepsilon_a \left[1 + \delta \cos\left(\frac{2\pi x}{d}\right)\right] = \varepsilon_0 \kappa \left[1 + \delta \cos\left(\frac{2\pi x}{d}\right)\right], \qquad (8.33)$$

where
   d is the spatial period of the structure
   δ is the parameter that determines the modulation depth of the medium's dielectric permittivity and is assumed to be small ($\delta \ll 1$)

In this case, the equation for a linearly polarized wave, which propagates along x-axis, in the region $x \geq 0$ can be written as

$$\frac{\partial^2 E}{\partial x^2} - \frac{1}{v^2}\left[1 + \delta \cos(2Kx)\right]\frac{\partial^2 E}{\partial t^2} = 0, \qquad (8.34)$$

where
   $E = E_z$ is the electric field of the wave
   $K = \pi/d$ and $v = c/\sqrt{\kappa}$

We assume solutions in the form of a monochromatic wave with amplitude that is a function of x:

$$E(x,t) = A(x)\exp(i\omega t). \tag{8.35}$$

Substituting this equation into Equation 8.34, we obtain the equation for the amplitude $A(x)$:

$$\frac{d^2A}{dx^2} + k^2\left[1 + \delta\cos(2Kx)\right]A = 0, \tag{8.36}$$

where $k = \omega/v$ is the wave number of a wave that propagates in the periodic medium. If δ is equal to zero, the solution of Equation 8.36 is well known:

$$A(x) = A^+\exp(-ikx) + A^-\exp(ikx), \tag{8.37}$$

where A^+ and A^- are the coordinate-independent amplitudes of waves traveling in the positive and negative directions of x, that is, amplitudes of forward- and backward-traveling waves. In the case of $\delta \ll 1$, the multiplication factor in the square brackets of Equation 8.36 has small periodic deviation from unity. So, for slow varying $\varepsilon(x)$, we can seek solution of Equation 8.36 in the form (8.37) with A^+ and A^- being slow varying functions of x, that is, $A^+ = A^+(x)$ and $A^- = A^-(x)$. To find the dependences of amplitudes A^+ and A^- on x, we need to substitute Equation 8.37 into Equation 8.36 and take into account that due to small variation of A^+ and A^-, we can neglect their second-order derivatives in comparison with the first-order derivatives, since

$$\left|\frac{d^2A^\pm}{dx^2}\right| \sim 2K\delta\left|\frac{dA^\pm}{dx}\right| \ll 2k\left|\frac{dA^\pm}{dx}\right|.$$

The most interesting case is the one in which the wave number k appears to be close to K. For $k = K$, the spatial period $d = \pi/k$ is equal to a half wavelength of the wave propagating in the medium ($d = \lambda/2$).

The condition $\lambda = 2d$ is a special case of the general condition of **Bragg's reflection** (8.32) for $\theta = 90°$. When this condition is satisfied, each wave out of a set of waves reflected from consecutive variations in ε has a path difference $m\lambda$ ($m = 1, 2, 3, \ldots$), and the waves interfere with the same phase. This means that the wave traveling in the forward direction is efficiently reflected from the periodic variations in ε, and its energy is transferred to the wave traveling backward. In turn, the backward wave is also reflected and returns part of its energy to an initial wave. Thus, when $k \approx K$ (spatial resonance), the two counterpropagating waves are related with each other.

Let us substitute in Equation 8.36 $K = k - \xi$, where ξ is a small deviation from the precise resonance condition. Using simple mathematical manipulations, it is possible to arrive at the following system of related equations for the amplitudes of the two waves:

$$\frac{dA^+}{dx} = -i\left(\frac{k\delta}{4}\right)A^-\exp(2i\xi x),$$

$$\frac{dA^-}{dx} = i\left(\frac{k\delta}{4}\right)A^+\exp(-2i\xi x). \tag{8.38}$$

To solve this system, we express from the first equation the amplitude $A^-(x)$ and substitute it into the second equation. In that way, we arrive at the following simple second-order differential equation:

$$\frac{d^2A^+}{dx^2} - 2i\xi\frac{dA^+}{dx} - \left(\frac{k\delta}{4}\right)^2 A^+ = 0. \tag{8.39}$$

The general solution of this equation can be written as

$$A^+(x) = \left[C_1 \exp\left(-\sqrt{\left(\frac{k\delta}{4}\right)^2 - \xi^2}\, x \right) + C_2 \exp\left(\sqrt{\left(\frac{k\delta}{4}\right)^2 - \xi^2}\, x \right) \right] \exp(i\xi x), \qquad (8.40)$$

where the constants of integration C_1 and C_2 are determined from the boundary conditions.

We assume that a wave with an amplitude A_0 is incident on the periodic structure from the half-space $x < 0$. Therefore, the amplitude A_0 is equal to that of a forward wave at $x = 0$: $A^+(0) = A_0$. For the periodic medium occupying the half-space $x > 0$, it is necessary to satisfy this condition: For $x \to \infty$, the amplitude $A^+(x) \to 0$. From these conditions, it follows that $C_2 = 0$ and $C_1 = A_0$. As a result, the solution of Equation 8.39 takes the following form:

$$A^+(x) = A_0 \exp\left(-\sqrt{\left(\frac{k\delta}{4}\right)^2 - \xi^2}\, x \right) \exp(i\xi x). \qquad (8.41)$$

Substituting this solution into the first equation of the system (8.38), we find an expression for the amplitude of the backward wave:

$$A^-(x) = \frac{4iA_0}{k\delta}\left(i\xi - \sqrt{\left(\frac{k\delta}{4}\right)^2 - \xi^2} \right) \exp\left(-\sqrt{\left(\frac{k\delta}{4}\right)^2 - \xi^2}\, x \right) \exp(-i\xi x). \qquad (8.42)$$

From these expressions, it follows that for $|\xi| < k\delta/4$, that is, for a small deviation from Bragg's resonance, there is total reflection of the incident wave. This wave transfers all its energy to the backward wave, whose amplitude increases from value $A^- = 0$ at $x \to \infty$ to $A^- = A_0$ at $x = 0$. It follows from Equations 8.41 and 8.42 that the moduli of amplitudes of forward and backward waves are equal to an arbitrary value of x, that is,

$$\left| A^+(x) \right| = \left| A^-(x) \right| = A_0 \exp\left(-\sqrt{\left(\frac{k\delta}{4}\right)^2 - \xi^2}\, x \right). \qquad (8.43)$$

In the region for which $|\xi| > k\delta/4$, the structure is transparent to the wave propagating forward. Thus, there exists a band of wave numbers $k = \omega/v$ (and, respectively, frequencies) for which the incident waves are efficiently reflected by the periodic structure. This forbidden band is determined by the following inequalities:

$$K\left(1 - \frac{\delta}{4}\right) < k < K\left(1 + \frac{\delta}{4}\right),$$
$$Kv\left(1 - \frac{\delta}{4}\right) < \omega < Kv\left(1 + \frac{\delta}{4}\right). \qquad (8.44)$$

Exercise 8.5

Using the system of Equation 8.38 for the amplitudes of *counterpropagating waves* $A^{\pm}$, show that difference $\left|A^{+}\right|^{2}-\left|A^{-}\right|^{2}$ is a constant.

Solution. We multiply the first equation of the system (8.38) by $(A^{+})^{*}$ and the second equation of this system by $(A^{-})^{*}$:

$$(A^{+})^{*}\cdot\frac{dA^{+}}{dx}=-i\left(\frac{k\delta}{4}\right)(A^{+})^{*}\cdot A^{-}\exp(2i\xi x),$$

$$(A^{-})^{*}\cdot\frac{dA^{-}}{dx}=i\left(\frac{k\delta}{4}\right)(A^{-})^{*}\cdot A^{+}\exp(-2i\xi x).$$

Let us write the system that is the complex conjugate to the given system:

$$A^{+}\cdot\frac{d(A^{+})^{*}}{dx}=i\left(\frac{k\delta}{4}\right)A^{+}\cdot(A^{-})^{*}\exp(-2i\xi x),$$

$$A^{-}\cdot\frac{d(A^{-})^{*}}{dx}=-i\left(\frac{k\delta}{4}\right)A^{-}\cdot(A^{+})^{*}\exp(2i\xi x).$$

Let us add the first and second equations of these systems and then subtract:

$$(A^{+})^{*}\cdot\frac{dA^{+}}{dx}+A^{+}\cdot\frac{d(A^{+})^{*}}{dx}=-i\frac{k\delta}{4}(A^{+})^{*}\cdot A^{-}\exp(2i\xi x)+i\frac{k\delta}{4}A^{+}\cdot(A^{-})^{*}\exp(-2i\xi x),$$

$$A^{-}\cdot\frac{d(A^{-})^{*}}{dx}+(A^{-})^{*}\cdot\frac{dA^{-}}{dx}=-i\frac{k\delta}{4}(A^{+})^{*}\cdot A^{-}\exp(2i\xi x)+i\frac{k\delta}{4}A^{+}\cdot(A^{-})^{*}\exp(-2i\xi x).$$

Now let us subtract the second equation from the first. Taking into account that the right-hand sides of these equations are the same, we get

$$(A^{+})^{*}\cdot\frac{dA^{+}}{dx}+A^{+}\cdot\frac{d(A^{+})^{*}}{dx}-\left(A^{-}\cdot\frac{d(A^{-})^{*}}{dx}+(A^{-})^{*}\cdot\frac{dA^{-}}{dx}\right)=0$$

or

$$\frac{d}{dx}\left(\left|A^{+}\right|^{2}-\left|A^{-}\right|^{2}\right)=0,$$

from which it follows that for any coordinate x, we get

$$\left|A^{+}\right|^{2}-\left|A^{-}\right|^{2}=\text{const.}$$

The magnitudes $\left|A^{+}\right|^{2}$ and $\left|A^{-}\right|^{2}$ give the energy of the forward- and backward-propagating waves. Since the backward wave occurs as a result of multiple reflections of the direct wave, then in this case the difference of the quantities mentioned earlier is conserved.

8.6   WAVES IN PLANAR LAYERED PERIODIC STRUCTURES

Let us now consider a planar layered structure consisting of alternating layers of two nonmagnetic dielectrics ($\kappa_{m1,2} = 1$) with thicknesses d_1 and d_2 and different relative dielectric permittivities κ_1 and κ_2 (Figure 8.14). Two adjacent layers form the structure period $d = d_1 + d_2$.

Consider a linearly polarized plane wave propagating along the axis of the structure's periodicity (x-axis) normal to the interfaces. We assume that in the layer planes the wave fields do not depend on the coordinates y and z. Thus, in each layer, the electric field of the wave can be described by means of the following 1D wave equation:

$$\frac{\partial^2 E}{\partial x^2} - \frac{\kappa_j(x)}{c^2}\frac{\partial^2 E}{\partial t^2} = 0, \quad j = 1, 2. \tag{8.45}$$

For each layer, we seek solutions of these equations in the form

$$E(t, x) = E_j \exp(i\omega t - k_j x). \tag{8.46}$$

Following substitution of these solutions into Equation 8.45, we obtain the **Hill equation**:

$$\frac{d^2 E_j}{dx^2} + k_j^2(x)E_j = 0, \tag{8.47}$$

where the wave number $k_j(x)$ is a periodic function of the coordinate x:

$$k_j(x) = \begin{cases} k_1, & nd < x < nd + d_1, \\ k_2, & nd + d_1 < x < (n+1)d, \end{cases} \tag{8.48}$$

where
$$k_j = k_0\sqrt{\kappa_j}$$
$$n = 0, 1, 2, \ldots$$

In this case, $k(x)$ is a step function for which the solution of the Hill equation can be obtained in the following way. For each of the two adjacent layers, we write the solution of Equation 8.47 as a superposition of two counterpropagating waves:

$$\begin{aligned}
E_1(x) &= A_{11} \exp(-ik_1 x) + A_{12} \exp(ik_1 x), \quad nd < x < nd + d_1, \\
E_2(x) &= A_{21} \exp(-ik_2 x) + A_{22} \exp(ik_2 x), \quad nd + d_1 < x < (n+1)d.
\end{aligned} \tag{8.49}$$

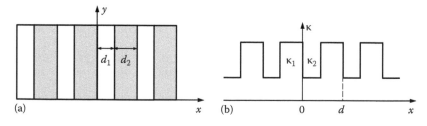

FIGURE 8.14   1D periodic crystal: alternating layers of two dielectrics (a) and dependence of relative dielectric permittivity on coordinate (b).

To determine the four constants A_{jl}, one has to take into account the continuity of the tangential components of electric and magnetic fields, that is, of the field E and its derivative dE/dx at the interfaces. Thus, at the boundary $x = d_1$, we have

$$E_1(x) = E_2(x), \quad \frac{dE_1(x)}{dx} = \frac{dE_2(x)}{dx}. \tag{8.50}$$

In addition to these boundary conditions, the solutions must satisfy the condition of periodicity, that is, the fields at two points separated by one period (e.g., at $x = 0$ and $x = d$) can differ only by a phase factor

$$E_1(0) = E_2(d)\exp(-ik_B d), \tag{8.51}$$

where the wave number k_B is called the **Bloch wave vector**. Equation 8.51 is called the **Floquet–Bloch theorem** (this theorem was used by Bloch for constructing the electron wave functions in a periodic crystal lattice).

By substituting the fields given by Equation 8.49 into Equations 8.50 and 8.51, we obtain

$$A_{11}e^{-ik_1 d_1} + A_{12}e^{ik_1 d_1} = A_{21}e^{-ik_2 d_1} + A_{22}e^{ik_2 d_1},$$

$$k_1\left(A_{11}e^{-ik_1 d_1} - A_{12}e^{ik_1 d_1}\right) = k_2\left(A_{21}e^{-ik_2 d_1} - A_{22}e^{ik_2 d_1}\right),$$

$$A_{11} + A_{12} = e^{-ik_B d}\left(A_{21}e^{-ik_2 d} + A_{22}e^{ik_2 d}\right),$$

$$k_1(A_{11} - A_{12}) = e^{-ik_B d}k_2\left(A_{21}e^{-ik_2 d} - A_{22}e^{ik_2 d}\right). \tag{8.52}$$

This is a system of four linear homogeneous equations for the unknown amplitudes of the waves propagating in the forward and backward direction: A_{11}, A_{12}, A_{21}, and A_{22}. A nontrivial (nonzero) solution of this system exists only if its determinant is equal to zero, that is,

$$\begin{vmatrix} e^{-ik_1 d_1} & e^{ik_1 d_1} & -e^{-ik_2 d_1} & -e^{ik_2 d_1} \\ k_1 e^{-ik_1 d_1} & -k_1 e^{ik_1 d_1} & -k_2 e^{-ik_2 d_1} & k_2 e^{ik_2 d_1} \\ 1 & 1 & -ye^{-ik_2 d} & -ye^{ik_2 d} \\ k_1 & -k_1 & -k_2 ye^{-ik_2 d} & k_2 ye^{ik_2 d} \end{vmatrix} = 0, \tag{8.53}$$

where the parameter $y = \exp(-ik_B d)$ is introduced. Expanding this determinant, we obtain the quadratic equation

$$y^2 - 2Fy + 1 = 0, \tag{8.54}$$

where

$$F = \cos(k_1 d_1)\cos(k_2 d_2) - \frac{1}{2}\left(\frac{k_1}{k_2} + \frac{k_2}{k_1}\right)\sin(k_1 d_1)\sin(k_2 d_2).$$

For the roots y_1 and y_2 of Equation 8.54, we have $y_1 y_2 = 1$, that is, $y_1 = \exp(ik_B d)$, $y_2 = \exp(-ik_B d)$, and $y_1 + y_2 = 2\cos(k_B d) = 2F$. As a result, we obtain the dispersion equation of the structure under consideration, which relates the Bloch wave number k_B with the known quantities k_1 and k_2:

$$\cos(k_B d) = \cos(k_1 d_1)\cos(k_2 d_2) - \frac{1}{2}\left(\frac{k_1}{k_2} + \frac{k_2}{k_1}\right)\sin(k_1 d_1)\sin(k_2 d_2). \qquad (8.55)$$

The case $|\cos(k_B d)| \leq 1$ corresponds to a transparency band of the structure. If $|\cos(k_B d)| > 1$, we find ourselves in an opacity band (the wave number k_B is imaginary, and the wave decays exponentially).

The analysis of the dispersion equation (8.55) is rather complicated, so we consider a special case, assuming $k_1 d_1 = k_2 d_2$. This means that the phase of the propagating wave has changed in each layer by the same amount. Then Equation 8.55 takes the form

$$\cos(k_B d) = F(k_1 d_1), \quad \text{here } F(k_1 d_1) = 1 - \frac{(k_1 + k_2)^2}{2k_1 k_2}\sin^2(k_1 d_1). \qquad (8.56)$$

In Figure 8.15a, the right-hand side of Equation 8.56, that is, the function $F(k_1 d_1)$, is presented. The points of intersection of this function with the line $F(k_1 d_1) = -1$ correspond to the boundaries of the nontransparency bands, which are shaded in this figure. In these ranges, $\cos(k_B d) < -1$, and thus, the effective Bloch wave number is imaginary that indicates the opacity of the periodic structure. Figure 8.15b illustrates the dependence of the wave number k_B on the magnitude of $k_1 d_1$. In fact, here, a type of the dispersion law is represented, that is, the dependence of k_B on the angular frequency ω, since $k_1 = (\omega/c_0)\sqrt{\kappa_1}$.

Exercise 8.6

A wave with amplitude A_0 is incident normally from region $x < 0$ on a 1D periodic structure whose thickness is $L \gg d$. Using Equations 8.40 and 8.38, find the coefficients of reflection and transmission of the structure.

Solution. The general solution of Equations 8.40 and 8.38 for the amplitudes of the forward and backward waves can be written in the form

$$A^+(x) = \left[C_1 \exp(-\beta x) + C_2 \exp(\beta x)\right]\exp(i\xi x),$$

$$A^-(x) = \left(\frac{4i}{k\delta}\right)\left[-C_1 \exp(-\beta x) + C_2 \exp(\beta x)\right]\exp(-i\xi x),$$

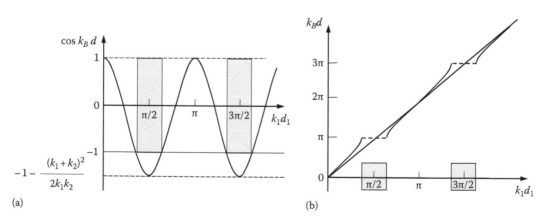

(a)   (b)

FIGURE 8.15   Dependence of function $F(k_1 d_1)$ (a) and the wave number k_B (b) on the magnitude of $k_1 d_1$.

where $\beta = \sqrt{(k\delta/4)^2 - \xi^2}$ and the constants of integration $C_{1,2}$ are determined from the boundary conditions. Since from region $x < 0$, the wave with amplitude A_0 is incident on periodic structure, then at $x = 0$, the amplitude of the forward wave is equal to $A^+(0) = A_0$.

In order to find the amplitude of the backward wave, it is necessary that at $x = L$, the condition $A^-(L) = 0$ must be satisfied. From these conditions, we get the following system:

$$A^+(0) = C_1 + C_2 = A_0,$$

$$A^-(L) = -C_1 \exp(-\beta L) + C_2 \exp(\beta L) = 0.$$

The solution of this system leads to the following expressions for the constants of integration:

$$C_1 = \frac{\exp(\beta L)}{2\cosh(\beta L)} A_0, \quad C_2 = \frac{\exp(-\beta L)}{2\cosh(\beta L)} A_0.$$

After the substitution of these expressions into the solutions of Equation 8.38, we get

$$A^+(x) = A_0 \exp(i\xi x) \frac{\cosh\left[\beta(L-x)\right]}{\cosh(\beta L)},$$

$$A^-(x) = \frac{4A_0}{k\delta} \exp(-i\xi x) \frac{\sinh\left[\beta(L-x)\right]}{\cosh(\beta L)}.$$

For the coefficients of transmission and reflection, we get

$$T = \left|\frac{A^+(L)}{A_0}\right| = \frac{1}{\cosh(\beta L)}, \quad R = \left|\frac{A^-(0)}{A_0}\right| = \frac{4}{k\delta} \frac{\sinh(\beta L)}{\cosh(\beta L)}.$$

8.7 PHOTONIC CRYSTALS

Historically, the theory of scattering of electromagnetic waves by 3D lattices began to develop rapidly in the wavelength range $\lambda \sim 0.1-1$ nm. In 1986, Eli Yablonovich from the University of California, Los Angeles, proposed the idea of the creation of a 3D dielectric structure, similar to normal crystals, in which electromagnetic waves in a particular spectral band could not propagate.

Such systems were called *photonic structures* and have a forbidden band (*photonic bandgap*) in which electromagnetic waves cannot propagate. Within a few years, such structures were manufactured by drilling millimeter size holes in a material with high refractive index. This artificial crystal, which did not transmit radiation in the millimeter range, was the first realization of a photonic structure with a forbidden band.

As it was shown earlier, due to the periodic variation of the material parameters in certain structures, it is possible to have allowed and forbidden frequency intervals (bands) for electromagnetic waves. A similar effect occurs in crystalline materials where allowed and forbidden energy bands for electrons are observed. In view of this analogy, materials with a spatially periodic structure, which are characterized by a change in the refractive index on a scale comparable with the wavelengths of light in the visible and near-infrared range, were called *photonic crystals*.

The main property of photonic crystals is the presence of *photonic band gaps* in their reflection and transmission spectra. If a wave with frequency that corresponds to the band gap is incident on a photonic crystal, the wave cannot propagate in this crystal and is reflected back. If the frequency of the incident wave corresponds to an allowed band, then it may propagate through the photonic crystal. Thus, the photonic crystal acts as an optical filter (generating bright colors in natural and artificial photonic crystals).

Based on the spatial changes of the refractive index, photonic crystals can be divided into three main classes: 1D, 2D, and 3D photonic crystals, depending whether the refractive index varies periodically in one, two, or three spatial directions.

One-dimensional (**1D**) *photonic crystals* consist of parallel layers of various materials with different refractive indices. Schematically, such a crystal is presented in Figure 8.16a. The symbol d denotes the period of variation of refractive indices and n_1 and n_2 are the refractive indices of the two materials (in the general case, one period may contain a greater number of layers of different materials).

Figure 8.16b shows a photonic crystal that consists of rectangular areas with refractive index n_1, which are embedded in a medium with refractive index n_2. The areas with refractive index n_1 can be arranged in the following *2D* lattices: oblique, square, hexagonal, primitive rectangular, and centered rectangular. The shape areas with refractive index n_1 is not limited to rectangles but can have any shape (triangles, circles, etc.).

Three-dimensional (**3D**) *photonic crystals* are formed by arrays of volume areas (spheres, cubes, etc.) arranged in the form of a 3D lattice.

One of the first important practical applications of 1D photonic crystals is in coatings that reduce reflection. Photonic crystals are high-performance spectral filters that reduce reflection from optical elements in a desirable interval of frequencies.

Another well-known example of 1D photonic structures is semiconductor lasers with distributed feedback as well as optical waveguides with a periodic longitudinal modulation of the physical parameters (a profile of a refractive index).

Since photonic crystals have controllable photonic bands, they are often considered as the optical analogs of crystalline materials and in particular semiconductors. Photonic crystals may become the basis of new devices for optical transmission and information processing. Lasers based on photonic crystals allow for realizing small-signal laser generators, the so-called low-threshold and non-threshold lasers. Waveguides based on photonic crystals can be very compact and have low losses. By using photonic crystals, it is possible to create "left-hand" media possessing negative refractive indices, which allow focusing light to a point with size less than the wavelength.

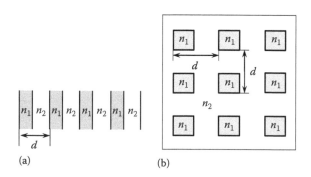

FIGURE 8.16   1D (a) and 2D (b) photonic crystals show their properties in the two spatial directions.

Exercise 8.7

Show that in an infinite 3D photonic crystal in the absence of absorption, the following relation is valid for the electric field of a wave that propagates through the crystal:

$$\mathbf{E}(\mathbf{r} + \mathbf{d}_n) = \mathbf{E}(\mathbf{r}) \exp(i\mathbf{k}_B \cdot \mathbf{d}_n),$$

where
$\mathbf{d}_n = n_1\mathbf{d}_1 + n_2\mathbf{d}_2 + n_3\mathbf{d}_3$ is a translation vector (which connects two identical points in different crystalline cells)
$\mathbf{d}_i$ are the periods of the structure along the coordinate axes
n_i are integers
$\mathbf{k}_B$ is Bloch's wave vector

Solution. In a photonic crystal, all material parameters are periodic. Thus, for the dielectric permittivity, we can write $\varepsilon(\mathbf{r} + \mathbf{d}_n) = \varepsilon(\mathbf{r})$, which means that points $\mathbf{r}$ and $\mathbf{r} + \mathbf{d}_n$ are physically equivalent. From the last statement, it follows that the electric field of a wave at these points, $\mathbf{E}(\mathbf{r})$ and $\mathbf{E}(\mathbf{r} + \mathbf{d}_n)$, must be proportional to each other, that is,

$$\mathbf{E}(\mathbf{r} + \mathbf{d}_n) = C\,\mathbf{E}(\mathbf{r}).$$

The constant of proportionality must depend on the translation vector $\mathbf{d}_n$ and at the same time must not change the amplitude of the field, that is,

$$C = C(\mathbf{d}_n), \quad |C(\mathbf{d}_n)| = 1.$$

Going from point $\mathbf{r}$ to point $\mathbf{r} + \mathbf{d}_n$ can be accomplished by a direct translation by vector $\mathbf{d}_n$ or by two consecutive translations by vectors $\mathbf{d}_m$ and $\mathbf{d}_l$, which satisfy the condition $\mathbf{d}_n = \mathbf{d}_m + \mathbf{d}_l$, that is,

$$\mathbf{E}(\mathbf{r} + \mathbf{d}_l) = C(\mathbf{d}_l)\mathbf{E}(\mathbf{r}),$$

$$\mathbf{E}(\mathbf{r} + \mathbf{d}_l + \mathbf{d}_m) = C(\mathbf{d}_m)\mathbf{E}(\mathbf{r} + \mathbf{d}_l) = C(\mathbf{d}_m)C(\mathbf{d}_l)\mathbf{E}(\mathbf{r}),$$

$$\mathbf{E}(\mathbf{r} + \mathbf{d}_n) = C(\mathbf{d}_m)C(\mathbf{d}_l)\mathbf{E}(\mathbf{r}) = C(\mathbf{d}_n)\mathbf{E}(\mathbf{r}).$$

From the last expressions, we get the coefficients C that satisfy the condition

$$C(\mathbf{d}_n) = C(\mathbf{d}_m)C(\mathbf{d}_l).$$

This property is characteristic for the exponential function $C(\mathbf{d}_n) = \exp(i\,\mathbf{b} \cdot \mathbf{d}_n)$, whose exponential coefficient must be a dimensionless quantity. This implies that vector $\mathbf{b}$ must have the dimension of inverse length (m^{-1}). As $k_B = 2\pi/\lambda$, the translation by vector $\mathbf{d}_n$ leads to the multiplication of function $\mathbf{E}(\mathbf{r})$ by the phase factor $\exp(i\mathbf{k}_B\mathbf{d}_n)$ in a photonic crystal.

PROBLEMS

8.1 A monochromatic plane wave propagates along the z-axis and is incident normally on an opaque screen with a circular aperture of radius r_0. The diffraction pattern is observed at a point P directly behind the aperture, at a distance L from the aperture on a line that is normal to the screen (see Figure 8.3). Determine the number of Fresnel zones, which fit the aperture. Show that the areas of all zones are equal. (*Answer*: The number of Fresnel zones is $m = r_0^2\lambda/L$.)

8.2 A plane wave is incident normally on a circular aperture of radius r_0 in an opaque screen. Determine the changes in the intensity of the light diffracted along the z-axis behind the screen and indicated regions along the z-axis where the Fresnel, Fraunhofer, and geometrical optics diffractions are realized. (*Part of the answer* is shown in Figure 8.17.)

8.3 A green light beam of wavelength $\lambda = 0.55$ µm is incident normally upon a narrow slit of width $b = 40.0$ µm. Determine the angles of the first two bright diffraction fringes that are observed on the screen with respect to the original direction of the light. (*Answer:* $\varphi_1 = \pm 1.20°$, $\varphi_2 = \pm 2.00°$.)

8.4 A beam of light with wavelength of $\lambda = 0.50$ µm is incident normally upon a slit of width $b = 100$ µm. A diffraction pattern is observed on a screen that is parallel to the slit plane. Determine the distance l between the slit and the screen, if the width of the central diffraction maximum d is equal to 1.00 cm. (*Answer:* $l = 100$ cm.)

8.5 A monochromatic light beam of wavelength $\lambda = 0.50$ µm is incident normally upon a diffraction grating. A screen is located parallel to the grating at a distance $L = 1.00$ m from the grating (see Figure 8.18). The distance Δx between the first-order diffraction maxima is equal to 10.0 cm. Determine the period of the grating d and the total number of main diffraction maxima obtained by this grating. (*Answer:* $d = 10^{-5}$ m, $N = 41$.)

8.6 A monochromatic light beam of wavelength of $\lambda = 0.60$ µm is incident normally upon a diffraction grating. Determine the highest-order diffraction maxima, which can be obtained by this grating if its grating period d is equal to 2.00 µm. (*Answer:* $n_{max} = 3$.)

8.7 Determine the distance d between atomic planes in a rock salt crystal if the first-order X-ray diffraction maximum is observed at an angle $\varphi = 15°12'$ to the crystal's surface. The X-rays' wavelength λ is equal to 0.147 nm. (*Answer:* $d = 0.282$ nm.)

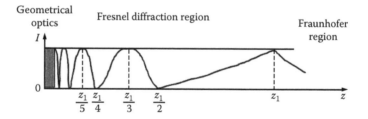

FIGURE 8.17 Dependence of the diffracted light intensity of a wave on the distance from the aperture on the z-axis.

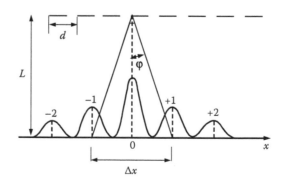

FIGURE 8.18 Schematic of the diffraction pattern of monochromatic light.

8.8 A narrow beam of neutrons that have the same energy is reflected from a natural facet of an aluminum single crystal at a glancing angle $\theta = 5°$ (see Figure 8.12). The distance between the atomic planes parallel of this facet of the single crystal is equal to $d = 0.20$ nm. What are the energy and the velocity of the neutrons that generate a first-order maximum? *Hint*: Use the notion from quantum physics that particles have wave properties. The behavior of such particles is described by their de Broglie wavelength given by

$$\lambda_B = \frac{h}{mv} = \frac{h}{\sqrt{2mE_{kin}}},$$

where $h = 6.62 \times 10^{-34}$ J s is Planck's constant, $m = 1.67 \times 10^{-27}$ kg is the neutron mass, and v is the neutron velocity. (*Answer*: $E_{kin} = 1.08 \times 10^{-19}$ J, $v = 1.13 \times 10^4$ m/s.)

8.9 A monochromatic plane wave is incident upon a 1D periodic structure with a period d, which occupies a region $0 < x < L$. Obtain expressions for the reflectance and transmittance coefficients with the assumption that the phase-matching conditions are satisfied between the forward and backward waves in the medium and that in the inlet boundary layer ($x = 0$) the amplitude of the wave is equal to A_0.

8.10 Using relation (8.55) for a plane periodic structure, obtain an expression for the effective refractive index of the structure for the case $\lambda \gg d$, where λ is the wavelength of an electromagnetic wave and d is the period of the structure. This approach is sometimes called the effective medium approximation. *Hint*: As $d \ll \lambda$, then for the layer thicknesses d_1 and d_2, we have $d_1 \ll \lambda$ and $d_2 \ll \lambda$.

$$\left(Answer: (k_B d)^2 = (k_1 d_1)^2 + (k_2 d_2)^2 + \left(\frac{k_1}{k_2} + \frac{k_2}{k_1} \right)(k_1 d_1)(k_2 d_2), \, k_1, \, k_2 \text{ are wave numbers in each} \right.$$

$$\left. \text{of the media and } n_{eff} = \frac{n_1 d_1 + n_2 d_2}{d_1 + d_2} = \frac{n_1 + n_2 \theta}{1 + \theta}. \right)$$

9 Waves in Guiding Structures

Guided waves in contrast to freely propagating waves in space may exist only in guided structures. The set of guiding elements forms a guided system. Guided systems are also called lines for energy transmission. A structure is considered *guided* if it provides local transmission of electromagnetic field energy in a desired direction. Such structure is also called a *transmission line* or a *waveguide*. Most of super-high-frequency (SHF) elements and units of radio equipment are based on segments of transmission lines. Such structures are used as channels to direct and deliver electromagnetic energy from the source to a device (e.g., from a generator to a transmitting antenna, from a receiving antenna to an amplifier). A rectangular guided structure with constant cross section is named a regular guided structure. If one of these properties is not satisfied (e.g., if the cross section is not rectangular or the cross section is variable), the transmission line is called irregular.

Widely used transmission lines such as cable and optical fibers are also considered to be guided structures. Cables consisting of twisted pairs of copper wires, coaxial cables, and optical fibers are heavily used in computers and computer networks. An *optical fiber* consists of a flexible glass fiber—that is, optical waveguide (its diameter for different applications is in the range of $5 - 100$ μm)—and is used to transmit optical signals. An optical fiber is the fastest transmission line (10 Gb/s and higher), and it has the highest quality of data transmission as the signal is protected from external noise. Further, we will consider several types of guided structures used from SHF range to optical range.

9.1 TYPES OF GUIDING STRUCTURES

For waves of different frequency ranges, there are various types of transmission lines.
The main ones are

- Rectangular and cylindrical hollow waveguides
- Two-wire, stripline, and coaxial transmission lines
- Planar and fiber optical waveguides

The centimeter waves—wavelength from 1 to 10 cm (or frequency from 3 to 30 GHz)—are assigned to the range of ultrahigh frequencies (UHFs). However, in practice, it is accepted to extend the microwave range to wider spectrum part, which includes decimeter, centimeter, and millimeter waves.

The spectrum of electromagnetic wave is given in Figure 9.1, and it is specified what types of transmission lines are reasonable to apply on its different intervals.

Further, we will pay our attention mainly to the study of waves in ideal transmission lines filled with homogeneous medium with constant (independent on coordinates) permittivity ε and permeability μ. Generally, in such lines, waves of the following three types can propagate independently from each other:

- Transverse electromagnetic (TEM) waves, in which the longitudinal components of the electric and magnetic fields are equal to zero and the propagation constant coincides with the wave number in the medium, that is, $\beta = \omega\sqrt{\kappa\varepsilon_0\kappa_m\mu_0} = (\omega/c)\sqrt{\kappa\kappa_m}$
- Transverse electric (TE) waves, in which the electric field is perpendicular to the direction of propagation and the magnetic field has both transverse and longitudinal components
- Transverse magnetic (TM) waves, in which only the electric field vector has a longitudinal component

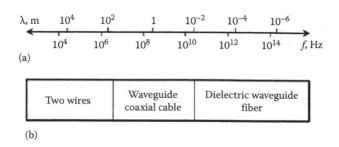

(a)

| Two wires | Waveguide coaxial cable | Dielectric waveguide fiber |

(b)

FIGURE 9.1   The spectrum of electromagnetic waves (a) and types of transmission lines (b) used in different spectral intervals.

In transmission lines with inhomogeneous filling (with parameters ε and μ depending on coordinate), only **hybrid waves** can exist. For these waves, the longitudinal components of both fields are nonzero. In general, transmission lines are systems of parallel metallic or dielectric layers, wires, rods, or tubes. In the radio and microwave ranges of wavelengths (with $\lambda > 1$ mm), homogeneous lines are most widely used. The space between conductors can be empty or filled with a medium.

By the nature of spatial localization of the electromagnetic field, the **transmission lines** are divided into **open** and **closed**. In open lines, there is no external metal shell, which confines the area of the field. The lines, in which the area of nonzero field is confined by an external closed metal shell, are called closed or screened. This shell is a metal pipe that the fields do not penetrate. Examples of such lines are metal rectangular and cylindrical waveguides and coaxial lines—cylindrical waveguide with a metal rod enclosed in it.

Cross sections of applied transmission lines are given in Figure 9.2. The simplest open transmission lines are single-wire, two-wire, and stripline transmission lines. Such lines are used in meter (MW) and even longer wavelength ranges. At shorter wavelengths, the radiation losses significantly increase in two-wire lines. Therefore, in the decimeter range, closed transmission lines are used.

The guiding of energy at ultrahigh frequencies (UHFs) by the usual two-wire transmission lines is almost impossible due to the following:

1. The transverse dimensions of the two-wire transmission lines are comparable with the wavelength. Thus, the lines play the role of antennas—instead of transmitting energy, they emit it in the space.
2. Due to the pronounced skin effect, the ohmic resistance of the line wires in the microwave range is so large that a considerable part of the energy is dissipated by heating the wires.

Striplines are widely used in microelectronics because they reduce the sizes and weight of the transmission lines significantly although the heat and radiation losses somewhat increase.

In the decimeter and centimeter wavelength ranges, the guiding of radio waves is carried out by means of waveguides and coaxial cables. Waveguide is a hollow metal pipe with constant cross section filled with an ideal dielectric or air. Waveguides allow the transmission of high power with low radiation and heat losses. For decimeter and waves of longer wavelengths, it is reasonable to use coaxial transmission lines (coaxial cables). The use of coaxial cable for guiding UHF waves is unprofitable. Though in this case no energy is emitted in the surrounding (as a cable shell at the same time is also a screen), the power losses in the core and dielectric washers (by means of which the core is attached in a cable) are great. In the millimeter wavelength range, dielectric waveguides made from a solid dielectric, such as polystyrene, without metal walls are used. The use of waveguides at longer wavelengths is impractical because of the large transverse dimensions of the waveguides, as the latter are always commensurable with the wavelength.

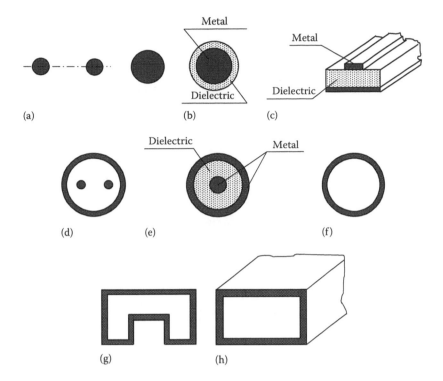

FIGURE 9.2 Cross sections of transmission lines: two-wire line (a), dielectric waveguide (b), stripline (c), screened two-wire line (d), coaxial line (e), single-wire line (f), shaped waveguide (g), and rectangular waveguide (h).

The main requirements for a transmission line are high efficiency, low losses in the conductor and insulator, small reflections, sufficient electric strength for transmitting power, broadband response, absence of noticeable amplitude and phase distortion in the operating frequency range, and absence of radiation into the surroundings. In addition, the line must have small dimensions in combination with simplicity of design and operation and also if it is necessary to possess a sufficient rigidity and vibration resistance.

Exercise 9.1

Determine the energy transferred by a plane monochromatic wave during time T (one period) through an area A, which is perpendicular to the direction of wave propagation. The wave propagates in free space (vacuum). The amplitude of the wave's electric field is equal to E_0 and its angular frequency is ω.

Solution. The power density, that is, the amount of energy transported by a wave during unit time through a unit area that is perpendicular to the direction of the wave propagation, is given by the Poynting vector $\mathbf{S} = \mathbf{E} \times \mathbf{H}$, where $\mathbf{E}$ and $\mathbf{H}$ are the electric field and magnetic field intensity, respectively. Taking into account that $\mathbf{E} \perp \mathbf{H}$, we get for the magnitude of the Poynting vector $S = E \cdot H$. Since the magnitudes of vectors $\mathbf{E}$ and $\mathbf{H}$ at each point of electromagnetic wave vary with time and the two fields have the same phase, the instantaneous value of S is equal to

$$S(t) = E_0 \sin \omega t \cdot H_0 \sin \omega t = E_0 H_0 \sin^2 \omega t.$$

Thus, we have

$$S(t) = \frac{1}{A} \frac{dW}{dt},$$

where dW is the energy transported by the wave through the area A during time dt. From the aforementioned relationships, it follows that

$$dW = SA\, dt = E_0 H_0 \sin^2 \omega t \cdot A\, dt.$$

In order to find dW, it is necessary to know the magnitude H_0, which we can find from the relationship

$$\frac{\kappa \varepsilon_0 E_0^2}{2} = \frac{\kappa_m \mu_0 H_0^2}{2}, \quad H_0 = \sqrt{\frac{\kappa \varepsilon_0}{\kappa_m \mu_0}} E_0.$$

For vacuum $\kappa = \kappa_m = 1$, and therefore, $H_0 = \sqrt{\varepsilon_0/\mu_0}\, E_0$. As a result, we get

$$dW = \sqrt{\frac{\varepsilon_0}{\mu_0}}\, E_0^2 \sin^2 \omega t \cdot A\, dt.$$

The energy transferred by the wave during time T is equal to

$$W = \sqrt{\frac{\varepsilon_0}{\mu_0}} A E_0^2 \int_0^T \sin^2 \omega t \cdot dt = \sqrt{\frac{\varepsilon_0}{\mu_0}} A E_0^2 \left(\frac{T}{2} - \frac{\sin 2\omega T}{4\omega} \right) = \sqrt{\frac{\varepsilon_0}{\mu_0}} A E_0^2 \frac{T}{2}$$

since $\omega T = 2\pi$ and $\sin 4\pi = 0$.

9.2 FIELD STRUCTURE OVER THE CONDUCTING PLANE

In transmission lines, the fields of monochromatic waves are usually represented as the real parts of complex expressions of the forms

$$\mathbf{E}(\mathbf{r}_\perp, z, t) = \mathbf{E}(\mathbf{r}_\perp) \exp[i(\omega t - \beta z)],$$

$$\mathbf{H}(\mathbf{r}_\perp, z, t) = \mathbf{H}(\mathbf{r}_\perp) \exp[i(\omega t - \beta z)],$$

(9.1)

where
   z is the longitudinal coordinate along the transmission line
   $\mathbf{r}_\perp$ is the 2D radius vector in the plane of cross section at $z = \text{const}$
   ω and β are the angular frequency and wave propagation constant

The propagation constant is a real number if there is no absorption of the wave by the transmission line. In this case, waves described by Equations 9.1 satisfy Maxwell's equations and the boundary conditions on the walls of the transmission line. These waves are called *modes of the waveguide*.

Guiding structures are characterized by permittivity $\varepsilon_a = \varepsilon_0 \kappa$ and permeability $\mu_a = \mu_0 \kappa_m$ of its material. Here, $\varepsilon_0 = (1/36\pi) \times 10^{-9}$ F/m and $\mu_0 = 4\pi \times 10^{-7}$ H/m are the electric and magnetic constants, and κ and κ_m are the relative dielectric permittivity and magnetic permeability of the medium. Let us remind that constants ε_0 and μ_0 determine two important parameters: the speed of light in vacuum, $c = 1/\sqrt{\varepsilon_0 \mu_0} = 3 \times 10^8$ m/s, and the impedances of vacuum, $Z_0 = E/H = \sqrt{\mu_0/\varepsilon_0} = 120\,\pi\,\Omega \approx 377\,\Omega$ and $Z_a = \sqrt{\mu_a/\varepsilon_a} = Z_0 \sqrt{\kappa_m/\kappa}$.

To localize the electromagnetic field and to guide waves, metal planes of various configurations are used. Consider the field pattern arising near a highly conducting metal plane when a plane wave is incident on it. When a plane wave is incident on an ideal (not absorbing) metal surface, there is its total reflection. Since the fields of the incident and reflected waves at the same point in space have different phases and different directions, the resultant field has a rather complex structure that differs considerably from the structure of the incident and reflected waves. If a plane homogeneous wave is incident obliquely on the interface, it is totally reflected, and the reflected wave carries the same energy as the incident wave.

Figure 9.3 shows a vector diagram, in which the Poynting vectors of the incident and reflected waves ($\mathbf{S}^i$ and $\mathbf{S}^r$) are decomposed onto normal ($\mathbf{S}_n^i$, $\mathbf{S}_n^r$) and tangential ($\mathbf{S}_t^i$, $\mathbf{S}_t^r$) components relative to the interface. Since the tangential components of the incident and reflected waves are added and the normal components cancel each other, the flow of energy occurs along the interface. Thus, a wave process guided by a metal surface is realized.

Depending on polarization, the electric vector of the incident wave $\mathbf{E}^i$ can be variously oriented relative to a plane of incidence. Let us consider two cases of polarization of the incident wave: (1) vector $\mathbf{E}^i$ is perpendicular to the plane of incidence (s-polarization) and (2) vector $\mathbf{E}^i$ lies in the plane of incidence (p-polarization).

For s-polarization of the incident and reflected waves, the instantaneous picture of the distribution of electric and magnetic field lines near a metal surface is shown in Figure 9.4. The picture of field lines moves along the metal plane from left to right with a wave phase velocity v_{ph}. The electric field $\mathbf{E}$ moves along the interface and possesses only the components, which are transverse in relation to the propagation direction. Field $\mathbf{H}$ possesses a longitudinal component ($H_z \neq 0$, $E_z = 0$). Such wave is called the **wave of magnetic type** and is designated as "TE" or "H."

Figure 9.5 shows a diagram of the intersection of the fronts of the incident and reflected waves. If we choose the section of the front of the incident plane wave, which is equal to the wavelength, from the drawing, it is possible to find the characteristic sizes determining the height and width of the contours of the closed field lines:

$$l = \frac{\lambda}{2\cos\theta_i}, \quad d = \frac{\lambda}{2\sin\theta_i}. \tag{9.2}$$

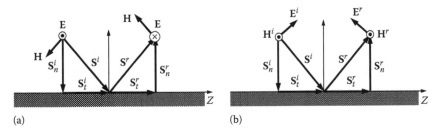

(a)   (b)

FIGURE 9.3 Energy flows near a metal surface: vector $\mathbf{E}^i$ is perpendicular to the plane of incidence (s-polarization) (a) and vector $\mathbf{E}^i$ lies in the plane of incidence (p-polarization) (b).

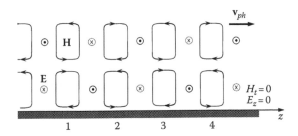

FIGURE 9.4 Character of magnetic field intensity and electric field lines near a metal surface with s-polarization of wave.

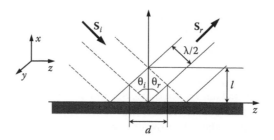

FIGURE 9.5 Crossing of the fronts of incident and reflected waves.

Thus, the size of contours of the electric and magnetic field lines is determined by the wavelength λ and the angle incidence θ_i of the plane wave with the metal plane.

Figure 9.6 shows the character of the electric field and magnetic field intensity lines near a metal surface for p-polarization of the wave. In this case, the vector **E** lies in a plane of incidence and the vector **H** is perpendicular to this plane. Thus, the boundary condition $E_t = 0$ has to be satisfied.

The wave field moves along the interface and has only a component, which is transverse in relation to the propagation direction of the magnetic field **H**. The electric field **E** has both transverse and longitudinal components coinciding in direction with the vector of phase velocity ($H_z = 0$, $E_z \neq 0$). Such wave is called *wave of the electric type* and designate either "TM" or "E."

We note that the field in the direction of the normal to the interface has the structure of a standing wave with nodes and antinodes for the electric and magnetic fields. Therefore, transfer of the field energy in this direction is absent and takes place only along the boundary.

Exercise 9.2

A plane monochromatic wave with s-polarization is incident at an angle θ on the plane surface of a metal with conductivity σ. Find the specific power loss, that is, the part of the wave's energy that is absorbed by the metal per unit surface.

Solution. To find the absorbed energy, it is necessary to calculate the average value of the Poynting vector directed into the metal. If the electric and magnetic fields at the metal surface are known, then

$$< \mathbf{S} > = \frac{1}{2} \mathrm{Re}(\mathbf{E}_t \times \mathbf{H}_t^*),$$

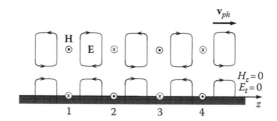

FIGURE 9.6 Character of magnetic field intensity and electric field lines near metal surface for p-polarization of wave.

where $\mathbf{E}_t$ and $\mathbf{H}_t$ are tangential components of the fields at the metal's surface. Let us use Leontovich boundary condition (7.24)

$$\mathbf{E}_t = Z_m(\mathbf{H}_t \times \mathbf{n}_0),$$

where the metal complex impedance has the form

$$Z_m = \sqrt{\frac{\kappa_m \mu_0}{\tilde{\kappa}\varepsilon_0}} = \sqrt{-\frac{\kappa_m \mu_0 \omega}{i\sigma}} = (1+i)\sqrt{\frac{\kappa_m \mu_0 \omega}{2\sigma}},$$

and $\mathbf{n}_0$ is a unit vector normal to the metal surface (directed toward the metal). After substitution of these relationships, we get

$$<\mathbf{S}> = \frac{1}{2}\operatorname{Re}\left(Z_m(\mathbf{H}_t \times \mathbf{n}_0) \times \mathbf{H}_t^*\right) = \frac{\mathbf{n}_0}{2}|\mathbf{H}_t|^2 \operatorname{Re} Z_m.$$

Since $\operatorname{Re} Z_m = \sqrt{\kappa_m \mu_0 \omega / 2\sigma}$, then for specific power losses, we get

$$P = \frac{1}{2}\sqrt{\frac{\kappa_m \mu_0 \omega}{2\sigma}}|\mathbf{H}_t|^2.$$

As $H_t = H_0 \cos\theta$, where H_0 is the amplitude of the incident wave, then

$$P = \sqrt{\frac{\kappa_m \mu_0 \omega}{8\sigma}}|\mathbf{H}_0|^2 \cos^2\theta.$$

For a nonmagnetic metal, the magnetic permeability κ_m is equal to unity. For an ideal metal, $\sigma \to \infty$, so the power losses in the case of ideal metal become zero: $P \to 0$.

9.3 FIELD BETWEEN TWO PARALLEL METAL PLANES

Propagation of electromagnetic energy is also possible between two parallel metal planes. Now, we explore the conditions under which such propagation occurs. Let us place above the first metal plane a second ideal metal plane so as not to disturb the existing field pattern. For this purpose, the second plane should be placed at a distance $a = nl$, where $n = 1, 2, 3, \ldots$. Thus, the field pattern between the two planes is the same as in the case of a single plane. The wave field $\mathbf{E}$ will be oriented normally to the metal planes and the field $\mathbf{H}$ tangentially.

A wave of this kind possesses only one variation of a field in the direction perpendicular to the planes. In the case of s-polarization and $n = 1$, the wave is called a wave H_1, and in the case of p-polarization, a wave E_1. The picture of field lines of these waves is shown in Figure 9.7.

Equation 9.2 related the distances between the metal planes and the angle of incidence θ_i:

$$a = nl = \frac{n\lambda}{2\cos\theta_i}. \tag{9.3}$$

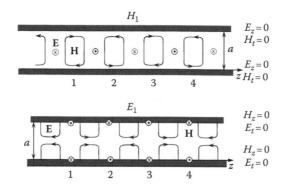

FIGURE 9.7   Field picture for waves of types H_1 and E_1.

For the existence of a higher type of wave, for example, with $n = 2$ at a fixed angle θ_i, it is necessary to increase the distance between the planes by a factor of two. The minimum distance, a, at which a wave with $n = 1$ is possible, is equal to $\lambda/2$. Thus, a uniqueness condition of a wave with $n = 1$ is $\lambda/2 < a < \lambda$. Waves with larger indices cannot propagate in this case between two planes. For $\lambda < a < 3\lambda/2$, between the planes, there is a wave with $n = 2$, and the wave with $n = 3$ does not exist. However, at the same time, a wave with $n = 1$, which is called the ***fundamental wave***, can exist. By suitable choice of the distance between the planes, it is possible to provide the conditions for a single-wave distribution mode for a wave of the fundamental type.

    The field in a two-plane waveguide can be considered as the result of the addition of plane uniform waves called ***partial waves*** reflected multiply from its boundary surfaces. The propagation constant, phase, and group velocities and the wavelength of such waveguide (as well as for the free space) are related to each other: $\beta = \omega\sqrt{\varepsilon_a\mu_a} = \omega\sqrt{\varepsilon_0\mu_0\kappa\kappa_m} = (\omega/c)\sqrt{\kappa\kappa_m} = \omega/v_0$.

    For an incident wave, that is, a wave in free space with propagation direction along the axis z_1, this relationship has the following form:

$$v_{ph} = \frac{\omega}{\beta} = v_0, \quad v_{gr} = \frac{\partial\omega}{\partial\beta} = v_0, \quad \lambda = \frac{2\pi}{\beta} = \frac{v_0}{f} = \frac{2\pi v_0}{\omega}, \tag{9.4}$$

where $v_0 = c/\sqrt{\kappa\kappa_m}$ is the propagation velocity of the wave in the medium (Figure 9.8).

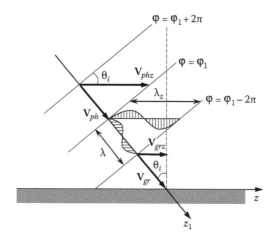

FIGURE 9.8   Phase and group velocities and a wavelength.

In a waveguide, the parameters of a plane wave propagating along the waveguide axis (i.e., along the z-axis) are given by

$$\beta_z = \beta \sin \theta_i, \quad v_{phz} = \frac{v_0}{\sin \theta_i}, \quad v_{grz} = v_0 \sin \theta_i, \quad \lambda_z = \frac{\lambda}{\sin \theta_i} = \frac{v_{phz}}{f}. \tag{9.5}$$

From this, it follows that the wavelength in the waveguide $\lambda_w = \lambda_z$ is greater than λ in free space for the same oscillation frequency and $v_{ph} > v_0$, $v_{gr} < v_0$.

Let us write the condition of wave propagation in two-plane waveguide. At constant values of the parameters a and n (n is the number of half-waves that fit the size of a), the angle of incidence of the partial wave depends on λ:

$$\cos \theta_i = \frac{\omega_{cr}}{\omega} = \frac{\lambda}{\lambda_{cr}} = \frac{n\lambda}{2a}. \tag{9.6}$$

If $\lambda \ll \lambda_{cr} = 2a/n$, then the angle θ_i is close to 90° and the partial waves upon reflection move along the waveguide walls (Figure 9.9). With increasing λ, the angle θ_i decreases and becomes equal to zero, that is, the wave propagation stops. Thus, the transverse dimension of the waveguide restricts the wavelength range of waves that can propagate in it by the following relation:

$$\lambda \leq \lambda_{cr} = \frac{2a}{n}. \tag{9.7}$$

The upper limit of the range in which a wave of a given type can propagate in a waveguide is called the ***critical wavelength*** λ_{cr}. In a waveguide, propagation is possible only for waves with wavelength $\lambda < \lambda_{cr}$, frequency $\omega > \omega_{cr}$, and angle of incidence

$$\sin \theta_i < \sqrt{1 - \left(\frac{\omega_{cr}}{\omega}\right)^2} = \sqrt{1 - \left(\frac{\lambda}{\lambda_{cr}}\right)^2}. \tag{9.8}$$

Note that at $\lambda = \lambda_{cr}$ the group velocity of the wave in the waveguide tends to zero. This shows that in a waveguide a wave with $\lambda > \lambda_{cr}$ does not propagate.

Exercise 9.3

A wave of mode H_1 with frequency $f = 6.00$ GHz is propagating in a double-plate metal waveguide with the distance between plates, a, equal to 3.00 cm. The waveguide is filled by a nonabsorbing dielectric with relative dielectric permittivity $\kappa = 2.5$. Find the phase and group velocities, the critical wavelength, and the wavelength of the wave in the waveguide.

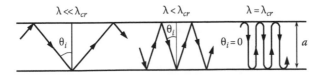

FIGURE 9.9 Beam picture of process of wave propagation in a two-plane waveguide for different wavelengths.

Solution. The wavelength of the wave in the space filled by dielectric is equal to

$$\lambda = \frac{c}{f\sqrt{\kappa}} = 3.16 \text{ cm}.$$

For the given mode of the wave ($n = 1$), the critical wavelength in the structure is equal to

$$\lambda_{cr} = \frac{2a}{n} = 2a = 6.00 \text{ cm}.$$

The propagation angle of the wave in the waveguide is equal to

$$\cos\theta = \frac{\lambda}{\lambda_{cr}} \approx 0.53, \quad \sin\theta = \sqrt{1 - \cos^2\theta} = 0.85, \quad \theta = 58°.$$

Thus, the wavelength of the wave in the waveguide is equal to

$$\lambda_z = \frac{\lambda}{\sin\theta} = \frac{c}{f\sqrt{\kappa}\sin\theta} = \frac{3.16}{0.85} = 3.72 \text{ cm}.$$

The phase and group velocities of the wave in the waveguide are equal to

$$v_{phz} = \frac{c}{\sin\theta} \approx 3.53 \times 10^8 \text{ m/s}, \quad v_{grz} = c \times \sin\theta \approx 2.55 \times 10^8 \text{ m/s}.$$

9.4 FIELDS IN A RECTANGULAR WAVEGUIDE

Theoretical consideration of electromagnetic wave propagation in a waveguide has to be based on the solution of Maxwell's equations. It is necessary to find such solution of these equations, which will satisfy boundary conditions at the waveguide walls and describe the propagation of electromagnetic waves along the waveguide axis.

We assume that the waveguide walls are made of an ideal metal (i.e., conductivity $\sigma \to \infty$). Under this condition, the electric field at the waveguide walls should be equal to zero (to have finite current density $\mathbf{j} = \sigma\mathbf{E}$ on the walls, it is necessary to require $\mathbf{E} \to 0$). The waveguide cavity can be filled with a homogeneous dielectric (with permittivity ε_a).

The sizes of a rectangular waveguide cavity along x- and y-directions are equal to a and b, respectively. The waveguide length in z-direction is infinite (Figure 9.10).

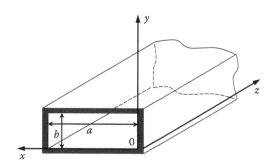

FIGURE 9.10   Rectangular metal waveguide.

Electromagnetic waves propagating in this waveguide are running along the z-axis of the wave-guide waves and standing in the other two transverse directions. Standing waves are formed due to multiple reflections of the waves from the waveguide walls.

The fields for a monochromatic wave field in a waveguide, Equations 9.1, in this case can be written in the following form:

$$\mathbf{E}(x,y,z,t) = \mathbf{E}(x,y)\exp[i(\omega t - \beta z)],$$

$$\mathbf{H}(x,y,z,t) = \mathbf{H}(x,y)\exp[i(\omega t - \beta z)],$$

(9.9)

where ω and β are the angular frequency and propagation constant of the natural wave, respectively.

In rectangular waveguides, an infinite number of type H and type E waves, which differ from each other by the field structure, can exist. The waves are described by two integer indices: m and n. To each combination of these numbers, there corresponds a particular structure of waves of type H_{mn} and E_{mn}. Indices m and n specify the number of half-waves that fits along x- and y-axes, respectively. Usually, the x-axis is directed along the longer side of a waveguide.

For a type H_{mn} wave, the solution of Maxwell's equations together with the boundary conditions mentioned earlier leads to the following coordinate dependence of the components of the electric field and magnetic fields intensity:

$$E_x = i\omega\mu_a \frac{\pi n}{g^2 b} H_0 \cos\left(\frac{\pi m x}{a}\right)\sin\left(\frac{\pi n y}{b}\right),$$

$$E_y = -i\omega\mu_a \frac{\pi m}{g^2 a} H_0 \sin\left(\frac{\pi m x}{a}\right)\cos\left(\frac{\pi n y}{b}\right), \quad E_z = 0,$$

$$H_x = i\frac{\beta\pi m}{g^2 a} H_0 \sin\left(\frac{\pi m x}{a}\right)\cos\left(\frac{\pi n y}{b}\right),$$

(9.10)

$$H_y = i\frac{\beta\pi n}{g^2 b} H_0 \cos\left(\frac{\pi m x}{a}\right)\sin\left(\frac{\pi n y}{b}\right),$$

$$H_z = H_0 \cos\left(\frac{\pi m x}{a}\right)\cos\left(\frac{\pi n y}{b}\right),$$

where $g = \sqrt{k_0^2 \kappa\kappa_m - \beta^2}$ is the transverse wave number. For the corresponding components of a wave of type E_{mn}, we get

$$E_x = -i\frac{\beta\pi m}{g^2 a} E_0 \cos\left(\frac{\pi m x}{a}\right)\sin\left(\frac{\pi n y}{b}\right),$$

$$E_y = -i\frac{\beta\pi n}{g^2 b} E_0 \sin\left(\frac{\pi m x}{a}\right)\cos\left(\frac{\pi n y}{b}\right),$$

$$E_z = E_0 \sin\left(\frac{\pi m x}{a}\right)\sin\left(\frac{\pi n y}{b}\right),$$

(9.11)

$$H_x = i\omega\varepsilon_a \frac{\pi m}{g^2 b} E_0 \sin\left(\frac{\pi m x}{a}\right)\cos\left(\frac{\pi n y}{b}\right),$$

$$H_y = -i\omega\varepsilon_a \frac{\pi n}{g^2 a} E_0 \cos\left(\frac{\pi m x}{a}\right)\sin\left(\frac{\pi n y}{b}\right), \quad H_z = 0.$$

For each set of indices m and n, a quantity called the critical frequency is determined from the dimensions of a waveguide and the properties of the dielectric filling the waveguide. This frequency is given by the equation

$$\omega_{cr} = \pi v_0 \sqrt{\left(\frac{m}{a}\right)^2 + \left(\frac{n}{b}\right)^2}, \tag{9.12}$$

where $v_0 = 1/\sqrt{\varepsilon_a \mu_a} = c/\sqrt{\kappa \kappa_m}$ is the velocity of the wave in a medium with parameters ε_a and μ_a. For each critical frequency, it is possible to calculate a critical wavelength corresponding to it:

$$\lambda_{cr} = \frac{2\pi v_0}{\omega_{cr}} = \frac{2}{\sqrt{(m/a)^2 + (n/b)^2}}. \tag{9.13}$$

We note that the critical wavelength does not depend on dielectric parameters and depends only on the type of a wave and the waveguide sizes. If $\omega > \omega_{cr}$ or $\lambda < \lambda_{cr}$, types E_{mn} and H_{mn} waves with indices m and n can propagate in the line.

The critical frequency (or critical wavelength) is involved in all expressions for the wave parameters in a waveguide. Thus, the transverse wave number and the distribution constant β_{mn} are given by

$$g = \frac{2\pi}{\lambda_{cr}}$$

$$\beta_{mn} = k\sqrt{1 - \left(\frac{\omega_{cr}}{\omega}\right)^2} = k\sqrt{1 - \left(\frac{\lambda}{\lambda_{cr}}\right)^2}. \tag{9.14}$$

The latter is always smaller than the wave number of the dielectric in the waveguide, $k = 2\pi/\lambda = \omega\sqrt{\varepsilon_a\mu_a}$. Wavelength λ_{in} in a waveguide is larger than the wavelength in the dielectric, $\lambda = 2\pi v_0/\omega$. The former is given by the expression

$$\lambda_{in} = \frac{\lambda}{\sqrt{1 - (\omega_{cr}/\omega)^2}} = \frac{\lambda}{\sqrt{1 - (\lambda/\lambda_{cr})^2}}. \tag{9.15}$$

The phase velocity of a wave in a waveguide is given by the expression

$$v_{phz} = \frac{\omega}{\beta_{mn}} = \frac{v_0}{\sqrt{1 - (\omega_{cr}/\omega)^2}}, \tag{9.16}$$

from which it follows that $v_{phz} > v_0$. Therefore, in a waveguide without filling, the phase velocity is higher than the speed of light in vacuum ($v_{phz} > c$). We note that it is the group velocity v_{grz} in a waveguide that determines the rate of energy transfer and v_{grz} is smaller than the speed of light:

$$v_{grz} = \frac{c^2}{v_{phz}} = c\sqrt{1 - \left(\frac{\lambda}{\lambda_{cr}}\right)^2}. \tag{9.17}$$

H_{10} waves are used in microwave application more often than any other wave. This is the reason why the H_{10} wave is called the *fundamental wave*. This wave has the largest value of the critical wavelength, which is given by Equation 9.13, if we substitute the values of the indices $m = 1$, $n = 0$ we get $\lambda_{cr}^{10} = 2a$.

For H_{01} wave, the critical wavelength is $\lambda_{cr}^{01} = 2b$. For waves H_{20} and H_{02}, the critical wavelengths are $\lambda_{cr}^{20} = a$ and $\lambda_{cr}^{02} = b$, respectively. The critical wavelengths for certain types of waves can coincide. Such waves are called degenerate. Critical lengths are identical for H_{11} and E_{11} waves:

$$\lambda_{cr}^{11} = 2ab/\sqrt{a^2 + b^2}.$$

The fundamental wave of a rectangular waveguide H_{10} contains the following nonzero components of electric and magnetic fields:

$$E_y = -i\frac{\omega\mu_a a}{\pi} H_0 \sin\left(\frac{\pi x}{a}\right),$$

$$H_x = i\frac{\beta a}{\pi} H_0 \sin\left(\frac{\pi x}{a}\right), \tag{9.18}$$

$$H_z = H_0 \cos\left(\frac{\pi x}{a}\right).$$

Exercise 9.4

What is the length l of a closed resonator with rectangular cross section area $a \times b = 6 \times 3$ cm^2 if it is known that at the resonance frequency $f_{res} = 6.00$ GHz, only three standing waves can fit along the waveguide's axis. Find the wavelength of the standing wave in the resonator.

Solution. Consider a segment of the rectangular waveguide with cross section $a \times b$, bounded by two metal end surfaces at $z = 0$ and $z = l$ (Figure 9.11).

Such a closed metallic structure is called a rectangular cavity resonator. In the following, we study one of the modes of natural oscillations in this resonator. In the cavity of the resonator, a standing wave is formed along the z-axis. The standing wave is a sum of forward and backward propagating waves due to reflections at the ends of the cavity. Assume that we have a mode H_{10} wave which has a y-component of the electric field vector with complex amplitude:

$$E_y^+(x,z) = E_{max} \sin\left(\frac{\pi x}{a}\right)\exp(-i\beta z),$$

$$E_y^-(x,z) = GE_{max} \sin\left(\frac{\pi x}{a}\right)\exp(i\beta z),$$

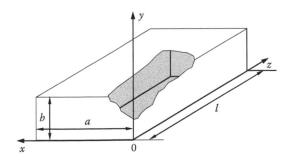

FIGURE 9.11   Rectangular resonator.

where G is an unknown amplitude coefficient. At $z = 0$, the total electric field $E_y = E_y^+ + E_y^-$ must be equal to zero (because of the boundary condition for an ideal conductor). Therefore, the boundary condition at $z = 0$ requires that $G = -1$. Using Euler's equation for the sum of two exponential functions with imaginary exponents, we get

$$E_y(x,z) = E_{max} \sin\left(\frac{\pi x}{a}\right)(e^{-i\beta z} - e^{i\beta z}) = -2iE_{max} \sin\left(\frac{\pi x}{a}\right)\sin(\beta z).$$

This expression describes a standing wave along the x- and z-axes (along the y-axis, the electric field does not change). The wavelength of the standing wave along z-axis can be determined from the longitudinal wave number β. At the end surface of the resonator $z = l$, we have the boundary condition $E_y = 0$, from which it follows that

$$\beta l = p\pi,$$

where p is a positive integer, known as "longitudinal index." If $p = 1$, half of the wavelength fits into the resonator. Three standing waves fit along the waveguide's axis if $p = 6$. The value of longitudinal wave number β, which satisfies the obtained equality, is called resonant wave number:

$$\beta_{res} = \frac{p\pi}{l}.$$

In general, in a closed rectangular resonator, certain oscillation modes of the electromagnetic field can form standing waves, which are denoted as H_{mnp} or E_{mnp}. The resonant frequencies of such oscillations are given by the expression

$$\omega_{res} = 2\pi f_{res} = c\sqrt{\left(\frac{m\pi}{a}\right)^2 + \left(\frac{n\pi}{b}\right)^2 + \left(\frac{p\pi}{l}\right)^2},$$

where a, b, l are the dimensions of the resonator. Here, we assumed that the relative permittivity and permeability are both equal to unity for the empty cavity of the resonator. The main oscillation modes having the lowest resonant frequencies are H_{101}, H_{011}, and E_{110}. At $b < a$ and $b < l$, the main mode of oscillations is H_{101}.

Using the relationship for ω_{res}, we can calculate the value of the resonant wavelength in an empty resonator λ_{0res}:

$$\lambda_{0res} = \frac{2\pi c}{\omega_{res}} = \frac{c}{f_{res}} = \frac{2}{\sqrt{(1/a)^2 + (p/l)^2}} = 5.00 \text{ cm.}$$

From this expression, we find the resonator length

$$l = \frac{p}{\sqrt{(f_{res}/c)^2 - (1/a)^2}} = \frac{3}{\sqrt{(1/5)^2 - (1/6)^2}} \approx 27.3 \text{ cm.}$$

For a resonator length l, we get the following value of the resonant wavelength inside the resonator:

$$\lambda_{res} = \frac{2\pi}{\beta_{res}} = \frac{2l}{p} \approx 18.3 \text{ cm.}$$

From this analysis, we conclude that by filling the cavity with a dielectric and for a given generator frequency, the dimensions of the cavity can be made smaller. This property may be used to decrease the size and mass of radio equipment operating in SHF range.

9.5   WAVEGUIDE OPERATING CONDITIONS

The higher the frequency, the more the types of waves can propagate simultaneously in a waveguide. There are two **waveguide operating conditions** on the number of simultaneously propagating waves (modes) of different types—**single mode** and **multimode**. In the former case, only one type of wave propagates in the waveguide and in the second several types of waves can propagate simultaneously. At single-mode operation, all power is transferred by a wave H_{10} and at multimode operation by all propagating types of waves. Usually, devices coupled to a waveguide at its output are designed for a particular type of wave. Therefore, part of the power is lost. In practice, the single-mode regime with a H_{10} wave type is used as a rule.

Three regimes of propagation are possible in a waveguide in relation to each type of a wave H_{mn} and E_{mn}: **subcritical**, **critical**, and **supercritical**. For a given type of a wave, the ratio of the wave frequency and the critical frequency determines what regime is realized. The main regime of propagation is the **subcritical regime** that is realized at $\omega > \omega_{cr}$. In a subcritical operating regime of a waveguide, the distribution constant β is a real number:

$$\beta^2 = k^2 \left(1 - \frac{\omega_{cr}^2}{\omega^2} \right) > 0. \tag{9.19}$$

The propagation constant in a waveguide determines the phase shift of the wave per unit of the waveguide length. The presence of the phase shift suggests that the electromagnetic wave propagates through the waveguide. For fixed x- and y-coordinates, the time-average field is constant along the waveguide.

In the subcritical regime, the transverse field components $E_\perp$ and $H_\perp$ (in a rectangular waveguide, they are components E_x, E_y, H_x, H_y) are in phase (the phase of each of fields is $\varphi_{E,H} = \omega t - \beta z$). Therefore, the average over a period of the Poynting vector is different from zero and is maximum compared to other waveguides.

If a matched load is included in the line at the end of the waveguide, the traveling wave regime is realized, that is, in the waveguide the subcritical mode propagation of electromagnetic wave energy is possible.

Figure 9.12a shows the structure of the wave field H_{10} in a rectangular waveguide in the subcritical regime. It can be seen that the transverse fields $E_\perp$ and $H_\perp$ are mutually perpendicular and the location of the maxima of these components coincides along the waveguide. Instants of the times at which these fields reach their maximum values also coincide.

If for this type of a wave $\omega = \omega_{cr}$, the regime is called **critical**. In this regime, the propagation constant β in the waveguide is equal to zero. This means that there is no phase shift along the waveguide length. Lack of a phase shift shows that the electromagnetic wave does not propagate through the waveguide. In this case, all components of the field do not have a dependence on the z-coordinate along the direction of wave propagation. At all points, both electric and magnetic fields have an identical phase, independent of the x-, y-, and z- coordinates. In other words, in this regime, the phase and amplitude of a field do not depend on the longitudinal coordinate z.

When $\beta = 0$, the phases of the transverse field components $E_\perp$ and $H_\perp$ are mutually shifted by $\pi/2$, that is, $\varphi_E - \varphi_H = \pi/2$.

From this, it follows that the field has a **reactive** character, and the transmission of energy along the waveguide is absent: the Poynting vector and velocity of field energy movement are equal to zero.

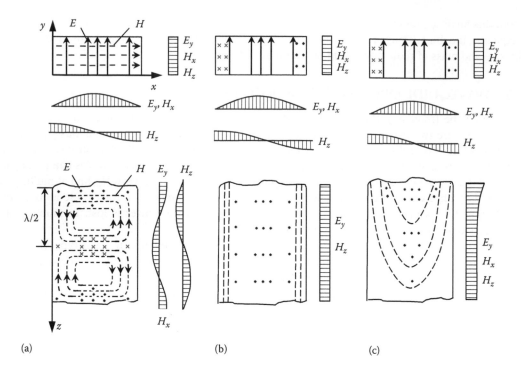

FIGURE 9.12 The structure of the wave field H_{10} in a rectangular waveguide in three regimes: subcritical $(\omega > \omega_{cr})$ (a), critical $(\omega = \omega_{cr})$ (b), and supercritical $(\omega < \omega_{cr})$ (c).

The magnetic field intensity lines are drawn along the walls of the waveguide (Figure 9.12b) and conduction currents uniformly flow through the walls. The presence in the wall of these currents results in large thermal losses.

If for the given type of a wave in a waveguide $\omega < \omega_{cr}$, the regime is called **_supercritical_**. For this regime, the propagation constant is purely imaginary, that is, $\beta^2 < 0$ and $\beta = -i\beta''$. Thus, the magnetic and electric fields in all points of the waveguide have constant phases φ_H and φ_E, differing by $\pi/2$. This means that the electromagnetic field in the waveguide in this regime is reactive, and thus, there is no transmission of energy along the waveguide.

For $\omega < \omega_{cr}$, in contrast to the critical regime, the field amplitude along the waveguide is not constant and decreases exponentially. Therefore, the electric field of the wave is given by

$$E(x, y, z, t) = E(x, y) \exp(-\beta'' z) \exp(i\omega t). \tag{9.20}$$

The field decreases even in the absence of losses in the walls and the dielectric. The decrease in field is caused by interference effects, instead of dissipation. The field pattern for a wave H_{10} in the supercritical regime is shown in Figure 9.12c.

Exercise 9.5

A rectangular waveguide is filled with air and has cross-sectional area $a \times b = 10 \times 5$ cm^2. Identify all wave modes that can exist in the waveguide at a frequency of 5 GHz. Find the critical wavelength of the waveguide and the phase and group velocities for the fundamental wave and for the wave with the highest values of m and n.

Solution. Let us find wavelength in free space:

$$\lambda = \frac{c}{f} = 6 \text{ cm.}$$

The critical wavelength in the waveguide can be found using the equation

$$\lambda_{cr} = \frac{2}{\sqrt{(m/a)^2 + (n/b)^2}} = \frac{2}{\sqrt{(m/10)^2 + (n/5)^2}} \text{ cm.}$$

Inside the waveguide only the wave that satisfies the following conditions can propagate:

$$\lambda < \lambda_{cr}.$$

For different integers m and n, we can find the wave modes. Thus, for $m = 1$ and $n = 0$, we get $\lambda_{cr,10} = 20$ cm, that is, $\lambda < \lambda_{cr}$, and a TE_{10} mode wave may exist in the waveguide for which the inequality $\lambda < \lambda_{cr,10}$ is valid. Analogously, we can find that waves of modes TE_{01}, TE_{20}, TE_{30}, TE_{11}, and TM_{11} may propagate. Other modes of waves cannot propagate in the waveguide. Indeed, let $m = 2$ and $n = 1$. Then, we get

$$\lambda_{cr,21} = 2.8 \text{ cm} < \lambda = 6 \text{ cm,}$$

that is, the required inequality is not satisfied.

We will now determine the wavelength of the wave in the waveguide and the corresponding phase and group velocities for mode TE_{10} (or H_{10}):

$$\lambda_{in} = \frac{\lambda}{\sqrt{1 - (\lambda/\lambda_{cr,10})^2}} = 6.3 \text{ cm,}$$

$$v_{phz} = \lambda_{in} \times f = 3.15 \times 10^8 \text{ m/s,} \quad v_{grz} = \frac{c^2}{v_{phz}} = 2.86 \times 10^8 \text{ m/s.}$$

For the wave of mode TE_{30}, we find $\lambda_{cr} = 6.7$ cm, $\lambda_{in} = 13.4$ cm, $v_{phz} = 6.7 \times 10^8$ m/s, and $v_{grz} = 1.35 \times 10^8$ m/s.

9.6 DAMPING OF WAVES IN WAVEGUIDES

In real waveguide structures, the propagation of waves is always accompanied by damping caused by losses in the metallic walls that have a finite conductivity σ as well as by losses in the dielectric material inside the waveguide. Here, we consider the last one as it is easy to take into account the absorption of energy in a homogeneous dielectric medium using the real and imaginary parts of the longitudinal wave number $\tilde{\beta} = \beta' - i\beta''$ for a medium with complex electrical permittivity $\kappa = \kappa' + i\kappa''$ and magnetic permeability $\kappa_m = \kappa'_m + i\kappa''_m$. In an absorbing medium the sign before the exponential that contains parameter β'' must always be negative which leads to attenuation of the wave in the propagation direction. The dependence of the wave's electric field on z and time can be written in the form

$$\mathbf{E}(z,t) = \mathbf{E}_0 \exp(-\beta'' z) \exp[i(\omega t - \beta' z)]. \tag{9.21}$$

Very often in order to write electric fields in different absorbing structures, the attenuation coefficient $\alpha = \beta''$ and the real constant of propagation $\beta = \beta'$ are introduced, that is,

$$\mathbf{E} = \mathbf{E}_0 \exp(-\alpha z)\exp[i(\omega t - \beta z)]. \tag{9.22}$$

Since only the real part of a parameter has physical meaning, the following equation derived from Equation 9.22 should be used:

$$\mathrm{Re}\,\mathbf{E} = |\mathbf{E}_0|\exp(-\alpha z)\cos(\omega t - \beta z + \varphi). \tag{9.23}$$

Here, the phase $(\omega t - \beta z + \varphi)$ describes oscillations of the field in time and space, which is a characteristic of a monochromatic wave with real longitudinal wave number β. The exponential factor gives the decrease in amplitude of the wave in the direction of the wave propagation. Usually, the parameter α is used as a measure of the damping that is called attenuation constant or damping coefficient. Sometimes, the inverse of α is used instead, $l_\alpha = 1/\alpha$, which is called the damping length, and it is equal to the distance for which the wave amplitude is attenuated by a factor equal to e.

In the presence of losses in the walls and in the dielectric in the subcritical regime, when the Poynting vector, averaged over a period, is not zero, the field amplitude decreases along the waveguide exponentially, that is, proportionally $\exp(-\alpha z)$ (Figure 9.13).

The rate of decrease of the field is determined by the losses, that is, by the attenuation coefficient α; the greater the coefficient α, the faster the field amplitude decreases.

For waves of various types, attenuation coefficient due to losses in the conductors is not identical as it depends on the form of the lines of surface current density on the walls of the waveguide, which is different for various types of waves. To reduce losses in the walls, nonmagnetic materials with high conductivity should be used, such as brass, copper, and aluminum, and also the walls should be covered with materials that are good conductors and do not corrode.

The thickness of these cover layers has to be equal to several penetration depths and, therefore, has to depend on the operating frequency. To avoid losses in waveguides, it is necessary that the surfaces of the internal walls are to be well polished, free from dust, and not be covered with films of poorly conductive oxides.

For given values of the imaginary parts of κ and κ_m, the parameters α and l_α depend on the deviation of the wave frequency from the critical value. The damping coefficient abruptly increases as ω approaches ω_{cr}. For $\omega \gg \omega_{cr}$, the ratio of the imaginary and the real parts of the propagation constant is proportional to the parameter known as loss tangent, that is, $\tan\delta = 2\alpha/\beta$. The energy losses in the metallic walls of the waveguide can be taken into account using Leontovich boundary conditions

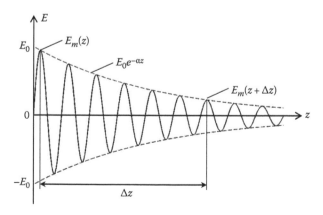

FIGURE 9.13 Change of amplitude of damped oscillations along a direction of wave propagation in the presence of losses.

that establish the relationship between the tangential components of the fields at the boundary of the conductor when $\sigma \gg \varepsilon_0 \kappa_s \omega$ and the electric field that is concentrated practically on the surface of the conductor:

$$\mathbf{E}_t = Z_s[\mathbf{H} \times \mathbf{n}], \quad Z_s = Z_0 \sqrt{\frac{\kappa_{ms}}{\kappa_s}}, \tag{9.24}$$

where

  $\mathbf{n}$ is the inward-pointing normal to the boundary with conductor
  $Z_0 = \sqrt{\mu_0/\varepsilon_0}$ is the wave impedance of vacuum
  Z_s is the surface impedance of the conductor that is determined by its complex dielectric constant
    κ_s and relative magnetic permeability κ_{ms}

In conductors that are typically used in transmission lines (in radio and SHF ranges), $\kappa_{ms} \approx 1$ and $\kappa_s \approx -i\sigma/\omega\varepsilon_0$. For a wide range of frequencies in metals, we have $|\kappa_s| \gg 1$; this leads to small impedance of metals compared to the impedance of vacuum. Relation (9.24) between the electric and the magnetic fields allows us to write an expression for the linear power losses in conductors (the average energy flux onto the wall per unit length of line) in the following form:

$$P_s = \frac{1}{2} \operatorname{Re} Z_s \oint |\mathbf{H}|^2 dl, \tag{9.25}$$

where the integral is taken over the boundary contour of cross section of the line. It follows from energy conservation that the power loss P_s and the rate of change of wave's energy flux are related as follows:

$$\frac{dP_w}{dz} = -P_s. \tag{9.26}$$

If we take into account the quadratic dependence of P_w on the field amplitudes and the exponential dependence of these amplitudes on coordinate z, we get

$$P_w = P_0 \exp(-2\alpha z), \quad \frac{dP_w}{dz} = -2\alpha P_w. \tag{9.27}$$

These relationships allow us to calculate wave's attenuation coefficient:

$$\alpha = \frac{P_s}{2P_w} = \frac{\operatorname{Re} Z_s \oint |\mathbf{H}|^2 \, dl}{2 \operatorname{Re} Z_\perp \iint |\mathbf{H}_\perp|^2 \, dA} \tag{9.28}$$

(the surface integral in the denominator is taken across the line's cross section). In the case of relatively weak damping ($\alpha \ll \beta$), instead of field $\mathbf{H}$ in this equation, we can use field $\mathbf{H}_0(x,y)$, which is calculated for the ideal transmission line (at $Z_s = 0$). For each mode of the wave, the value of the attenuation coefficient depends on the dimensions of the line, the conductivity of walls, and the frequency. The study of these dependences on the basis of the general expressions for the fields of TE, TM, and TEM waves in ideal transmission lines gives us the results that will be discussed now.

In a rectangular waveguide for the lowest TE_{10} mode, which is most frequently used for the transmission of centimeter range waves, the attenuation coefficient using the relationships obtained earlier can be written as

$$\alpha = \sqrt{\frac{2\omega_{cr}\varepsilon_0}{\sigma}} \cdot \frac{(a/2b)u^2 + 1}{a\sqrt{u(u^2-1)}},$$ (9.29)

where
$$u = \omega/\omega_{cr}$$
$$\omega_{cr} = \pi c/a\sqrt{\kappa\kappa_m}$$

Curve A in Figure 9.14 shows the dependence of αD on u for a rectangular waveguide with an aspect ratio $a/b = 2$. Here $D = a\sqrt{\sigma/2\omega_{cr}\varepsilon_0}$. Minimum absorption is achieved in this case when $u \approx 2.42$, but in reality usually smaller values of this parameter ($u < 2$) are used, since at $u > 2$ the waves of higher order, TE_{01} and TE_{20}, begin propagating, which complicates the situation in communication systems.

The absolute value of the attenuation coefficient decreases when the waveguide size increases. Close to the absorption minimum on curve A, we get $\alpha \propto 1/a^{3/2}$ because ω_{cr} is proportional to $1/a^{1/2}$ (see Equation 9.29). In the region of frequencies, where ω is substantially larger than ω_{cr} (i.e., $u = \omega/\omega_{cr} \gg 1$), we get from Equation 9.29 that $\alpha \propto 1/a$. In particular, for a copper (or brass) waveguide with conductivity $\sigma \approx 5.50 \times 10^7$ S/m and dimensions $a = 2$ cm and $b = 1$ cm at $\lambda = 2.5$ cm ($u = 1.7$), we find $\alpha \approx 1.4 \times 10^{-2}$ 1/m, that is, the attenuation length is equal to $l_\alpha \approx 70$ m. Due to oxidation of the surface layers of the waveguide's walls, which results in the reduction of effective value of conductivity, and to the presence on the wall of small irregularities leading to scattering of the wave, the real attenuation coefficient is somewhat higher than the calculated value. Thus, for energy transmission over distances of more than 10 m, rectangular waveguides in the SHF range are not commonly used.

Other modes of waves in rectangular and other types of waveguides have similar behavior and approximately the same value of attenuation coefficients and analogous dependence of α on frequency. The only exception is the TE_{0n} wave mode in a circular waveguide. The magnetic field intensity tangential component of the TE_{0n} mode on the wall, which defines the energy flux into

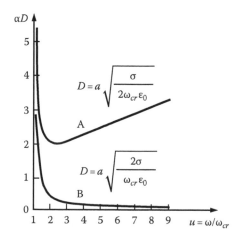

FIGURE 9.14 Attenuation coefficients as functions of frequency for the wave TE_{11} of rectangular waveguide with $a/b = 2$ (curve A) and wave TE_{01} of circular waveguide (curve B).

metal (the line integral in Equation 9.12), coincides with the longitudinal component H_z, which tends to zero at $\omega/\omega_{cr} \to \infty$ (since the wave approaches the transverse wave, $H_z/H_r \approx \omega_{cr}/\omega$). As a result, the attenuation coefficient with increasing frequency tends to zero as $\omega^{-3/2}$. The frequency dependence of the attenuation coefficient of TE_{01} mode is given by the expression

$$\alpha = \sqrt{\frac{\omega_{cr}\varepsilon_0}{2\sigma}} \cdot \frac{1}{a\sqrt{u(u^2-1)}},$$ (9.30)

where a is the radius of the waveguide. This dependence is shown in Figure 9.14 (curve B). Note that here $D = a\sqrt{2\sigma/\omega_{cr}\varepsilon_0}$. Unfortunately, wave TE_{01} is not the lowest mode wave of a circular waveguide. Therefore, despite the weak attenuation, it has limited applications in communication lines.

The attenuation coefficient depends on the surface current density that the wave generates on the walls of a waveguide. This is why the energy losses for different modes of waves have different values. For E_{mn} waves with $m \neq 0$ and $n \neq 0$, the attenuation coefficient is equal to

$$\alpha_{mn}^E = \frac{2R_s}{bZ_0\sqrt{1-(\omega_{cr}/\omega)^2}} \cdot \frac{m^2(b/a)^3+n^2}{m^2(b/a)^2+n^2},$$ (9.31)

where

$R_s = \sqrt{\omega\mu_a/2\sigma}$ is the surface resistance of the conductor

$Z_0 = 120\pi$ is the wave impedance of vacuum

For H_{mn} waves with $m \neq 0$ and $n \neq 0$, the attenuation coefficient is equal to

$$\alpha_{mn}^H = \frac{2R_s}{bZ_0\sqrt{1-(\omega_{cr}/\omega)^2}}\left[\frac{m^2(b/a)^2+n^2(b/a)}{m^2(b/a)^2+n^2}+\left(\frac{\omega_{cr}}{\omega}\right)^2\frac{m^2(b/a)^3+n^2}{m^2(b/a)^2+n^2}\right].$$ (9.32)

For H_{mn} waves with the fundamental mode $m = 1$ and $n = 0$ that are not included in this equation,

$$\alpha_{10}^H = \frac{2R_s}{bZ_0\sqrt{1-(\omega_{cr}/\omega)^2}}\cdot\left[1+\frac{b}{a}\left(\frac{\omega_{cr}}{\omega}\right)^2\right].$$ (9.33)

9.7 REFLECTIONS IN TRANSMISSION LINES AND NEED OF THEIR MATCHING

In any transmission line, it is necessary to match the line with the load. This means that the load of the line has to be active and has conductivity equal to the wave conductivity of the line. In this case, there is no reflected wave from the load and the traveling wave regime is set, in which (if no losses are present) the time-averaged density of the wave energy flow along the line is constant.

In a mismatched line, there are reflections. If the amplitude of the reflected wave is equal to the incident wave amplitude (the reflection coefficient is equal to the unit), we have a standing wave. In this case, z-field structure is established along the waveguide, for which the field maxima (antinodes) alternate with zero values (nodes). Nodes and antinodes of the field are time independent. The field phase does not change from node to node. Upon transition through a node, the field phase changes abruptly by π. Energy transmission along the waveguide is absent though the excitation frequency is higher than the critical value.

If the amplitude of the reflected wave is smaller than that of the incident wave, we have the following situation: the field maxima alternate with field minima at which the field does not reduce to zero and the phase changes smoothly.

To characterize the degree of matching of the transmission line, three interconnected values are used: the reflection coefficient (Γ), the standing wave ratio (SWR), and the traveling wave ratio (TWR). The reflection coefficient is equal to the ratio of the complex amplitudes of the reflected and the incident waves. The standing wave coefficient is equal to the ratio of the time-averaged maximum and minimum values of the field measured at the corresponding sections of the line (SWR $\geq$ 1). The value reciprocal to the SWR is called the TWR (TWR $\leq$ 1). The relationship between these parameters is

$$\text{SWR} = \frac{1+|\Gamma|}{1-|\Gamma|} = \frac{1}{\text{TWR}} \quad \text{TWR} = \frac{1-|\Gamma|}{1+|\Gamma|} = \frac{1}{\text{SWR}}. \tag{9.34}$$

Values of the parameters characterizing a degree of matching of a transmission line at various regimes of its operation are as follows:

Regime	Parameters Characterizing a Degree of Matching		
	Γ	SWR	TWR
Traveling wave	0	1	1
Standing wave	1	$\to\infty$	0
Mixed wave	<1	>1	<1

Reflections in a transmission line can arise not only from a load but also from any irregularities of the tract.

The mismatch of the line leads to a power decrease in the load and a reduction of the maximum permissible power transferred to the load. As a result, losses increase and the line efficiency decreases. Thus, matching of lines and realization of a situation close to that of a traveling wave in most of the transmission line length are of great practical importance.

To achieve matching, special matching elements are introduced in the transmission line. In microwave transmission lines, the value of SWR = (1.1–1.3) is considered as good enough. The reflected wave can be removed with the help of a valve—a special device introduced into the transmission line for these purposes.

For transmission lines, their *efficiency* has great practical importance. This is defined as the ratio of active power in a load to total active power on a line input.

If we assume that only reflections in the transmission take place from the load, in the presence of losses, the efficiency (E) is calculated as

$$E = (1-|\Gamma|^2)\exp(-2\alpha l), \tag{9.35}$$

where
   $|\Gamma|$ is the reflection coefficient module
   α is a damping factor
   l is the length of the line

One can see that in order to increase the transmission line efficiency, it is necessary to reduce, whenever possible, the length of the line, to minimize reflections, and to reduce losses in the walls of the waveguide and in the dielectric.

9.8 TWO-WIRE, COAXIAL, AND STRIPLINE TRANSMISSION LINES

1. A two-wire line is formed by two parallel conductors, surrounded by a dielectric with the parameters ε_a and μ_a (usually air). Figure 9.15 shows a symmetric two-wire transmission line. This line consists of two identical conductors with circular cross section of radius r. The distance b between the axes of the conductors is constant along the line. For such line, a normal wave is purely transverse wave (what we often designate as T-type wave). The picture of the field lines of this wave is a system of closed magnetic field lines around each conductor and a system of electric field lines, beginning on one conductor and ending on another.

This line is characterized by inductance and capacitance per unit of length of the line:

$$L_1 = \frac{\mu_a}{\pi} \ln\left(\frac{b}{r} - 1\right) \frac{H}{m}, \quad C_1 \simeq \pi\varepsilon_a / \ln\left(\frac{b}{r} - 1\right) \frac{F}{m}. \tag{9.36}$$

Channeling of electromagnetic energy along a symmetric two-wire line is possible only for large wavelengths, that is, when

$$\lambda \gg b. \tag{9.37}$$

Indeed, as the currents of frequency ω equal in magnitude and opposite in phase flow through the wires and the distance b between two wires of the line is small compared to the wavelength, the electromagnetic waves created by wires will appear almost the same in an antiphase and opposite in phase at all points of space, therefore these waves almost completely cancel each other. The resulting radiation field of the line is thus practically absent. If condition (9.37) is not satisfied, there are energy losses on radiation in the transmission line.

Some advantages of the symmetric transmission line are simplicity of design and low cost. The disadvantage is the absence of screening of the electromagnetic wave propagating along the lines. Therefore, the symmetric line is exposed to influence by external disturbances and has rather large power losses in metal objects located near it.

Symmetric two-wire lines are used as antenna feeders in the transmitting and receiving of radio signals for frequencies 30–50 MHz. Twisting the two wires with a separate insulation reduces radiation of the line, so such line can be used for frequencies of the order of hundreds of megahertz. The possibility of energy radiation into the surrounding space restricts the use of two-wire transmission lines in the microwave range.

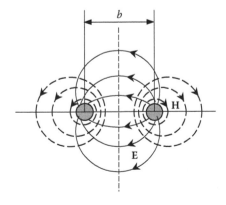

FIGURE 9.15 Picture of electric field and magnetic field intensity lines of a T wave in a symmetric two-wire transmission line.

2. From among transmission lines of T-type waves, in the microwave range, a coaxial transmission line is most often used. This line consists of two coaxial metal cylinders with diameters d and D, separated by a dielectric layer with permittivity ε_a and permeability μ_a (Figure 9.16).

The electric field distribution in the transverse plane of the coaxial transmission line, which operates at a type T wave, has a field structure similar to that of a cylindrical capacitor. The electric field vector is radial with magnitude E_r:

$$E_r = \frac{U}{r \cdot \ln(D/d)}, \quad d \leq r \leq D, \tag{9.38}$$

where U is the potential difference between the inner and outer conductors at $z = 0$. The complex amplitude of the electric vector of the traveling wave is written as follows:

$$\mathbf{E}(r,z) = \mathbf{E}_r \cdot \exp(-i\beta z). \tag{9.39}$$

The main feature of the coaxial line is that the current in it, traveling from the generator to the load on the inter cylinder, returns to the generator on the outer conductor. Because of this, in the space between the cylinders, the magnetic field intensity lines have the same form as in the case of current flow on a single cylindrical conductor, that is, they form concentric circles.

In a cylindrical coordinate system, the magnetic field intensity vector has the single azimuthal component H_φ. The amplitude of the magnetic field can be easily expressed through the wave impedance of the T wave:

$$H_\varphi(r) = \frac{E_r}{Z} = \sqrt{\frac{\varepsilon_a}{\mu_a}} \frac{2\pi E_r}{\ln(D/d)}, \tag{9.40}$$

where

$$Z \approx \sqrt{\frac{L_1}{C_1}} = \frac{1}{2\pi} \sqrt{\frac{\mu_a}{\varepsilon_a}} \ln \frac{D}{d}.$$

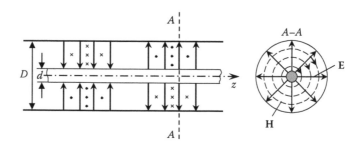

FIGURE 9.16 Structure of an electromagnetic field of T-type wave in a coaxial transmission line.

The running parameters of a coaxial transmission line are

$$L_1 = \frac{\mu_a}{2\pi} \ln \frac{D}{d} \, \text{H/m}, \quad C_1 = \frac{2\pi\varepsilon_a}{\ln(D/d)} \, \text{F/m}. \tag{9.41}$$

To suppress losses due to higher harmonics that can arise in a waveguide, it is necessary that the frequency satisfies the inequality

$$\omega \leq \frac{4}{\sqrt{\varepsilon_a \mu_a}(d+D)}. \tag{9.42}$$

The main advantages of the coaxial transmission line are their broadband, the absence of radiation leak, and the possibility of being manufactured in the form of flexible coaxial cables.

The main drawbacks are large attenuation, low dielectric strength, and design complexity. Coaxial transmission lines are most often applied in the form of coaxial cables to connect the radio equipment blocks. Coaxial lines and elements of coaxial type are used at frequencies up to about 20 GHz.

3. In the microwave range, the guiding systems called stripline transmission lines are widely used as well. They are especially convenient in printed and integrated microwave circuits.

In Figure 9.17, stripline transmission lines of the asymmetric and symmetric types are shown. These lines are filled with different types of dielectrics.

The rigorous theory of striplines is quite complex. In striplines, the T waves cannot propagate in pure form. However, what is known as quasi-T wave can exist in these lines, if the width of the current-carrying conductor and the distance between it and the grounded plate are less than a half of the wavelength in the transmission line. In this case, the electric and magnetic fields are concentrated mostly in the space between the conductor and the grounded plate. The picture of electric field lines is similar to that of the static field in a plane capacitor.

The asymmetrical line is constructed most simply. However, it has an essential drawback: part of the wave propagates in air and causes undesirable interference with other nearby elements. The symmetric line is almost completely screened. Rather big attenuation and small values of transmitting power are the other shortcomings of striplines.

Striplines are used predominantly in the 3–10 GHz range for the transmission of low power. Their advantages in comparison with hollow waveguides consist of the simplicity of their manufacturing, their compactness, and low cost. The striplines manufactured by printed circuits offer especially great advantages in the manufacture of small functional units for the microwave region. In miniature microwave integrated circuits, asymmetrical microstriplines, which have technological advantages, are used.

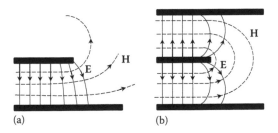

(a)   (b)

FIGURE 9.17 Structure of an electromagnetic field in stripline transmission lines: the asymmetric (a) and the symmetric (b).

Exercise 9.6

An open two-wire line with air insulation is formed by two parallel wires separated by a distance $D = 16.0$ mm. The wire radius $d = 2.00$ mm. Find the inductance and capacitance per unit length of the line.

Solution. 1. First, we find the inductance per unit length of the line. Since for air $\kappa_m = 1$ and $\mu_0 = 4\pi \times 10^{-7}$ H/m, then for air line Equation 9.41 takes the form

$$L_1 = \frac{\mu_0}{\pi} \cdot \ln\left(\frac{D}{d}\right) = 0.92 \cdot \ln\left(\frac{D}{d}\right) \ \mu H/m,$$

for the given case

$$L_1 = \frac{\mu}{\pi} \cdot \ln\left(\frac{D}{d}\right) = 0.92 \cdot \ln\left(\frac{D}{d}\right) = 0.92 \cdot 0.90 = 0.83 \ \mu H/m.$$

2. Next, we find the linear capacitance per unit length of the line. Since for air

$$\varepsilon_a = \kappa \varepsilon_0 = \frac{10^{-9}}{36\pi} \ F/m,$$

the capacitance per unit length is equal to

$$C_1 = \frac{\pi \varepsilon_a}{\ln(D/d)} = \frac{12.1}{\ln(D/d)} = \frac{12.1}{\ln(16/2)} = 13.4 \ pF/m.$$

If the line is filled by dielectric with $\varepsilon_a \neq \varepsilon_0$, then we need to introduce the dielectric constant κ.

9.9 OPTICAL WAVEGUIDES (LIGHTGUIDES)

The operation of waveguides is based on the phenomenon of ***total internal reflection*** of light. The basic elements of an integrated optics and fiber optics are planar and fiber optical waveguides, which we will discuss next.

1. The ***planar waveguide*** consists of a thin dielectric layer on a dielectric substrate with a lower refractive index. The light wave in such layer can propagate due to total internal reflection at the "layer–substrate" and "layer–covering medium" interfaces (Figure 9.18). The thickness of the optical lightguide layers is comparable with the wavelength of light and is, usually, of $h = 0.3$–3 µm.

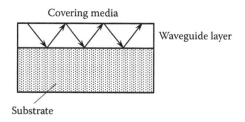

FIGURE 9.18   Planar optical waveguide.

The materials suitable for the manufacture of waveguide structures are a variety of glasses, polymers, electrooptic crystals, active dielectrics, and semiconductors.

In a dielectric waveguide, modes at frequency ω are the solutions of the wave equation that satisfy the boundary conditions. In optical waveguides, the existence of both radiative and directed (waveguide) modes is possible. Waveguide mode represents a wave, which is traveling in the propagation direction but forms a standing wave in the direction normal to the plane of the waveguide.

Assuming that the direction of propagation coincides with the z-axis of a Cartesian coordinate system and the direction of the normal is along x-axis, the expression for the field mode can be written as

$$\mathbf{E}(t, x, z) = \mathbf{E}(x)\exp[i(\omega t - \beta z)], \tag{9.43}$$

where
  $\mathbf{E}$ is the electric field of the light wave
  $\beta = \omega/v_{ph}$ is the propagation constant

From solutions of Maxwell's equations, it follows that for an isotropic optical waveguide and also for waveguides based on uniaxial crystals (when the light propagates along the principal axes), the waveguide modes have either purely TE or TM structure. From the form of the refractive index distribution over the cross section (along the x-axis), the planar optical waveguides are divided into homogeneous, for which a change of the refractive index is constant inside of the waveguide and has a steplike change on its boundary, and inhomogeneous (or graded), with a smooth change of the refractive index. The first type of waveguide is often named a waveguide with step discontinuity or even called using the shorter term step waveguide. In step waveguides, the formation of a guided mode can be regarded as the result of wave propagation in a homogeneous film in the zigzag fashion shown in Figure 9.19. To realize the waveguide regime in such a waveguide, it is necessary to satisfy the condition

$$n_1 > n_0, \quad n_1 > n_2, \tag{9.44}$$

where n_0, n_1, and n_2 are the refractive indices of the covering medium (often, it is simply air), the waveguide layer, and the substrate, respectively.

In a gradient waveguide, there is no explicit interface of the substrate with the waveguide layer. In this case, the direction of wave propagation is changed due to the inhomogeneity of the material

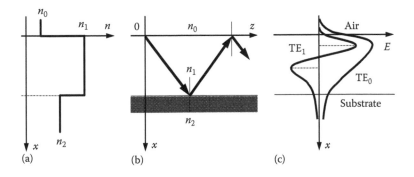

FIGURE 9.19   Refractive index profile (a), beam propagation model (b), and field distribution of modes TE_0 and TE_1 (c) in step lightguide.

refractive index. As a result, the light beam in the waveguide region propagates by refraction along curvilinear paths (Figure 9.20).

The profile $n(x)$ is usually defined by the application and in particular by the wavelength of the wave to be transmitted. The electric field in the interval $0 < x < x_t$, where x_t is the turning point, is an oscillating function. Outside this region, the mode field decays exponentially. The turning point is defined as the coordinate at which the tangent to the path of the light beam, which propagates in the gradient waveguide by refraction, becomes parallel to the waveguide surface.

Energy carried by the guided modes is localized in the region of the waveguide layer. Radiative modes carry out the energy from the waveguide layer to the coating layer or (and) the substrate.

2. The main element in optical fiber lines is a fiber lightguide. The simplest lightguide is an optically transparent circular dielectric rod, called the core, surrounded by an optically transparent dielectric shell (Figure 9.21).

The refractive index of a core n_1 is higher than the refractive index of a shell n_2. Depending on fiber manufacturing technology, a refractive index can change either abruptly or smoothly. Also in the case of planar lightguides, in the former case, fibers are called step fiber, and in the second, they are called gradient fiber. The shell of fiber lightguides is fabricated from clear quartz glass SiO_2, and the increase of the core refractive index is achieved by doping it with germanium or phosphorus.

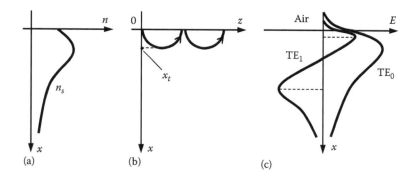

FIGURE 9.20 Refractive index profile (a), beam model of light wave propagation (b), and field distribution of modes TE_0 and TE_1 (c) in gradient lightguide.

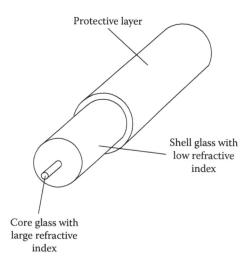

FIGURE 9.21 Structure of an optical fiber.

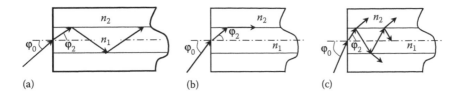

(a)   (b)   (c)

FIGURE 9.22   Light propagation in a cylindrical lightguide for different angles of wave entry into the waveguide: $\varphi_0 < \varphi_{cr}$ (a), $\varphi_0 = \varphi_{cr}$ (b), and $\varphi_0 > \varphi_{cr}$ (c).

In recent years, optical fibers from multicomponent glasses and polymers are applied more and more widely. They are used in the cheap mass production of optical fibers of average quality for communication over short distances. The outside of the fiber is coated with polyethylene to protect it from mechanical damage. The fiber ends are polished, and a source of radiation is attached to one of them. Light-emitting diodes and semiconductor lasers are used as sources of radiation in optical communication lines.

Figure 9.22 shows the propagation of radiation in a cylindrical lightguide at different angles of incidence. For the transmission of light by optical fibers, the principle of total internal reflection is used. This is possible if $n_1 > n_2$, and the angle of incidence at the interface between two media (core–shell) has such value that the angle of refraction in the second media approaches 90° (critical angle). If the light beam enters the fiber at an angle, $\varphi_0 < \varphi_{cr}$, relative to the optical axis (Figure 9.22a), the refracted beam will experience total internal reflection at the interface between the core and the cladding layer (shell). Increasing the angle of incidence also increases the angle of refraction and, at some critical angle φ_{cr}, the refracted beam will not be reflected, but it will propagate along the boundary between the core and the cladding (Figure 9.22b). A further increase in the angle of incidence will result in the radiation penetrating into the shell (Figure 9.22c). The angle of incidence from which all the energy is reflected from the interface between the core and the cladding is called *aperture*. To have total internal reflection, it is necessary to provide a difference in the refractive indices near the outer surface of the core and that of the cladding. In the case of a steplike change of the refractive index, the optimal situation is when the refractive index of the cladding differs from that of core by a factor of 1.4.

For gradient lightguides, a parabolic dependence of the refractive index along the radial direction with a maximum value on the lightguide axis has special importance. Glass and quartz fibers with a quadratic function $n(r)$ are widely used in communication for transmitting signals and the formation of optical images. This is due to the fact that light pulses, propagating in multimode optical fibers, experience minimal temporary broadening and thus distortion, when fibers with a parabolic transverse dependence of the refractive index are used.

PROBLEMS

9.1   A plane monochromatic wave with s-polarization and frequency $f = 5.00$ GHz is incident from a nonmagnetic dielectric medium with permittivity $\varepsilon = 1.50\varepsilon_0$ onto the surface of a nonmagnetic metal at an angle $\theta = 30°$ with respect to normal to the metal surface. Determine the ratio of the wavelengths and phase and group velocities of the wave in the dielectric to the guided wave at the metal surface. (*Answer:* $\lambda/\lambda_w = 0.50$, $v_{ph}/v_{phw} = 0.50$, $v_{gr}/v_{grw} = 2.00$)

9.2   Consider a cylindrical coaxial cable that consists of a central conductor of radius r_1 and an outer shell (see Figure 9.23). The space between the central conductor and the shell (i.e., from r_1 to r_2) is filled with a dielectric with relative permittivity κ. An alternating current voltage V is applied between the central conductor and the outer shell; the current in the cable is equal to I. Neglecting the resistance of conductors, find the dependence of the power transmitted in the dielectric as a function of radius r (r is an arbitrary point between r_1 and r_2; see Figure 9.23). Prove that alternating current energy is transmitted only through the dielectric. $\left(Answer: S = \dfrac{VI}{2\pi r^2 \ln(r_2/r_1)} \cdot \right)$

9.3 Power P is transmitted along a two-wire transmission line. The voltage applied between the two wires is equal to V and a current in the wires is equal to I; the wires are surrounded by air. By neglecting the resistance of the wires whose radius is equal to r_0 (see Figure 9.24), find the dependence of the energy flux density (the Poynting vector) on the coordinate x at $y = 0$. $\left(Answer: S_z = \dfrac{VId^2}{4\pi x^2 (d-x)^2 \ln(d/r_0)} \cdot \right)$

9.4 The width b of the smaller wall of a rectangular waveguide is equal to 3.00 cm (see Figure 9.10). Determine the minimum size a of a larger wall of the waveguide if it is filled with air, where a wave of type TE_{11} with frequency $f = 6.00 \times 10^9$ Hz can propagate. What is the critical wavelength? (*Answer*: $\lambda_{cr} = 5.00$ cm.)

9.5 Prove that in a rectangular waveguide, the phase velocity of a TE_{10} wave (or H_{10} wave) is greater than the speed of light.

9.6 A TE_{11} (or H_{11}) wave with frequency $f = 6.00$ GHz is propagating in a rectangular waveguide with wall sides $a = 5.00$ cm and $b = 3.00$ cm. The waveguide is filled with a dielectric with dielectric permittivity $\varepsilon = 2.25\,\varepsilon_0$. Find the critical wavelength, the wavelength in the waveguide, and the phase velocity. (*Answer*: $\lambda_{cr} = 5.15$ cm, $\lambda_{in} = 4.38$ cm, $v_{ph} = 2.65 \times 10^{10}$ cm/s.)

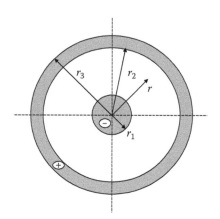

FIGURE 9.23   Cross section of a coaxial cable.

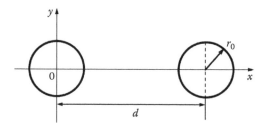

FIGURE 9.24   Two-wire transmission line.

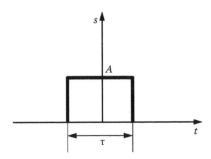

FIGURE 9.25   A rectangular wave pulse.

9.7 Find the spectral distribution of a rectangular pulse

$$s(t) = \begin{cases} A, & |t| < \tau/2, \\ 0, & |t| > \tau/2, \end{cases}$$

whose duration is equal to τ and its amplitude is equal to A (see Figure 9.25).

9.8 Pulses with a rectangular envelope are propagating in a rectangular waveguide with wall dimensions of $a = 2.30$ cm and $b = 1.00$ cm. The pulse duration $\tau = 10.0$ ns, the carrier (central) frequency $f_0 = 9.00$ GHz, and the waveguide length $l = 100$ cm. Determine the distortion of the pulse when it reaches the end of the waveguide. *Hint*: The front and the back of the rectangular pulse propagate with different group velocities through the waveguide.

10 Emission of Electromagnetic Waves

The electromagnetic field of fixed or uniformly moving charged particles is inseparably linked to these particles. Accelerated motion of charged particles, on the other hand, results in an electromagnetic field, which is "detached" from the charges and exists independently from them in the form of a wave propagating in space. Maxwell predicted the existence of electromagnetic waves based on his equations; experimental proof was obtained by H. Hertz in 1886 after Maxwell's death.

The simplest emitters of electromagnetic waves are accelerated electric charges and oscillating electric dipoles in the form of a line segment or conductor through which an alternating current exists. The electric charges in the conductor follow an oscillatory motion, so the alternating current carrying conductor acts as an ***antenna***. An antenna transforms the oscillations of charges into free waves that are radiated in certain directions that depend on the geometry of the antenna. Different types of antennas will be discussed in this chapter.

The structure and nature of electromagnetic field distribution in space depend on the distance from the emitter to the observation zone, which is subdivided into near- and far-field zones. The ***near-field zone*** extends from a source to distances that have the same order of magnitude with the wavelength of the electromagnetic wave generated by the source. In this zone, the field strength usually decreases sharply with distance from the source, and the fields continue to exist when the current frequency tends to zero (i.e., the time-independent fields can exist in the near-field zone).

The ***far-field zone*** is located at a distance much greater than the wavelength of the emitted wave, and the field strength decreases inversely proportional to a distance from the source. Here, the field propagates as a plane wave whose energy is divided equally between its electric and magnetic components. In free space, the relation between strengths of electric field and magnetic field intensity is $E = Z_0 H$, where the impedance of vacuum is $Z_0 = \sqrt{\mu_0/\varepsilon_0} = 377\ \Omega$.

10.1 RADIATION EMITTED BY AN ACCELERATED MOVING CHARGE

A description of the electromagnetic field generated by an accelerating charge can be directly obtained from Maxwell's equations. However, in the following text, we present a less rigorous but more physical derivation based on the continuity of the electric field lines in vacuum. Consider the motion of a point charge q, which was at rest for $t < 0$ and at time $t = 0$ begins moving from its rest point $z = 0$ in the positive direction of the z-axis with a constant acceleration a.

After a short time interval Δt, the acceleration stops and the charge continues to move with constant velocity $v = a\Delta t$ (Figure 10.1).

Consider now the picture of the electric field lines generated by a charge that moves as it is shown in Figure 10.1a. At time t ($t \gg \Delta t$), information about changes in the motion of charge q has not reached the points that are outside the sphere of radius $r = ct$, where c is the speed of light in vacuum. Therefore, the electric field in this area coincides with the field of a point charge q at rest at the origin of coordinates: $\mathbf{E}(\mathbf{r}) = (k_e q/r^3)\mathbf{r}$.

Inside the sphere of radius $r = c(t-\Delta t)$, there is a field of a charge moving with uniform velocity. For $v \ll c$, the field of the charge at time t coincides with the field of point charge q, which is in a point $z = \Delta v \cdot t$. Thus, for time $t \gg \Delta t$, the displacement of the charge during the acceleration can be neglected. Taking into account the continuity of the electric field lines, these lines have the form shown in Figure 10.1b.

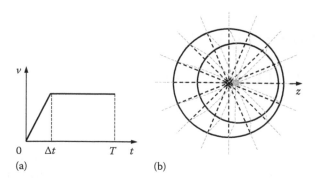

(a)   (b)

FIGURE 10.1 Time dependence of the velocity of a charged particle (a) and the distribution of the electric field lines of the accelerated charge (b).

The change in shape of the electric field lines between spheres $r = c(t - \Delta t)$ and $r = ct$ is related with the change in the charge velocity $v(t)$ shown in detail in Figure 10.2. Consider one of the field lines that passes through the observation point O', which is at a distance r from the origin of coordinates. The direction vector of point O' forms an angle θ with the axis $0z$. Let us determine the ratio of tangential E_θ and radial E_r components of a vector of electric field $\mathbf{E}$ between spheres $r = c(t - \Delta t)$ and $r = ct$. As it is shown in Figure 10.2, the radial component BO' of vector $\mathbf{E}$ (its length is equal to AO') is proportional to $c\Delta t$ and the tangential component AB is proportional to $\Delta v \cdot t \sin \theta$, so their ratio is

$$\frac{E_\theta}{E_r} = \frac{\Delta v \cdot t \sin \theta}{c \Delta t}. \tag{10.1}$$

The radial component of the electric field $E_r = k_e q/r^2$ is a static component of the field of a point charge. Since $\Delta v = a \Delta t$ (here $a = dv/dt$ is acceleration of the charge) and $r = ct$, the electric component of the wave field is

$$E_\theta = \frac{\Delta v \cdot t \sin \theta}{c \Delta t} E_r = \frac{1}{4 \pi \varepsilon_0} \frac{q a \sin \theta}{r c^2}. \tag{10.2}$$

For the magnetic component of the wave field ($E = Z_0 H$, $Z_0 = \sqrt{\mu_0/\varepsilon_0} = 1/c\varepsilon_0$, $c = 1/\sqrt{\varepsilon_0 \mu_0}$), we obtain

$$H_\varphi = \varepsilon_0 c E_\theta = \frac{q a \sin \theta}{4 \pi r c}. \tag{10.3}$$

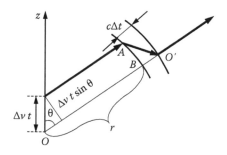

FIGURE 10.2 Derivation of formula (10.1).

In this case, the Poynting vector $\mathbf{S} = \mathbf{E} \times \mathbf{H}$ (the vector that determines the energy flow through a unit area per unit time) is directed radially outward along the position vector $\mathbf{r}$, and its magnitude is

$$S_r = E_\theta H_\varphi = Z_0 \left(\frac{qa \sin\theta}{4\pi rc} \right)^2, \quad Z_0 = \sqrt{\frac{\mu_0}{\varepsilon_0}}. \tag{10.4}$$

From this, it can be seen that the flux of the emitted energy is proportional to the square of the accelerated charge.

The spatial distribution of the energy flow depends on the direction of angle θ: the flow reaches its maximum value for $\theta = \pi/2$ and vanishes at $\theta = 0$ and $\theta = \pi$.

In the following text, we calculate the power radiated by an accelerated charge. To do this, it is necessary to integrate $\mathbf{S}$ over the surface of a sphere of radius r:

$$P = \oint \mathbf{S} d\mathbf{A} = Z_0 \int_0^\pi \int_0^{2\pi} \left(\frac{qa \sin\theta}{4\pi rc} \right)^2 r^2 \sin\theta \, d\theta d\varphi = \frac{1}{6\pi\varepsilon_0} \frac{q^2 a^2}{c^3}. \tag{10.5}$$

Thus, the total energy flow through the sphere does not depend on the distance r of the charge. If the charge follows a simple harmonic oscillation described by $z(t) = z_0 \cos \omega t$, its acceleration is defined as $a = d^2z/dt^2 = -z_0\omega^2 \cos \omega t$. Thus,

$$(qa)^2 = (qz_0\omega^2 \cos \omega t)^2 = (p_0\omega^2 \cos \omega t)^2, \tag{10.6}$$

where $p_0 = qz_0$ is the magnitude of the electric dipole moment created by the oscillating charge. In this case, the Poynting vector can be written as

$$\mathbf{S} = \frac{p_0^2\omega^4 \sin^2\theta \cos^2[\omega(t - r/c)]}{16\pi^2\varepsilon_0 r^2 c^3} \times \frac{\mathbf{r}}{r}. \tag{10.7}$$

The electric field lines that correspond to E_θ lie in a plane passing through an axis of motion of a charge, and those for the magnetic field intensity H_φ lie in a plane parallel the plane $x0y$ (Figure 10.3).

The propagation direction of the emitted waves is radially outward and coincides with the radius vector $\mathbf{r}$. Such field pattern is the feature of transverse electromagnetic wave (T-type).

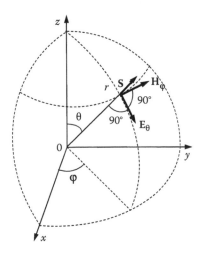

FIGURE 10.3 Electric field and magnetic field intensity vectors of an accelerated moving charge in far-field zone.

Note that the wave arrives at a distance r from the accelerating charge in time r/c. For this reason, in the expressions for the fields and energy flows, it is necessary to take into account retardation and to replace t with $t - r/c$. Thus, the phase of an emitted wave in a region far away from the emitter (far-field zone) becomes $\omega(t - r/c) = \omega t - k_0 r$. For higher frequencies used in practical applications ($\omega \sim 10^8 - 10^{16}$ s^{-1}), it makes sense to speak only about energy flow averaged over time. Since $< \cos^2[\omega(t - r/c)] > = 1/2$, we obtain

$$< \mathbf{S} > = \frac{p_0^2 \omega^4 \sin^2 \theta}{32\pi^2 \varepsilon_0 r^2 c^3} \times \frac{\mathbf{r}}{r}. \tag{10.8}$$

From this, it follows that the averaged over time radiation intensity is proportional to ω^4.

If we integrate Equation 10.8 over the surface of a sphere of radius r, we obtain the mean power radiated by the oscillating charge:

$$< P > = \frac{\omega^4 p_0^2}{12\pi\varepsilon_0 c^3}. \tag{10.9}$$

Taking into account the relation $qa = \partial^2 p/\partial t^2$ obtained in Equation 10.6, we can write Equations 10.2 and 10.3 for the wave field of an oscillating electric dipole in the far-field zone, that is, at distances $r \gg \lambda$ as

$$E_\theta(r,t) = \frac{\sin \theta}{4\pi\varepsilon_0 c^2 r} \frac{\partial^2 p(t - r/c)}{\partial t^2}, \quad H_\varphi(r,t) = \frac{\sin \theta}{4\pi c r} \frac{\partial^2 p(t - r/c)}{\partial t^2}, \tag{10.10}$$

where θ is the angle between the dipole axis and the line that connects the dipole center with the observation point.

Exercise 10.1

Because of energy emission, the accelerated electron loses energy and slows down. This means that the electron experiences a braking force called *radiative friction force* $\mathbf{F}_{rad}$. Find an expression and estimate the magnitude of this force for an electron oscillating at angular frequency $\omega = 4 \times 10^{16}$ s^{-1} with an amplitude $x_0 = 0.05$ nm. Assume that $x = x_0 \cos \omega t$.

Solution. For a periodic motion of an electron, the work done on the electron during one period by the radiation friction force is negative of the radiation power given by Equation 10.5 integrated over one period from 0 to $T = 2\pi/\omega$, i.e.

$$\int_0^T \mathbf{F}_{rad} \cdot \mathbf{v} dt = -\int_0^T \frac{e^2 a^2}{6\pi\varepsilon_0 c^3} dt = -\int_0^T \frac{e^2}{6\pi\varepsilon_0 c^3} \frac{d\mathbf{v}}{dt} \cdot \frac{d\mathbf{v}}{dt} dt,$$

Integrating the right hand part of the above equation by parts, we see that the boundary term in the integral by part equal zero for the periodic motion.

$$\int_0^T \mathbf{F}_{rad} \cdot \mathbf{v} dt = -\frac{e^2}{6\pi\varepsilon_0 c^3} \frac{d\mathbf{v}}{dt} \cdot \mathbf{v} \Big|_0^T + \int_0^T \frac{e^2}{6\pi\varepsilon_0 c^3} \frac{d^2\mathbf{v}}{dt^2} \cdot \mathbf{v} dt = -0 + \int_0^T \frac{e^2}{6\pi\varepsilon_0 c^3} \frac{d\mathbf{a}}{dt} \cdot \mathbf{v} dt.$$

It is obvious from the above that

$$\mathbf{F}_{rad} = \frac{e^2}{6\pi\varepsilon_0 c^3} \frac{d\mathbf{a}}{dt}.$$

For the periodic motion $x(t) = x_0 \cos \omega t$ we get

$$\mathbf{F}_{rad} = \frac{e^2 \omega^3 x_0}{6\pi\varepsilon_0 c^3} \sin \omega t = F_0 \sin \omega t,$$

where $F_0 = e^2 \omega^3 x_0 / 6\pi\varepsilon_0 c^3$. For the oscillation frequency $\omega = 4 \times 10^{16}$ s^{-1} and $x_0 = 0.05$ nm, we obtain the amplitude of radiation friction force equal to $F_0 \approx 1.82 \cdot 10^{-14}$ N.

10.2 RADIATION EMITTED BY AN ELECTRIC DIPOLE (HERTZ ANTENNA)

The **Hertz dipole** consists of a metal conductor of length l, through which flows an alternating current $I = I_0 \cos \omega t$. Using thin wires, Hertz constructed a symmetric antenna on the ends of which he installed metal spheres with large capacitance. Due to the concentration of electric charges on the Hertz dipole ends, the current amplitude varies slightly along the wire and the antenna behaves like a dipole with a length equal to the length of the metal conductor.

As we obtained earlier, the accelerated charge and the time-varying dipole have three field components—the radial component E_r of **E**, azimuthal component $H_\varphi(t)$ of **H**, and tangential (or polar) component $E_\theta(t)$ of **E**. The electric $E_\theta(t)$ and magnetic $H_\varphi(t)$ field components depend on time and on angle θ. The radial component E_r is time and angle independent for large distances from the source. Thus, we can conclude that the accelerated charge and the time-varying dipole create a spherical wave in space (they are the sources of electromagnetic waves). The criterion for determining the type of field zone is based on the value of $k_0 r$. If $k_0 r \ll 1$ (i.e., $r \ll \lambda$), we are in the near-field zone; if on the other hand $k_0 r \gg 1$ (i.e., $r \gg \lambda$), we are in the far-field zone.

In the **near-field zone**, the wave field amplitude decreases very rapidly with increasing distance from the emitter. We do not consider this zone in details. Electric and magnetic fields differ by a factor $i = \sqrt{-1}$, that is, they are shifted in phase by $\pi/2$. Therefore, if at some time t the magnetic field intensity is minimum, then the electric field is maximum. The Poynting vector $\mathbf{S} = \mathbf{E} \times \mathbf{H}$ has two components: S_r and S_θ. The instantaneous values of these components can be expressed as

$$S_r = E_{\theta 0} \sin \omega t \cdot H_{\varphi 0} \cos \omega t = \frac{1}{2} E_{\theta 0} \cdot H_{\varphi 0} \sin 2\omega t,$$

$$S_\theta = E_{r0} \sin \omega t \cdot H_{\varphi 0} \cos \omega t = \frac{1}{2} E_{r0} \cdot H_{\varphi 0} \sin 2\omega t. \tag{10.11}$$

From these equations, it is seen that both components vary with time as $\sin 2\omega t$. The average value of each components of the Poynting vector over one period is zero, that is,

$$< S_{r,\theta} > = \frac{1}{T} \int_0^T S_{r,\theta}(t)dt = 0. \tag{10.12}$$

Thus, in the near-field zone, there is no net transmission of energy but the movement of the field energy has an oscillatory character instead.

In the **far-field zone**, the wave field components H_φ and E_θ of a wave emitted by a Hertz dipole are nonzero: the radial component E_r is time independent, and so it can be neglected in the consideration of the wave process. The instantaneous values of the components E_θ and H_φ at the distance r from the dipole can be written as

$$E_\theta = E_0(r)\sin(\omega t - k_0 r), \quad H_\varphi = H_0(r)\sin(\omega t - k_0 r), \tag{10.13}$$

where the amplitudes of the corresponding fields decrease as r^{-1} and are given by

$$E_0(r) = Z_0 \frac{I_0 l \omega}{4\pi c r} \sin\theta, \quad H_0(r) = \frac{I_0 l \omega}{4\pi c r} \sin\theta. \tag{10.14}$$

In the far-field zone, the Poynting vector has only a radial component $S = S_r = E_\theta \cdot H_\varphi$. The instantaneous value of this component has the form

$$S = E_0 \sin(\omega t - k_0 r) \cdot H_0 \sin(\omega t - k_0 r) = E_0 H_0 \sin^2(\omega t - k_0 r). \tag{10.15}$$

The instantaneous value of the Poynting vector is always positive. This means that the wave field energy moves only in one direction from the emitter outward and represents the energy of the emitted electromagnetic wave. Thus, the average value of the energy flux density over one period is

$$<S> = \frac{E_0}{\sqrt{2}} \frac{H_0}{\sqrt{2}} = \frac{Z_0}{2} \left(\frac{I_0 l \omega}{4\pi r c} \sin\theta \right)^2. \tag{10.16}$$

The average energy flux density over one period determines the emitted power density, which can be represented as

$$P_{rad} = \frac{I_0^2 l^2 k_0^3}{12\pi\omega\varepsilon_0}. \tag{10.17}$$

According to this expression, the radiation power P_{rad} is proportional to the square of the amplitude of the alternating current flowing through the emitter. There is analogy of Equation 10.17 for radiation power P_{rad} with the conventional equation for power dissipated on an active resistor R that carries an alternating current $I = I_0 \cos \omega t$, that is, with $P = I_0^2 R/2$. Therefore, the power of radiation can be expressed as

$$P_{rad} = \frac{1}{2} I_0^2 R_{rad}, \tag{10.18}$$

where $R_{rad} = l^2 k_0^3/6\pi\omega\varepsilon_0$ is the resistance of radiation, which in the theory of antennas is widely used for the estimation of the resistance of antennas.

The normalized directional characteristic of an emitter is the function

$$F(\theta, \varphi) = \sin\theta, \tag{10.19}$$

whose graphical representation in space looks like a donut. This function shows that in the direction $\theta = \pi/2$ and $\theta = 3\pi/2$ radiation is maximum and in the directions $\theta = 0, \pi$ radiation vanishes.

Since $<S>$ does not depend on the azimuthal angle φ, in the plane perpendicular to the dipole axis, the Hertz dipole is an isotropic emitter.

In summary, we will show in Figure 10.4 a simplified scheme of generation of electromagnetic waves by a Hertz dipole. The electric current of the dipole with a harmonic time dependence forms closed field lines for magnetic fields. In turn, the alternating magnetic field leads to the appearance

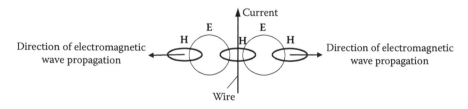

FIGURE 10.4 Simplified picture of the generation of an electromagnetic wave by the Hertz dipole.

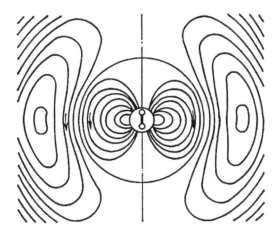

FIGURE 10.5   Picture of the electric field lines near a dipole at a fixed instant of time.

of closed electric field lines. The electromagnetic waves propagating from the dipole axis in all directions consist of the combination of time-dependent electric and magnetic fields.

The pattern of the electric field lines near the dipole obtained from the relations (10.10) at a fixed time is shown in Figure 10.5. The magnetic field intensity lines at each point in space are orthogonal to the electric field lines. Note that many simple antennas used in radio bands of long, medium, and short waves have a distribution of the wave field in space similar to a field distribution of the Hertz dipole.

Exercise 10.2

A simple emitter of length $l = 10$ cm is excited by a current with amplitude $I_0 = 3$ A and frequency $f = 100$ MHz. Determine the amplitudes of the electric field and magnetic fields intensity created by this emitter at a point that is located at a distance of $r = 4$ km from the emitter at an angle $\theta = \pi/4$ to the dipole axis.

Solution. The wavelength of the emitted waves is given by

$$\lambda = \frac{v}{f} = \frac{c}{f\sqrt{\kappa\kappa_m}}.$$

For air, $\kappa = 1$, $\kappa_m = 1$, and $\lambda = 3$ m, which is much larger than the linear size of the antenna $l = 10$ cm. Therefore, it is possible to consider that the current along the length of the emitter has a constant phase.

Since $r \gg \lambda$, we can assume that the observation point is in the far-field zone of the emitter. Using formulas (10.14), valid for the far field, we find the amplitudes of the magnetic and electric fields at a given distance from the emitter:

$$H_0(R) = \frac{I_0 l \omega}{4\pi c r} \sin\theta = \frac{3 \times 0.1 \times 2\pi \times 10^8}{4\pi \times 3 \times 10^8 \times 4 \times 10^3} \sin\left(\frac{\pi}{4}\right) \approx 8.75 \ \mu\text{A/m},$$

$$E_0(r) = Z_0 H_0(r) = 120\pi \times 8.75 \times 10^{-6} \approx 3.3 \ \text{mV/m}.$$

10.3   RADIATION OF AN ELEMENTARY MAGNETIC DIPOLE

Another simple emitter of electromagnetic waves is a small wire loop that carries an alternating current. When the coil dimensions are small compared to the wavelength that corresponds to the frequency of the alternating current, ($\lambda = 2\pi c/\omega$), the amplitude and phase of the current are practically the same at all points of the coil. As a result, there is an alternating magnetic field that is perpendicular to the plane of the coil. So, the magnetic field of a horizontal coil is analogous to the electric field of a vertical electric dipole, and the electric field of the coil is identical to the magnetic field of an electric dipole. Therefore, a horizontal coil with alternating current can be treated as the equivalent of a *vertical magnetic dipole*.

At a distance r from the magnetic dipole, the instantaneous values of the wave fields in the far-field zone for an arbitrary direction are given by expressions analogous to Equation 10.13:

$$E_\varphi = E_0 \cos(\omega t - k_0 r), \quad H_\theta = H_0 \cos(\omega t - k_0 r), \tag{10.20}$$

where the amplitudes of corresponding fields are given by

$$E_0 = Z_0 \frac{I_0 l_m \omega}{4\pi cr} \sin\theta, \quad H_0 = \frac{I_0 l_m \omega}{4\pi cr} \sin\theta. \tag{10.21}$$

where

I_0 is the amplitude of current in the coil
the value $l_m = k_0 A$ is interpreted as the length of the magnetic dipole, where A is the area of the coil

Thus, the average value of the energy flux density of a wave emitted by the coil has a form similar to expression (10.16):

$$<S> = \frac{E_0}{\sqrt{2}} \frac{H_0}{\sqrt{2}} = \frac{Z_0}{2} \left(\frac{I_0 l_m \omega}{4\pi rc} \sin\theta \right)^2. \tag{10.22}$$

Note the following important point: the wave fields of the electric and magnetic dipoles, which have identical current phases, are shifted in phase by an angle $\pi/2$. This is indicated by the factors $\sin(\omega t - k_0 r)$ and $\cos(\omega t - k_0 r)$ in Equations 10.13 and 10.20.

For comparison, Figure 10.6 shows the directions of the electric field and magnetic field intensity, as well as the direction of the propagation of wave energy flux for the electric and magnetic dipoles. In both cases, the direction of the wave propagation coincides with the direction of the position vector **r** (in the figure, this vector lies in the drawing plane). The electric field lines of the dipole E_θ and magnetic field lines of the coil H_θ lie in the plane formed by axis z and vector **r**, and they are perpendicular to **r**. The electric field lines of the coil E_φ and the magnetic field intensity lines of the electric dipole H_φ are directed perpendicularly to the plane formed by axis z and vector **r**.

The curves around each dipole show the relative change of the field strength amplitude in the plane of each dipole as functions of the polar angle θ of a cylindrical coordinate system (the angle θ is counted from the direction of the dipole axis). This change of field is determined by the factor $\sin\theta$.

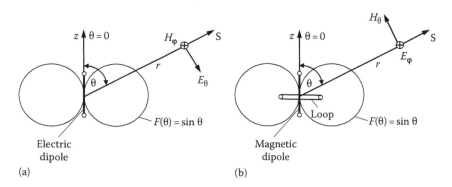

FIGURE 10.6   Wave structure in the far-field zone of electric (a) and magnetic (b) dipoles.

Along the axis of the coil (as well as along the axis of the electric dipole), no radiation is emitted, as for the direction $\sin\theta = 0$. The maximum of radiation lies in the coil plane, that is, in the equatorial plane of the magnetic dipole.

Exercise 10.3

Find the radiation power P_{rad} and the radiation resistance R_{rad} in the far-field zone of a magnetic dipole that has the form of a square loop. In the frame with sides $b = 10$ cm, we generate an alternating current with amplitude $I_0 = 2$ A and frequency $f = 150$ MHz. The magnetic dipole is placed in air. Express your result in terms of the wavelength of the emitted wave.

Solution. For an arbitrary direction of emission, at a distance r from the magnetic dipole, the instantaneous values of the wave fields in the far-field zone are determined by Equations 10.20. Since $\omega = 2\pi c/\lambda$, the amplitudes of these fields can be written as

$$E_0 = Z_0 \frac{I_0 l_m \omega}{4\pi c r}\sin\theta = \frac{Z_0 I_0 l_m}{2\lambda r}\sin\theta,$$

$$H_0 = \frac{I_0 l_m \omega}{4\pi c r}\sin\theta = \frac{I_0 l_m}{2\lambda r}\sin\theta,$$

where
$l_m = k_0 A = \omega A/c$ is the magnetic dipole length
$A = b^2$ is the loop area

The energy flux density, that is, the Poynting vector $\mathbf{S} = \mathbf{E}\times\mathbf{H}$, in a far-field zone has only a radial component S_r. Its average value over one period is

$$<S(\theta)> = \frac{1}{T}\int_0^T S_r(t)dt = \frac{E_0}{\sqrt{2}}\frac{H_0}{\sqrt{2}} = \frac{Z_0}{2}\left(\frac{I_0 l_m}{2\lambda r}\sin\theta\right)^2.$$

The total flux of the radiated power is obtained by integrating the flux density $<S(\theta)>$ on the surface of a sphere of radius r:

$$P_{rad} = \oint <S(\theta)>dA = \frac{Z_0}{2}\left(\frac{I_0 l_m}{2\lambda}\right)^2 \int_0^\pi\int_0^{2\pi}\left(\frac{\sin\theta}{r}\right)^2 r^2\sin\theta\, d\theta d\varphi$$

$$= \pi Z_0 \left(\frac{I_0 l_m}{2\lambda}\right)^2 \int_0^\pi \sin^3\theta\, d\theta = \frac{4\pi^3}{3}\left(\frac{b}{\lambda}\right)^4 I_0^2 Z_0.$$

Here, the area element is

$$dA = r^2\sin\theta\cdot d\theta\cdot d\varphi.$$

It is seen that the total energy flux through the chosen sphere does not depend on the distance to the loop. Since the surrounding medium is air, the wavelength of the emitted wave is $\lambda = c/f = 2$ m and $Z_0 = 377\ \Omega$, we get:

$$P_{rad} = 160\pi^4\left(\frac{b}{\lambda}\right)^4 I_0^2 = 0.39\ \text{W}.$$

Using the analogy with the expression for the alternating current power dissipated by a resistance R_{rad}, the radiation power can be represented in the following form:

$$P_{rad} = \frac{1}{2} I_0^2 R_{rad},$$

where the radiation resistance is given by

$$R_{rad} = \frac{8\pi^3}{3} \left(\frac{b}{\lambda}\right)^4 Z_0 = 0.19\ \Omega.$$

10.4   DIRECTIONAL DIAGRAM

In practice, for some situations, it is necessary to have an antenna emitting uniform radiation with equal intensity in all directions. In other situations, it is necessary to concentrate the radiated electromagnetic energy in certain directions, that is, to use directional antennas. A visual representation of the distribution of the radiated energy in space is given by the amplitude characteristic of the pattern that is determined by the dependence of the electric E or magnetic H fields, generated by the antenna, on azimuthal and polar angles φ and θ, respectively. The graphic representation of the directional characteristic of field is called a ***directional diagram***. The space directional diagram is represented in the form of a surface, $E(\theta, \varphi)$ or $H(\theta, \varphi)$. For convenience, we will use here only $E(\theta, \varphi)$ diagram unless we specifically indicate otherwise. Plotting such spatial diagram is cumbersome. This is why the directional diagrams are often plotted in a certain plane, so they become a function of only one angle, that is, $E(\varphi)$ or $E(\theta)$. This definition corresponds to the directional diagram of the field. In some cases, it is important to know the dependence of the power flux density as a function of angles, so the power flux density directional diagram is used. The power flux density is the power passing through a unit area perpendicular to the direction of wave propagation. Therefore, the directional diagram of power is proportional to $E^2(\theta, \varphi)$.

The directional diagram, for which the maximum value of the function $E(\theta, \varphi)$ is set to unity, is called the normalized directional diagram and is denoted as $F(\theta, \varphi)$. The $F(\theta, \varphi)$ diagram is obtained by dividing the values of the nonnormalized diagram by its maximum value, that is,

$$F(\theta, \varphi) = \frac{E(\theta, \varphi)}{E_{max}(\theta, \varphi)}. \tag{10.23}$$

We will discuss several directional diagrams that are often observed in practice. As it was shown earlier, the simplest emitters in the form of elementary electric and magnetic dipoles have the toroid-shaped directional diagram given in Figure 10.7. For this diagram, the function $F(\theta, \varphi) = F(\theta) = \sin\theta$ does not depend on the azimuthal angle. From this, it follows that radiation is uniformly distributed over the entire 0–2π azimuthal angle range.

Figure 10.8 shows an example of a unidirectional diagram that corresponds to electromagnetic waves produced by a directional antenna. Most of the radiation of the antenna with such directional diagram is concentrated within a small angle. Fan directional diagrams are also widely used. Such diagram is compressed in one plane (usually horizontal) and extended in another direction.

The angle $2\theta_{0.5}$ is introduced to estimate the directional effect of an antenna. The angle $2\theta_{0.5}$ is the angle between the directions, along which the field magnitude decreases by a factor of $\sqrt{2}$ with respect to the maximum of radiation at the main lobe center, and thus, the power flux is reduced by a factor of 2 with respect to the maximum (Figure 10.9). This angle is referred to as the ***width of diagram***.

Antennas also differ in their operating ranges. Operating wavelength range is the range within which the antenna retains its essential characteristics (directed action, the polarization structure of

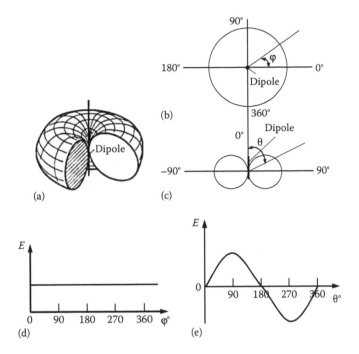

FIGURE 10.7   Toroidal directional diagram of a dipole (a) and details of dependence of field on angle φ (b) and (d) and on angle θ (c) and (d).

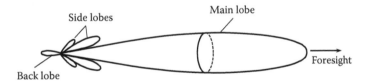

FIGURE 10.8   Unidirectional diagram of radiation by a directional antenna.

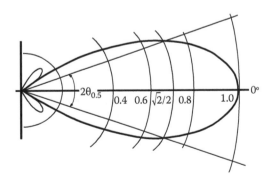

FIGURE 10.9   Definition of the directional diagram width.

the field) with a given accuracy. If the width of the operating range does not exceed a few percent of the middle wavelength, such an antenna is called narrow-range antenna; the antennas that have an operating range of a few tens of a percent of the middle wavelength and more are called *wide-range antennas.*

Note that the view of the antenna pattern is significantly affected by the Earth's surface, which has a rather high conductivity. Thus, if a horizontal antenna is located above the ground at a height

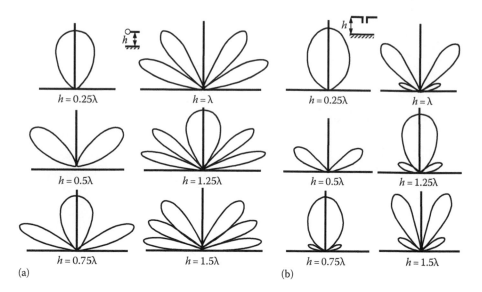

FIGURE 10.10   Directional diagrams of a half-wave Hertz dipole antenna located above the Earth in the planes perpendicular (a) and parallel (b) to its axis (insets in [a] and [b] show orientation of the antenna).

that is shorter compared to the wavelength, then there is practically no radiation as the antenna radiation is suppressed by its mirror image formed by the conducting ground. Increasing antenna's distance from the ground results in the increase of the distance between the antenna and its mirror image. When this distance becomes comparable to the wavelength, the two wave sources radiate in antiphase (i.e., their phases are shifted by $\pi/2$). The form of the directional diagram varies depending on the height of the antenna from the ground.

Directional diagrams of the half-wave Hertz dipole located at a height h above the ground are shown in Figure 10.10 in the planes perpendicular (a) and parallel (b) to the antenna axis. A half-wave symmetric Hertz dipole whose length is equal to a quarter wavelength (total length of the antenna is equal to half the wavelength) consists of two identical rectilinear conductors. The driving alternating current source is connected between the two conductors. The above diagrams show that with an increasing height, the number of lobes in the directional diagram and simultaneously the radiation directionality increase. The higher the antenna is suspended, the closer to the horizon a lower lobe will be.

Exercise 10.4

Plot directional diagrams of the symmetric antenna for different ratios of its length $2l$ to length of the emitted wave λ: $l/\lambda = 0.25$, 0.5, 0.75 and 1. The antenna (Figure 10.11) consists of two equally sized and shaped conductors, between which an alternating-current generator is

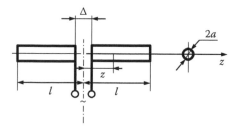

FIGURE 10.11   The symmetric dipole antenna.

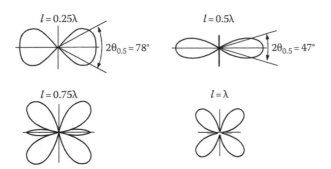

FIGURE 10.12   Directional diagram of antenna for different values of l.

connected. The current $I(z)$ has a sinusoidal dependence: $I(z) = I_0 \sin(k_0(l - |z|)) / \sin(k_0 l)$, where I_0 is the current in points of a power supply of the antenna (at $z = 0$). Use the following equation for the electric field component of the symmetric antenna in a point at a distance $r \gg \lambda$, located in the far field zone:

$$E(r, \theta, \varphi) = \frac{i 60 I_0}{r \sin(k_0 l)} f(\theta, \varphi) \exp(-i k_0 r),$$

where $k_0 = 2\pi/\lambda$ and the directional function of radiation has the form

$$f(\theta, \varphi) = f(\theta) = \frac{\cos(k_0 l \cos\theta) - \cos(k_0 l)}{\sin\theta}.$$

Solution. The symmetric antenna differs from the elementary dipole of Hertz by a nonuniform current distribution along the length. From the expression for the electric field it can be seen that the symmetric antenna radiates spherical waves and it does not possess directional properties in the plane, which is perpendicular to the dipole axis. Directional properties of the radiation occur only in a tangential plane (in the plane of electric field vector).

From the given relations it follows that the directional properties of the symmetric antenna, with a sinusoidal current propagation in it, are determined only by the ratio l/λ. We introduce the normalized directional diagram of the symmetric antenna:

$$F(\theta) = \frac{f(\theta)}{f(\pi/2)} = \frac{\cos(k_0 l \cos\theta) - \cos(k_0 l)}{(1 - \cos(k_0 l)) \sin\theta}.$$

The directional diagrams of the symmetric antenna with different ratios l/λ are represented in Figure 10.12.

10.5   TYPES OF ANTENNAS

A *transmitting antenna* is a device that transforms the energy of a high-frequency current into the energy of electromagnetic waves emitted in a certain direction. The transmitting antenna should not simply radiate electromagnetic waves but provide the most effective energy distribution in space. In this regard, the most important property of transmitting antennas is the directionality of a field radiated by them. Requirements to the directionality of antenna range within rather wide limits—from a weak directionality (for radio and television broadcasting) to sharply defined directionality (for distant cosmic radio communication, radiolocation, and radio astronomy). The directionality of an antenna allows increasing the field power radiated in a given direction without increasing the power of the transmitter. A proper orientation also reduces disturbances by radio systems promoting a solution of the electromagnetic compatibility problem. Directionality can be obtained only when the antenna dimensions significantly exceed the wavelength of the emitted wave.

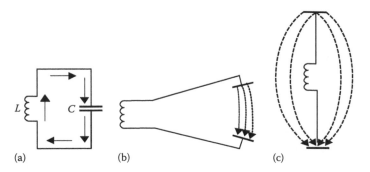

FIGURE 10.13 Transition from a closed circuit to an open circuit antenna: (a) closed circuit, (b) partially open circuit, and (c) open circuit antenna.

A *feeder* is the line that transmits power from the generator to the antenna (in transmission mode) or from the antenna to the receiver (in reception mode). The feeder plays an important role in the operation of antennas. The wave resistance of the feeder has to be matched with the resistance of the antenna to reduce losses.

For the understanding of the operating principle of most types of antennas, we consider the so-called ***open oscillatory circuit*** (Figure 10.13). Electromagnetic waves are radiated by conductors that carry a high-frequency current. If the current oscillations occur in a "closed" oscillatory circuit consisting of a coil and a capacitor, the alternating magnetic field is related to the coil, and the alternating electric field is concentrated in the space between the capacitor plates. A closed oscillatory circuit, which has an inductor whose size is much smaller than the operating wavelength, practically does not radiate electromagnetic waves into the surrounding space. In the closed oscillatory circuit, the magnetic field is concentrated inside the coil and the electric field between the capacitor plates.

If we change the geometry of the oscillatory circuit, by moving the capacitor plates apart, then electromagnetic waves created by the capacitor plates will propagate in space. Thus, transformation of the closed oscillatory circuit into an open oscillatory circuit turns it into an antenna. As the displacement current of an open oscillatory circuit covers large volume of space, efficient radiation of waves becomes possible. In the simplest case, an open oscillatory circuit is a rectilinear conductor with a small coil, which is inserted to connect with a generator or receiver. An antenna, obtained by such transformation, is characterized by geometrical symmetry and is called a symmetric antenna. It has a nonuniform distribution of current and voltage along its length. For the antenna length $l \approx \lambda/2$, the distribution of current and voltage has the form shown in Figure 10.14.

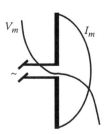

FIGURE 10.14   Current and voltage distribution in an antenna.

In practice, the asymmetric antenna, in which the Earth's surface is used as a lower half of a dipole, is also widely applied. This becomes possible due to the relatively good conductivity of the Earth's surface. Note also that each wire has its own inductance and capacitance distributed along its length. Therefore, each piece of a conductor is an oscillatory circuit. Approximately, one can assume that each meter of wire has a capacitance of $C \approx 5$ pF and an inductance of $L \approx 2 \times 10^{-6}$ H.

The *receiving antenna* transforms the energy of free electromagnetic waves, coming from a certain direction to the energy, which enter through the feeder to the receiver input. The directional properties in the receiving antennas increase the registered power, which enters into the receiver.

Owing to the reversibility of electromagnetic processes that occur in the antenna systems, it is possible in principle to use the same antenna both for transmission and for reception. However, this does not mean that transmitting and receiving antennas are identical in design, as each type of antennas has its own design features. It is usually assumed that the receiving antenna is oriented in space so that a reception is carried out from the direction of the maximal reception.

If the antenna is located in the area that has electromagnetic waves, an *emf* appears at the terminals of the antenna. If we connect a receiver to the antenna terminals, current flows in the antenna circuit, which generates a voltage at the receiver input. The *emf* across a receiving antenna E_a depends on the electric field strength E at reception point and also on the form and dimensions of the receiving antenna. Their relation is given by the expression $E_a = h_a E$, where h_a is called the *operating height of a receiving antenna*.

Antennas can be classified according to different criteria: wavelength ranges and application purposes (for radio, radio broadcasting, radio astronomy, television, etc.). It is most expedient to classify them by the type of emitting elements. On this basis, all antennas can be divided into the following basic groups.

Linear antennas include any radiating system of small cross-sectional dimension in which the alternating currents are flowing along the system axis. This can be either a thin metal wire, in which an alternating current is excited, or a narrow slit in a metal screen, between the edges of which an alternating voltage is applied. A characteristic feature of the linear antennas is that the current distribution along their axis does not depend on the wire configuration. Therefore, the linear antennas are not always straight lines but can also be curve, but the radius of the curvature should be small compared to the total length of the antenna and to the wavelength. These antennas can be symmetric and asymmetric antennas, frame antennas, wire antennas of traveling wave (including helical antennas), and thin slot antennas of standing and traveling waves.

Examples of *aperture antennas* are the pyramidal funnel-shaped antennas, the optical-type parabolic antennas, lens antennas, and the open-ended waveguide antennas.

Aperture antennas, for which it is possible to define some bounded surface, through which the all flow of radiated (or received) electromagnetic energy passes. This surface (an opening, an aperture) is often represented in the form of a plane. Dimensions of the aperture are usually much larger than the wavelength. Examples of aperture antennas are pyramidal horn antennas, optical-type parabolic antennas, lens antennas, and radiating open-ended waveguide antennas.

Ground-wave antennas are excited by electromagnetic waves propagating along the antenna and radiate predominantly in the direction of propagation, an example is the rod dielectric antenna.

All three types of antennas can be used as single antennas or can be grouped in multielement systems. In particular, *antenna lattices* are widely applied. Such antennas consist of several single-type emitters and are arranged in space in a certain way and excited by one generator or several coherent generators. Here, it is possible to receive both required spatial distribution of emitted energy and required control in time by this distribution. Typical antenna lattice is the VHF director antenna, which represents the linear array of half-wave symmetric antennas.

Exercise 10.5

A receiving antenna consists of a rectangular frame with dimensions $a = 30$ cm, $b = 15$ cm. The received signal frequency $f = 10$ MHz. The magnetic field is perpendicular to the plane of the frame, and the effective value of the magnetic field intensity of the signal at the location of the frame is $H = 80$ mA/m. Find the greatest *emf* induced by a signal in the frame. Determine the lowest frequency at which the signal can be received if the receiver sensitivity is 15 μV.

Solution. Let us direct the x-axis along one side of the frame and the y-axis along its other side. For a plane wave propagating along the x-axis, the expressions for the fields can be written as

$$E_y(x,t) = E_y(x)\exp(i\omega t), \quad H_z(x,t) = H_z(x)\exp(i\omega t),$$

where

$$E_y(x) = E_0 \exp(-ik_0 x), \quad H_z(x) = H_0 \exp(-ik_0 x).$$

The wave impedance of air is

$$Z_0 = \sqrt{\frac{\mu_0}{\varepsilon_0}} = 120 \ \pi\Omega.$$

We assume that in the frame plane, the magnetic field (at $x = 0$) is $H_z(0) = H_0 = 80$ μA/m, and the electric field is

$$E_y(0) = E_0 = Z_0 H_0 \approx 3 \times 10^{-2} \text{ V/m}.$$

The wavelength and wave propagation constant in air are

$$\lambda = \frac{c}{f} = 30 \text{ m}, \quad k_0 = \frac{\omega}{c} = \frac{2\pi f}{c} = \frac{2\pi}{\lambda} \approx 0.21 \text{ m}^{-1}.$$

When the wave propagates along the large side of the frame, the change of its phase is $k_0 a = 2\pi a/\lambda = 0.063$ rad. If you turn the frame in the plane by 90°, the change in phase of the oscillation will be $k_0 b = 0.0315$ rad. We calculate the *emf* induced in the frame by means of the expression

$$\mathcal{E}_i = \oint_L \mathbf{E} d\mathbf{l},$$

where an integration is carried out along the frame. The direction of a path tracing corresponds to the right-hand rule relative to the direction of vector **H**. For $z = a$, the field $E_y(a) = E_0\exp(-ik_0 a)$, so the induced *emf* in the frame is given by

$$\mathcal{E}_i = bE_0\exp(-ik_0 a) - bE_0 = bE_0\left[\exp(-ik_0 a) - 1\right].$$

In this example, $k_0 a = 0.063 \ll 1$. Therefore, to calculate the exponent, we use the first two terms of the expansion

$$\exp(-ik_0 a) \approx 1 - ik_0 a.$$

As a result, the *emf* induced in the antenna is determined by the expression

$$\mathcal{E}_i = -ik_0 abE_0 = -ik_0 abZ_0 H_0 = -i \times 0.063 \times 0.15 \times 120\pi \times 80 \times 10^{-6} = -i \times 0.285 \text{ mV}.$$

The complex value of *emf* means a phase shift of *emf* with respect to the current by

$$\Delta\varphi = \frac{\pi}{2} \quad \text{as} \quad \exp\left(\frac{-i\pi}{2}\right) = -i.$$

The induced *emf* is proportional to the frequency, that is,

$$\mathcal{E}_i = -ik_0 abZ_0 H_0 = -i2\pi f abZ_0 H_0 / c$$

$$\frac{\mathcal{E}_i}{f} = \frac{\mathcal{E}_{min}}{f_{min}} = -i\frac{2\pi abZ_0 H_0}{c} = \text{const.}$$

Therefore, the minimum frequency, at which this frame can be used as a receiving antenna for the receiver with sensitivity $\mathcal{E}_{min}$, is

$$f_{min} = \mathcal{E}_{min}\frac{f}{\mathcal{E}_i} = 0.53 \text{ MHz.}$$

10.6 HORN ANTENNAS

As an example of the aperture antenna, we will consider a waveguide horn antenna. Some advantages of these antennas are design simplicity, small losses, and well-controlled wave length. The efficiency of the horn antennas is very high (close to 100%). This results in their broad application in the millimeter, centimeter, and decimeter ranges. Such antennas can form the directional diagrams of a width range from 100° to 140° (when a special opening is used) to 10°–20° in pyramidal horns. Possibility of further narrowing of the horn diagram is limited by the necessity of sharp increase of its length.

A horn antenna is formed by a smooth increase in transverse dimensions of a waveguide. Since rectangular and circular waveguides are usually used, the horns formed from these waveguides are the most widely applied (Figure 10.15). If broadening of a rectangular waveguide occurs only in one plane, then the horn obtained in this way is called sectoral horn. Sectoral horns, obtained by broadening of a rectangular waveguide in one of the two planes (horizontal or vertical), include *H*-plane sectoral horn (a) and *E*-plane sectoral horn (b).

Sectoral horns allow narrowing the directional diagram of the horn antenna only in that plane in which broadening is carried out. In other plane, the directional diagram remains the same as that of the open end of the waveguide of which the horn was formed. Thus, the sectoral horns create fan-type directional diagrams.

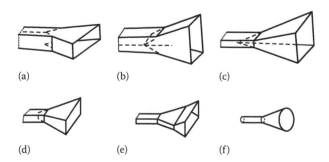

FIGURE 10.15 Types of horn antennas: (a) *H*-plane sectoral horn, (b) *E*-plane sectoral horn, (c) pointed pyramidal horn, (d) wedge pyramidal horn, (e) combined horn, and (f) conical horn.

In order to narrow the directional diagram in both planes, a pyramidal horn is applied. It is formed by broadening of the waveguide in both planes (c, d). For pointed pyramidal horn, the edges meet in one point (c), and for wedge horn, the edges meet in a line (d).

Broadening of the waveguide in both planes can be made not simultaneously but successively. The resulting horn is called a combined horn (e). This horn can be matched better than the pyramidal horn, but because of its design complexity, it is seldom used. Broadening of a circular waveguide forms a conical horn (f).

Since the horn is formed as a result of a smooth increase in the waveguide cross section, the oscillations that are excited in it are of the same type as in the entrance feeding waveguide. In Figure 10.16, the configuration of an electromagnetic field in a pyramidal horn antenna is shown; the antenna is fed from the waveguide excited by a wave H_{10}.

As seen from Figure 10.16, the oscillations excited in the horn are of the same type as that in the waveguide. However, the wave front at the transition from the waveguide to the horn is transformed from flat into spherical. The line where the wave front meets with the plane of the longitudinal section of the horn has the form of an arc of a circle, which has its center in the horn top O. In the plane *E*, the wave front meets with the horn walls at an angle of 90°, since the tangential component of electric field on a conducting surface is equal to zero.

An increase in the antenna aperture leads to a decrease in the main lobe width and an increase in the antenna directional effect. The greater is the angle of the horn aperture, the greater is the path difference between the central and peripheral rays and the greater the phase distortions $\varphi_{max} = k\Delta r$ at its edges. Violation of equiphase conditions of the radiating surface leads to distortions of a horn directional diagram. As a consequence, the main lobe of directional diagram widens, the intensity of the side lobes increases, and the zero between

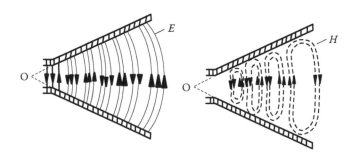

FIGURE 10.16 Electromagnetic field distribution in the pyramidal horn.

the lobes disappears. In the plane E (at an uniform amplitude distribution), the directional diagram is distorted more than in the plane H.

In the design of the horn antenna, the following initial basic data are optimized: wavelength λ or range of operating wavelengths λ_{min}, ..., λ_{max}, radiation power P, and directional diagram width at half-power level $2\theta_{0.5}$ and $2\varphi_{0.5}$.

Exercise 10.6

Determine the phase distribution in the aperture plane of the H-plane sectoral horn antenna. The width of the horn aperture is L, the size of wide wall of the waveguide is a, and the depth of the horn is R (Figure 10.17).

Solution. In the horn cavity in a plane E, the wave front is flat; therefore, for this direction, the phase is constant in the plane of the wave front. To determine the type of phase distribution in a plane H, we will consider the geometric relationships in the horn. At the horn edges, the wave front lags behind the front at the horn center. The maximum path difference can be determined from the following geometric relationships:

$$R^2 + \left(\frac{L}{2}\right)^2 = (R + \Delta r)^2 = R^2 + 2R\Delta r + (\Delta r)^2.$$

Since $\Delta r \ll R$ and $(\Delta r)^2 \ll 2R\Delta r$, then $(L/2)^2 \approx 2R\Delta r$. Therefore, the maximum path difference is

$$\Delta r_{max} \approx \frac{(L/2)^2}{2R}.$$

The maximum phase difference of aperture in the H plane is

$$\varphi_{max} = k_0 \Delta r_{max} = k_0 \frac{(L/2)^2}{2R} = \frac{\pi L^2}{4\lambda R}.$$

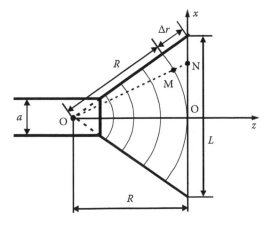

FIGURE 10.17 H-plane horn antenna.

The equation that gives the phase change of aperture in the H plane (phase distribution) is obtained if in this expression we substitute, instead of the maximum value of the coordinate $x_{max} = L$, the current coordinate $-L \leq x \leq L$:

$$\varphi(x) = \frac{\pi x^2}{4 \lambda R}.$$

Thus, along the x-axis, the phase is proportional to the square of x. The square law of phase change in the aperture widens the main lobe of the directional diagram compared with a uniform distribution.

PROBLEMS

10.1   According to the classical model of the hydrogen atom, the electron revolves around the nucleus (proton) in a circular orbit whose radius is $r_B = 0.53 \times 10^{-10}$ m. Determine the energy that an electron loses by radiation per unit time (radiant power). Estimate the time required for an electron to lose all its energy. (*Answer:* $|E| \approx 2.17 \times 10^{-12}$ J, $\tau \approx 4.70 \times 10^{-5}$ s.)

10.2   Compare the magnitude of the electrostatic field of a stationary proton with the wave electric field, which is created in the far-field (wave) zone by a rotating electron in its first Bohr orbit in a hydrogen atom. Estimate the values of these fields at a distance of 1.00 cm from the atom $(r \gg r_B = 5.30 \times 10^{-11}$ m) in a direction perpendicular to the orbital plane. (*Answer:* $E_p/E_{e0} \approx 10^{-4}$, $E_{e0} \approx 0.14$ V/m.)

10.3   An indoor radio antenna is a straight wire of length $l = 0.50$ m. The frequency f of the emitted signal is equal to 2.00 MHz. The amplitude of the electric field at the location of the antenna is $E = 5.00$ mV/m. Assume that the receiving antenna is in the far-field zone of the transmitter at a distance $R \gg \lambda$ and orient the antenna so that the induced *emf* in it is maximized. Calculate the amplitude of the induced *emf*. (*Answer:* $\mathcal{E}_{max} = 2.5$ mV.)

10.4   An antenna consists of a circular coil of diameter $d = 0.25$ m, which is placed in an electromagnetic field of frequency $f = 1.50$ MHz. The amplitude of the electric field in the center of the coil is $E_0 = 0.15$ mV/m. Locate the coil so that the induced *emf* in it is maximized. Determine the amplitude of this *emf*. (*Answer:* $\mathcal{E}_{max} = 0.24$ μV.)

10.5   Find the ratio of the electromagnetic wave powers emitted by electric dipoles, which have the same magnitude of charge and the same amplitude of charge oscillations if the oscillation frequency ratio of the two waves is equal to 10. (*Answer:* The ratio is equal to 10^4.)

10.6   A receiving antenna consists of a rectangle with sides $a = 0.40$ m and $b = 0.20$ m. The frequency f of the received signal is equal to 10.0 MHz; the amplitude H of the magnetic intensity at the location of the frame is equal to 0.10 mA/m. Find the lowest frequency at which the signal reception by this antenna is still possible if the receiver's sensitivity is equal to 20.0 μV. (*Answer:* $f_{min} = 0.32$ MHz.)

10.7   The amplitude of the electric field of a wave emitted by a Hertz dipole in the equatorial plane at a distance $r = 6.00$ km is $E_{max} = 2.00 \times 10^{-3}$ V/m. Find the total power, P_{rad}, of radiating dipole. (*Answer:* $P_{rad} = 1.60$ W.)

10.8   Find the current in the elementary electric dipole (Hertz dipole) of length equal to 15.0 cm if at the point with coordinates $r = 10.0$ km and $\theta = \pi/2$ the electric field amplitude is $E_0 = 10^{-4}$ V/m. The antenna is perpendicular to the wave front and the oscillation frequency is equal to 0.10 GHz.

10.9   Find the radiation resistance of an elementary electric dipole (Hertz dipole) of length $l = 10.0$ cm emitting a wave of wavelength $\lambda_0 = 1.50$ m. Find the radiation power if the amplitude of the current in the dipole is $I_0 = 1.00$ A.

10.10 Find the field components of an elementary electric dipole (Hertz dipole) of length $l = 10.0$ cm in the equatorial plane at a distance of $r = 100$ m if the oscillation frequency $f = 0.60$ GHz. The amplitude of the current in the emitter is $I_0 = 8.00$ A.

10.11 A square frame with sides equal to 15.0 cm emits electromagnetic waves for which the electric field in the equatorial plane has an amplitude $E_0 = 10^{-3}$ V/m at a distance $r = 5.00$ km. Determine the current in the frame if the wavelength of the emitted radiation $\lambda_0 = 3.00$ m. (*Answer*: $I_0 = 1.69$ A.)

10.12 Find the radiated power of an elementary magnetic dipole if at a distance $r = 2.00$ km the electric field of the emitted wave has the amplitude $E_{max} = 5.00$ mV/m in the equatorial plane. (*Answer*: $P_{rad} = 1.11$ W.)

Section IV

Advanced Topics in
Electromagnetics and Optics

11 Electromagnetic Waves in Gyrotropic Media

In the previous chapter, we considered the aspects of wave propagation in anisotropic media, for which the dielectric permittivity tensor is symmetric. Such tensor in the coordinate system that coincides with the principal axes is transformed to a diagonal tensor. Anisotropic media, in which tensors of dielectric and/or magnetic permeability are antisymmetric and in the absence of absorption have purely imaginary nondiagonal components, are known as *gyrotropic*. Such media have the ability to rotate the polarization plane of linearly polarized electromagnetic waves, which propagate through it.

A medium, which exhibits gyrotropic properties under the influence of a constant external magnetic field, is called *magnetoactive*. Examples of such media are magnetized plasma and magnetized ferrites. The magnetoactive plasma is an example of a *gyroelectric* medium for which the tensor of dielectric permittivity is antisymmetric and the magnetic permeability is scalar. For a magnetized ferrite, the magnetic permeability is an antisymmetric tensor. In this case, the medium is called *gyromagnetic*.

11.1 DIELECTRIC PERMITTIVITY TENSOR OF MAGNETOACTIVE PLASMA

Plasma is an ionized gas consisting of a mix of neutral and charged particles. Usually, the charged particles are positively charged ions and negatively charged free electrons. In general, plasma is electrically neutral. Later, while considering the motion of plasma particles in external fields, we will neglect the movement of positively charged ions, since their mass M is much larger than the electronic mass m. The proton and the neutron are approximately 1836 times heavier than the electron, and atoms contain several protons and neutrons.

We will not consider the collision of charged particles with each other, that is, we neglect the energy losses during these collisions. As a result, it is possible to neglect conduction currents in comparison with displacement currents in Maxwell's equations.

First, let us consider a nonmagnetized plasma (magnetic field is not applied). Due to the chaotic thermal motion of the particles, such plasma should be isotropic. Assume that the plasma is placed in an electric field of a wave propagating through the plasma. This field is expressed by the form

$$\mathbf{E}(t) = \mathbf{E}_0 \exp(i\omega t), \tag{11.1}$$

where ω is the wave angular frequency. Under the influence of this electric field, the free electrons of the plasma oscillate about their equilibrium positions. The equation of motion of a free electron can be written as

$$m \frac{d^2\mathbf{r}}{dt^2} = -e\mathbf{E}, \tag{11.2}$$

where
  e is the elementary charge
  $\mathbf{r}$ is the electron displacement from its equilibrium position

Taking into account Equation 11.1, Equation 11.2 is easily integrated, and one gets

$$\mathbf{r}(t) = \mathbf{r}_0 \exp(i\omega t), \quad \mathbf{r}_0 = \frac{e\mathbf{E}_0}{m\omega^2}. \tag{11.3}$$

The displacement $\mathbf{r}$ of the electron from its equilibrium position results in the appearance of an electric dipole moment $\mathbf{p} = -e\mathbf{r}$. Note that in the definition of dipole by Equation 1.13, the vector $\mathbf{l}$ is directed from the negative to the positive charge, but $\mathbf{r}$ is the coordinate of the negative charge; this is why there is a negative sign (–) in the definition of $\mathbf{p}$ given earlier. If there are N free electrons per unit volume, which have the same displacement under the influence of the field, the total dipole moment per unit volume (***polarization***) taking into account Equation 11.3 is

$$\mathbf{P}(t) = -Ne\mathbf{r}(t) = -\frac{Ne^2\mathbf{E}(t)}{m\omega^2}. \tag{11.4}$$

Using the relation of the polarization vector and the electric field, $\mathbf{P} = \varepsilon_0(\kappa - 1)\mathbf{E}$, for the dielectric permittivity of the plasma, we obtain the expression

$$\kappa = 1 - \frac{Ne^2}{\varepsilon_0 m\omega^2} = 1 - \frac{\omega_p^2}{\omega^2}, \tag{11.5}$$

where we have introduced the ***plasma frequency*** $\omega_p = \sqrt{Ne^2/\varepsilon_0 m}$. This frequency plays the role of frequency "cutoff" below which an electromagnetic wave cannot propagate in an isotropic plasma. When $\omega < \omega_p$, the dielectric permittivity is negative, and the refractive index of the plasma is $\tilde{n} = \sqrt{\kappa}$, which is an imaginary value. From Equation 11.5, it also follows that the dielectric permittivity of a nonmagnetized plasma is scalar with strong frequency dispersion.

Now we consider magnetized plasma. Plasma is magnetized if charged particles of the plasma change their motion with the magnetic field applied to the plasma. We suppose that along with the field of Equation 11.1, a constant external magnetic field $\mathbf{B}_0$ (magnetic field intensity $\mathbf{H}_0 = \mathbf{B}_0/\mu_0$) is applied along the z-axis.

Thus, in the plasma, there is a well-defined direction given by the external magnetic field. Under the influence of the Lorentz force, $\mathbf{F}_m = -e(\mathbf{v} \times \mathbf{B}_0)$, the free electrons of the plasma move along helical paths, revolving around the external magnetic field with angular frequency

$$\omega_c = \frac{eB_0}{m} = \frac{e\mu_0 H_0}{m}. \tag{11.6}$$

The quantity ω_c is known as the ***cyclotron frequency*** and it does not depend either on the particle velocity or on the radius of its orbit. In the presence of an alternating electric field $\mathbf{E}$, the electron trajectories will be more complicated, and their forms depend on the angle between vectors $\mathbf{E}$ and $\mathbf{B}_0$. The polarization vector $\mathbf{P}$ of the plasma will depend on the direction of the vector $\mathbf{B}_0$, so the plasma is magnetized and exhibits the properties of an anisotropic medium.

Let $\mathbf{E}(t)$, described by Equation 11.1, be the alternating electric field of an electromagnetic wave propagating in the plasma. We make the following simplifying approximations: plasma ions can be regarded as immobile when $\omega \gg \omega_{ci}$, where ω_{ci} is the ion cyclotron frequency. If, in addition, $\omega \gg \omega_{col}$, where ω_{col} is a frequency of electron collisions with ions and neutral particles, conduction currents can be neglected compared with displacement currents (actually, we neglect losses in the plasma). Under these assumptions, we determine the dielectric permittivity of a magnetized plasma.

Since the wave electric field $\mathbf{E}$ has a harmonic time dependence, the plasma polarization vector will also possess the same time dependence, that is, $\mathbf{P}(t) = \mathbf{P}_0 \exp(i\omega t)$. Therefore, according to Equation 11.4, we have $d\mathbf{P}/dt = i\omega\mathbf{P} = -Ne\mathbf{v}$, where $\mathbf{v}(t) = \mathbf{v}_0 \exp(i\omega t)$ is the electron velocity. And we get

$$\mathbf{D} = \varepsilon_0 \mathbf{E} + \mathbf{P} = \varepsilon_0 \mathbf{E} + \frac{ieN\mathbf{v}}{\omega}. \tag{11.7}$$

We will find the velocity $\mathbf{v}$ from the solution of the equation of motion for an electron:

$$m\frac{d\mathbf{v}}{dt} = -e\left(\mathbf{E} + \mathbf{v} \times \mathbf{B}_0\right). \tag{11.8}$$

Since the time dependence of the velocity $\mathbf{v}$ is harmonic, $d\mathbf{v}/dt = i\omega\mathbf{v}$. With this, the projections of vector Equation 11.8 on the Cartesian coordinate system axes are given by

$$v_x = i\frac{e}{m\omega}E_x + i\frac{\omega_c}{\omega}v_y,$$

$$v_y = i\frac{e}{m\omega}E_y - i\frac{\omega_c}{\omega}v_x, \tag{11.9}$$

$$v_z = i\frac{e}{m\omega}E_z,$$

where we have introduced the cyclotron frequency defined by Equation 11.6. Solving Equation 11.9 with respect to the velocity components v_x, v_y, v_z and substituting the obtained expressions in Equation 11.7, we arrive at the following expressions for the projections of electric displacement vector $\mathbf{D}$:

$$D_x = \varepsilon_0 \left(1 - \frac{\omega_p^2}{\omega^2 - \omega_c^2}\right) E_x + i\varepsilon_0 \frac{\omega_p^2 \omega_c}{\omega(\omega^2 - \omega_c^2)} E_y,$$

$$D_y = -i\varepsilon_0 \frac{\omega_p^2 \omega_c}{\omega(\omega^2 - \omega_c^2)} E_x + \varepsilon_0 \left(1 - \frac{\omega_p^2}{\omega^2 - \omega_c^2}\right) E_y, \tag{11.10}$$

$$D_z = \varepsilon_0 \left(1 - \frac{\omega_p^2}{\omega^2}\right) E_z.$$

In this equation, we used the expression for the plasma frequency $\omega_p = \sqrt{Ne^2/\varepsilon_0 m}$ that was introduced in Equation 11.5. From Equation 11.10, it follows that the dielectric permittivity of a magnetoactive plasma is an antisymmetric tensor with imaginary off-diagonal components:

$$\hat{\kappa} = \begin{pmatrix} \kappa & i\kappa_a & 0 \\ -i\kappa_a & \kappa & 0 \\ 0 & 0 & \kappa_z \end{pmatrix}, \tag{11.11}$$

where we have introduced the expressions

$$\kappa = 1 - \frac{\omega_p^2}{\omega^2 - \omega_c^2}, \quad \kappa_a = \frac{\omega_p^2 \omega_c}{\omega(\omega^2 - \omega_c^2)}, \quad \kappa_z = 1 - \frac{\omega_p^2}{\omega^2}.$$

Thus, from Equation 11.11, it follows that the components of the dielectric permittivity tensor of a plasma placed in a magnetic field depend on cyclotron frequency. As the frequency ω approaches ω_c (i.e., $\omega \to \omega_c$), both κ and κ_a tend to infinity.

This means that in the magnetoactive plasma, there is a resonance at the cyclotron frequency ω_c, and at this resonant frequency, the amount of energy that is absorbed by the plasma will reach its maximum. When $\omega \to \omega_c$, κ and κ_a tend to infinity. The presence of these singularities is due to the fact that dissipative processes, that is, plasma conductivity, were not taken into account in the derivation of Equation 11.11. If energy dissipation is taken into account, both κ and κ_a in the magnetoactive plasma remain finite at the resonant frequency.

Exercise 11.1

Write an expression for the refractive index $\tilde{n}$ of an isotropic plasma (in the absence of an external magnetic field) taking into account the motion of free electrons and the positively charged ions (the charge of ion is equal to e). Assume 1D motion for electrons and ions, for example, along x-axis. Also take into account energy losses by particles (the result of nonelastic collisions) by introducing dissipative force $\mathbf{F}_e = -\delta_e \mathbf{v}_e = -\delta_e(dx/dt)$ and $\mathbf{F}_i = -\delta_i \mathbf{v}_i = \delta_i(dX/dt)$ into the equations of motion for the electrons and the ions, respectively (here x and X are displacements of an electron and an ion).

Solution. The equation of motion for electrons and ions under the influence of the electric field of a monochromatic wave in the presence of dissipative forces can be written as

$$m\frac{d^2x}{dt^2} + \delta_e \frac{dx}{dt} = -eE_0 \exp(i\omega t),$$

$$M\frac{d^2X}{dt^2} + \delta_i \frac{dX}{dt} = eE_0 \exp(i\omega t),$$

where m and M are the electron and ion masses, respectively. Let us introduce the parameters $\gamma_e = \delta/m$, $\gamma_i = \delta_i/M$ and look for harmonic solutions of these equations:

$$x(t) = x_m \exp(i\omega t), \quad X(t) = X_m \exp(i\omega t).$$

For the electron and ion oscillation amplitudes, we get

$$x_m = \frac{eE_0}{m} \cdot \frac{1}{\omega^2 - i\gamma_e\omega}, \quad X_m = -\frac{eE_0}{M} \cdot \frac{1}{\omega^2 - i\gamma_i\omega}.$$

The electric dipole moment of an electron and an ion can be found by multiplying their displacements with their charges, that is,

$$p_e(t) = -ex(t), \quad p_i = eX(t).$$

The total dipole moments of electron and ion are

$$p(t) = p_e(t) + p_i(t) = -ex(t) + eX(t)$$

$$= -\left(\frac{e^2/m}{\omega^2 - i\gamma_e\omega} + \frac{e^2/M}{\omega^2 - i\gamma_i\omega} \right) E_0 \exp(i\omega t).$$

The dipole moment per unit volume, $P(t)$, can be found if we multiply $p(t)$ by the number of electrons and ions in the unit volume N (let us assume that $N_e = N_i = N$):

$$P(t) = -\left(\frac{Ne^2/m}{\omega^2 - i\gamma_e\omega} + \frac{Ne^2/M}{\omega^2 - i\gamma_i\omega} \right) E(t).$$

Now, let us take into account the relation between the polarization vector and the electric field vector: $\mathbf{P} = \varepsilon_0(\kappa - 1)\mathbf{E}$. As a result, we get the following expressions for the dielectric permittivity and the refractive index for the plasma:

$$\kappa = n^2 = 1 - \left(\frac{Ne^2/\varepsilon_0 m}{\omega^2 - i\gamma_e\omega} + \frac{Ne^2/\varepsilon_0 M}{\omega^2 - i\gamma_i\omega} \right).$$

Let us introduce the plasma frequency for electrons and ions,

$$\omega_{pe} = \sqrt{\frac{Ne^2}{\varepsilon_0 m}} \quad \text{and} \quad \omega_{pi} = \sqrt{\frac{Ne^2}{\varepsilon_0 M}},$$

and take into account that for plasma the value of n is close to unity.
   Therefore,

$$n^2 = 1 - u, \quad u \ll 1, \quad n = \sqrt{1-u} \approx 1 - \frac{u}{2}.$$

The refractive index of the plasma is

$$\tilde{n} = 1 - \frac{1}{2}\left(\frac{\omega_{pe}^2}{\omega^2 - i\gamma_e\omega} + \frac{\omega_{pi}^2}{\omega^2 - i\gamma_i\omega} \right).$$

The refractive index is complex because of the dissipation, that is, $\tilde{n} = n - i\kappa$. In this case, it is necessary to separate the real and imaginary parts of n:

$$n = 1 - \frac{1}{2}\left(\frac{\omega_{pe}^2}{\omega^2 + \gamma_e^2} + \frac{\omega_{pi}^2}{\omega^2 + \gamma_i^2} \right), \quad \kappa = \frac{1}{2}\left(\frac{\omega_{pe}^2}{\omega}\frac{\gamma_e}{\omega^2 + \gamma_e^2} + \frac{\omega_{pi}^2}{\omega}\frac{\gamma_i}{\omega^2 + \gamma_i^2} \right).$$

We note that the plasma frequency for the ions is much smaller than the plasma frequency for the electrons ($\omega_{pi} \ll \omega_{pe}$) since $M \gg m$.

11.2   ELECTROMAGNETIC WAVES IN MAGNETOACTIVE PLASMA

Let us now consider a plane monochromatic wave propagating in a magnetoactive plasma, magnetized in the direction of z-axis by the external field $\mathbf{H}_0$. Maxwell's equations for the field vectors $\mathbf{E}$ and $\mathbf{H}$ take the form

$$\nabla \times \mathbf{E} = -i\omega\mu_0\kappa_m\mathbf{H},$$
$$\nabla \times \mathbf{H} = i\omega\varepsilon_0\,\hat{\kappa}\,\mathbf{E}, \tag{11.12}$$

where the dielectric permittivity tensor $\hat{\kappa}$ is determined by Equation 11.11 and the magnetic permeability of the plasma is practically equal to unity. Let us write the projections of Equation 11.12 on the Cartesian coordinate axes. The first of these equations can be written as

$$\frac{\partial E_z}{\partial y} - \frac{\partial E_y}{\partial z} = -i\omega\mu_0\kappa_m H_x,$$

$$\frac{\partial E_x}{\partial z} - \frac{\partial E_z}{\partial x} = -i\omega\mu_0\kappa_m H_y, \tag{11.13}$$

$$\frac{\partial E_y}{\partial x} - \frac{\partial E_x}{\partial y} = -i\omega\mu_0\kappa_m H_z,$$

and the second equation as

$$\frac{\partial H_z}{\partial y} - \frac{\partial H_y}{\partial z} = i\omega\varepsilon_0(\kappa E_x + i\kappa_a E_y),$$

$$\frac{\partial H_x}{\partial z} - \frac{\partial H_z}{\partial x} = i\omega\varepsilon_0(-i\kappa_a E_x + \kappa E_y), \tag{11.14}$$

$$\frac{\partial H_y}{\partial x} - \frac{\partial H_x}{\partial y} = i\omega\varepsilon_0\kappa_z E_z.$$

In what follows we will consider the following two simple but, at the same time, important special cases.

11.2.1   LONGITUDINAL PROPAGATION

Here, we consider the propagation of an electromagnetic wave along the direction of a constant external magnetic field, $\mathbf{H}_0$, that is, along the z-axis. We seek a solution of the system of Equations 11.13 and 11.14 in the form

$$\mathbf{E}(z) = \mathbf{E}\exp(-ikz), \quad \mathbf{H}(z) = \mathbf{H}\exp(-ikz), \tag{11.15}$$

where
   $\mathbf{E}$ and $\mathbf{H}$ are constant amplitudes
   k is the wave vector of the wave

We then substitute these solutions into the system of Equations 11.13 and 11.14. In the case of a medium unbounded in the transverse direction, the partial derivatives of the field vectors of the plane wave on variables x and y are equal to zero. Then, from the last equation of the system of

Equations 11.13 and 11.14, we find that $E_z = 0$ and $H_z = 0$, that is, in this particular case, the wave is purely transverse relatively to both the electric and magnetic fields (TEM wave). The remaining four equations take the form

$$kE_y = -\omega\mu_0\kappa_m H_x,$$

$$kE_x = \omega\mu_0\kappa_m H_y,$$

$$kH_y = \omega\varepsilon_0(\kappa E_x + i\kappa_a E_y),$$ (11.16)

$$-kH_x = \omega\varepsilon_0(-i\kappa_a E_x + \kappa E_y).$$

By eliminating the components H_x, H_y, Equation 11.16 is reduced to a system of linear homogeneous equations containing the variables E_x, E_y:

$$\left(k^2 - \varepsilon_0\mu_0\kappa_m\kappa\omega^2\right)E_x - i\varepsilon_0\mu_0\kappa_m\kappa_a\omega^2 E_y = 0,$$

$$i\varepsilon_0\mu_0\kappa_m\kappa_a\omega^2 E_x + \left(k^2 - \varepsilon_0\mu_0\kappa_m\kappa\omega^2\right)E_y = 0.$$ (11.17)

By equating to zero the determinant of this system, we obtain the dispersion equation that gives the dependence of the frequency on the wave number:

$$\left(k^2 - \varepsilon_0\mu_0\kappa_m\kappa\omega^2\right)^2 - \left(\varepsilon_0\mu_0\kappa_m\kappa_a\omega^2\right)^2 = 0.$$ (11.18)

This equation has two positive roots that yield two possible values of k:

$$k_\pm = k_0\sqrt{\kappa_m\kappa_\pm},$$ (11.19)

where
$k_0 = \omega/c$ is the wave number in vacuum
$\kappa_\pm = \kappa \pm \kappa_a$ is the effective dielectric constants of the plasma for the longitudinal magnetization

Substituting the values $k_\pm$ in Equation 11.17, we obtain the relationship that connects the electric field components:

$$E_y = \mp iE_x.$$ (11.20)

This equation corresponds to waves of circular polarization with two opposite directions of rotation.

Thus, we have two independent waves with opposite circular polarizations, which can propagate in a magnetoactive plasma along the direction of the external magnetic field. The first wave, for which the wave vector is $\mathbf{k}_+$ and the electric vector components are connected by the relationship $E_y = -iE_x$, is a right circularly polarized wave. For this wave, the electric field

$$\mathbf{E}_+(z) = A_1(\mathbf{i} - i\mathbf{j})\exp(-ik_+z)$$ (11.21)

rotates in the (x, y) plane in the clockwise direction if we look toward the direction of the wave propagation. In a similar way, the wave with wave vector $\mathbf{k}_-$ has left circular polarization. In this case, the electric field vector

$$\mathbf{E}_-(z) = A_2(\mathbf{i} + i\mathbf{j})\exp(-ik_-z)$$ (11.22)

rotates counterclockwise if we look toward the direction of the wave propagation. If we use the opposite "line of sight," that is, look in the direction opposite to the wave vector of the propagating wave, the direction of rotation of the polarization vectors of the two waves will change rotation direction. In Equations 11.21 and 11.22, we introduced the unit vectors $\mathbf{i}$ and $\mathbf{j}$ that are directed along x- and y-axes, respectively.

Two waves that are described by Equations 11.21 and 11.22 have different wave vectors and phase velocities and they are referred to as ***normal waves*** (compare with Section 6.2 [Equation 6.19]). The difference of those velocities explains the rotation of the polarization plane of a wave propagating in a magnetoactive plasma. Let us assume that the amplitudes of two normal waves are identical, that is, $A_1 = A_2 = A$. For $z = 0$, their total field $\mathbf{E}(z) = \mathbf{E}_+(z) + \mathbf{E}_-(z)$ is $\mathbf{E}(0) = 2A\mathbf{i}$, that is, it corresponds to a wave linearly polarized along an x-axis. In the plane $z = l$, we have

$$\begin{aligned}
\mathbf{E}(l) &= A(\mathbf{i} + i\mathbf{j})\exp(-ik_+ l) + A(\mathbf{i} - i\mathbf{j})\exp(-ik_- l) \\
&= 2A\exp(-ikl)\big(\mathbf{i}\cos(\Delta kl) + \mathbf{j}\sin(\Delta kl)\big),
\end{aligned} \tag{11.23}$$

where $k = (k_+ + k_-)/2$, $\Delta k = (k_+ - k_-)/2$. The wave that is determined by Equation 11.23 is also linearly polarized, but $\mathbf{E}$ forms a nonzero angle θ_F with the x-axis of

$$\tan \theta_F = \frac{E_y}{E_x} = \tan(\Delta kl), \quad \theta_F = \Delta kl. \tag{11.24}$$

According to Equations 11.23 and 11.24, the plane of polarization of a linearly polarized wave rotates as the wave propagates along the magnetic field in the magnetoactive plasma. The rotation angle θ_F of the polarization plane increases linearly with the distance that the wave travels in the plasma. The phenomenon described earlier for the plasma is part of more general phenomena of the rotation of the plane of polarization known as the ***Faraday effect***.

11.2.2 Transverse Propagation

Assume now that the wave propagates along the y-axis, which is perpendicular to the external magnetic field, $\mathbf{H}_0$. In this case, the partial derivatives with respect to x and z in Equations 11.13 and 11.14 are equal to zero. We will seek a solution of these equations in the form

$$\mathbf{E}(y) = \mathbf{E}\exp(-iky), \quad \mathbf{H}(y) = \mathbf{H}\exp(-iky). \tag{11.25}$$

Substituting Equation 11.25 in the system of Equations 11.13 and 11.14, we find that Equations 11.13 and 11.14 split into two independent subsystems of equations corresponding to two normal (characteristic) waves. The first subsystem

$$\kappa E_z = \omega\mu_0\kappa_m H_x, \quad \kappa H_x = \omega\varepsilon_0\kappa_z E_z \tag{11.26}$$

gives a solution in the form of a linearly polarized plane wave with the electric field vector $\mathbf{E}$, which is parallel to the external field $\mathbf{H}_0$ and the magnetic field of the wave, $\mathbf{H}$, which is orthogonal to the external field $\mathbf{H}_0$. This wave is called ***ordinary***, and its wave vector corresponds to a wave propagating in an isotropic plasma:

$$k_1 = k_0\sqrt{\kappa_m\kappa_z}, \quad Z_1 = \sqrt{\frac{\mu_0\kappa_m}{\varepsilon_0\kappa_z}}, \tag{11.27}$$

$$E_z(y) = E_z\exp(-ik_1 y), \quad H_x(y) = H_x\exp(-ik_1 y).$$

The second subsystem, corresponding to the **extraordinary** normal wave, has the form

$$\kappa E_x = -\omega\mu_0\kappa_m H_z,$$

$$\kappa H_z = -\omega\varepsilon_0(\kappa E_x + i\kappa_a E_y), \tag{11.28}$$

$$0 = i\omega\varepsilon_0(-i\kappa_a E_x + \kappa E_y).$$

From Equation 11.28, it can be seen that the magnetic field of this wave is linearly polarized in the direction of the external field $\mathbf{H}_0$. The electric field is elliptically polarized in the (x, y) plane orthogonal to the external field, since its components are related as

$$E_y = i\frac{\kappa_a}{\kappa} E_x.$$

The wave number and impedance of the extraordinary wave are

$$k_2 = k_0\sqrt{\kappa_m\kappa_\perp}, \quad Z_2 = \sqrt{\frac{\mu_0\kappa_m}{\varepsilon_0\kappa_\perp}},$$

$$E_{x,y}(y) = E_{x,y}\exp(-ik_2 y), \quad H_z(y) = H_z\exp(-ik_2 y), \tag{11.29}$$

where $\kappa_\perp = \kappa - \kappa_a^2/\kappa$ is the effective dielectric permittivity of the medium for transverse magnetization. The difference between the k vectors of the waves described by Equations 11.27 and 11.29 means the existence of birefringence (compare with Section 6.3), that is, a wave propagating perpendicular to the magnetic field in magnetized plasma is a superposition of two normal waves propagating with different velocities.

Exercise 11.2

A plasma is placed in a uniform magnetic field B_0. A circularly polarized electromagnetic wave with frequency ω is propagating along the magnetic field. Show that cyclotron resonance absorption is possible only in the case of left circular polarization of the electromagnetic wave.

Solution. The cyclotron frequency for electrons in plasma is

$$\omega_c = \frac{eB_0}{m}.$$

The cyclotron resonance is defined by the maximum of the imaginary part of the dielectric permittivity $\kappa_\pm = \kappa \pm \kappa_a$. When dissipation is taken into account, the corresponding components of the dielectric permittivity tensor defined by Equation 11.11 for the plasma in magnetic field take the form

$$\kappa = 1 - \frac{\omega_p^2}{\omega^2 - \omega_c^2 - i\gamma\omega}, \quad \kappa_a = \frac{\omega_p^2\omega_c}{\omega(\omega^2 - \omega_c^2 - i\gamma\omega)}.$$

Let us find an expression for the κ_+ and κ_- components of electrical permittivity tensor:

$$\kappa_\pm = \kappa \pm \kappa_a = 1 - \left(1 \mp \frac{\omega_c}{\omega}\right)\frac{\omega_p^2}{\omega^2 - \omega_c^2 - i\gamma\omega}.$$

Let us separate the real and imaginary parts of this expression:

$$\mathrm{Re}(\kappa_\pm) = 1 - \left(1 \mp \frac{\omega_c}{\omega}\right) \frac{\omega_p^2(\omega^2 - \omega_c^2)}{(\omega^2 - \omega_c^2)^2 + \gamma^2\omega^2},$$

$$\mathrm{Im}(\kappa_\pm) = \left(1 \mp \frac{\omega_c}{\omega}\right) \frac{\omega_p^2\gamma\omega}{(\omega^2 - \omega_c^2)^2 + \gamma^2\omega^2}.$$

For a frequency, which coincides with frequency ω_c, we get $\mathrm{Re}(\kappa_\pm) = 1$, and

$$\mathrm{Im}(\kappa_\pm) = \left(1 \mp \frac{\omega_c}{\omega}\right) \frac{\omega_p^2}{\gamma\omega}.$$

From this expression, it follows that in the case of right circular polarization of the wave in a plasma, the magnitude $\mathrm{Im}(\kappa_+) = (\omega - \omega_c)\omega_p^2/\gamma\omega^2$ at $\omega = \omega_c$ becomes equal to zero, that is, in this case, there is no resonance. Cyclotron resonance takes place only in the case of left circular polarization of the electromagnetic wave propagating in the plasma. As one can see for $\gamma \to 0$ $\mathrm{Im}(\kappa_-) = 2\omega_p^2/\gamma\omega_c \to \infty$.

11.3 MAGNETIC PERMEABILITY OF FERRITES AND MAGNETIC RESONANCE IN FERRITES

Ferrites are magnetic semiconductors or dielectrics possessing rather small level of losses in the microwave frequency and range. These are crystalline materials, which are usually made in the form of ceramics. From the chemistry standpoint, the ferrites are compounds of iron oxide Fe_2O_3 with oxides of other divalent metals. The magnetic properties of ferrites are due to the existence in their crystal lattice of atoms or ions possessing electrons with uncompensated spin. A ferrite, magnetized by an external magnetic field, possesses a special type of anisotropy—a *magnetic gyrotropy*.

Consider a homogeneous unbounded ferrite, magnetized to saturation by a constant external magnetic field $\mathbf{H}_0$. We will assume that the field $\mathbf{H}_0$ is applied along the z-axis of a Cartesian coordinate system. We also assume that in the absence of a magnetizing field the ferrite is isotropic, that is, we neglect all types of the anisotropy in it, except that induced by the field $\mathbf{H}_0$.

A magnetized ferrite is characterized by its magnetization vector $\mathbf{M}$, that is, the magnetic moment per unit volume of the ferrite. Consider the time evolution of vector $\mathbf{M}$ in an external magnetic field $\mathbf{H} = \mathbf{H}_0 + \mathbf{h}(t)$, which consists of a dc field $\mathbf{H}_0$ and alternating one $\mathbf{h}(t)$. For a lossless medium, the motion of the vector $\mathbf{M}$ is described by the *Landau–Lifshitz equation*:

$$\frac{d\mathbf{M}}{dt} = -\gamma\mu_0\mathbf{M} \times \mathbf{H}, \tag{11.30}$$

where γ is the *gyromagnetic ratio*, for a free electron $\gamma = e/m = 1.76 \times 10^{11}$ C/kg.

Let us study in more detail the magnetization vector motion in the presence of a constant external field $\mathbf{H} = \mathbf{H}_0 = H_0\mathbf{k}$, magnetizing the ferrite to saturation along the z-axis (here, $\mathbf{k}$ is the unit vector directed along z-axis). If we project the magnetization vector on three Cartesian axes, Equation 11.30 becomes

$$\frac{dM_x}{dt} = -\gamma\mu_0H_0M_y, \quad \frac{dM_y}{dt} = \gamma\mu_0H_0M_x, \quad \frac{dM_z}{dt} = 0. \tag{11.31}$$

We express the component M_y from the first equation of (11.31) and substitute it into the second equation. As a result, we obtain

$$\frac{d^2 M_x}{dt^2} + \omega_H^2 M_x = 0,$$ (11.32)

where we have introduced the angular frequency $\omega_H = \gamma \mu_0 H_0$. We obtained a differential equation that describes harmonic oscillations with an angular frequency ω_H for the magnetization component M_x. A similar equation can be obtained for the component M_y.

Taking into account Equation 11.31, one gets the solutions

$$M_x = A\sin(\omega_H t + \alpha), \quad M_y = -A\cos(\omega_H t + \alpha),$$ (11.33)

where A and α are arbitrary constants. From the third equation of the system of Equation 11.31, it follows $M_z = \text{const}$. This constant is equal to the **saturation magnetization** M_0, which is determined experimentally and is one of the main characteristics of the ferrite.

From Equation 11.33, it follows that the tip of the vector $\mathbf{M}$ rotates at an angular velocity ω_H in the plane perpendicular to the vector $\mathbf{H}_0$. If we look toward the tip of vector $\mathbf{H}_0$, the rotation $\mathbf{M}$ is in the counterclockwise direction (Figure 11.1). This type of magnetization vector motion is called **precession**.

Equation 11.30 does not take into account losses in energy. In a real ferrite, the energy of magnetic oscillations dissipates, turning into heat. In this case, the trajectory of the vector $\mathbf{M}$ end (tip) is not a circle but a spiral with a gradually decreasing radius. During the **relaxation time** (of the order of 10^{-8} s), the precession of the magnetic moment decays, and the magnetic moments of the atoms are set along the field. To maintain the precession, it is necessary to apply an external alternating field $\mathbf{H}_0$.

Let us suppose that in addition to the constant magnetic field $\mathbf{H}_0 = H_0 \mathbf{z}_0$, an alternating magnetic field oriented perpendicular $\mathbf{H}_0$ is applied, which has the form

$$\mathbf{H} = \mathbf{H}_0 + \mathbf{h}\exp(i\omega t).$$ (11.34)

The magnetization vector $\mathbf{M}$ will undergo forced oscillations at the same frequency ω, that is, it also will have an alternating component:

$$\mathbf{M}(t) = \mathbf{M}_0 + \mathbf{m}\exp(i\omega t).$$ (11.35)

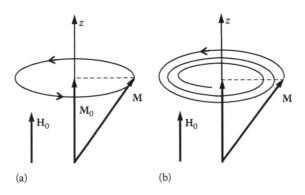

(a)   (b)

FIGURE 11.1 Precession of the magnetization vector in the magnetic field for the case without energy dissipation (a) and with the dissipation (b).

Substituting Equations 11.34 and 11.35 into Equation 11.30, we get

$$i\omega\mathbf{m} = -\gamma\mu_0\mathbf{m} \times \mathbf{H}_0 - \gamma\mu_0\mathbf{M}_0 \times \mathbf{h} - \gamma\mu_0(\mathbf{m} \times \mathbf{h})\exp(i\omega t). \tag{11.36}$$

Here, we took into account the fact that vector $\mathbf{M}_0$ is parallel to vector $\mathbf{H}_0$, and that is why $\mathbf{M}_0 \times \mathbf{H}_0 = 0$. The applied alternating field is assumed small compared to the time-independent field $\mathbf{H}_0$ so that the inequalities $|\mathbf{h}| \ll H_0$, $|\mathbf{m}| \ll M_0$ are satisfied. Then in the right-hand side of Equation 11.36, the third term can be neglected because it is much smaller than other terms.

We project Equation 11.36 on the axes of the coordinate system. Taking into account that $\mathbf{H}_0(0,0,H_0)$, $\mathbf{h}(h_x,h_y,0)$, $\mathbf{M}_0(0,0,M_0)$, $\mathbf{m}(m_x,m_y,m_z)$, we get

$$i\omega m_x = -\omega_H m_y + \omega_M h_y,$$

$$i\omega m_y = \omega_H m_x - \omega_M h_x, \tag{11.37}$$

$$i\omega m_z = 0,$$

where $\omega_M = \gamma\mu_0 M_0$. From here, we find the projections of high-frequency magnetization vector:

$$m_x = \frac{\omega_H \omega_M}{\omega_H^2 - \omega^2} h_x + i\frac{\omega\omega_M}{\omega_H^2 - \omega^2} h_y,$$

$$m_y = -i\frac{\omega\omega_M}{\omega_H^2 - \omega^2} h_x + \frac{\omega_H \omega_M}{\omega_H^2 - \omega^2} h_y, \tag{11.38}$$

$$m_z = 0.$$

The relationship between high-frequency components of the magnetization and the alternating magnetic field can be expressed in tensor form as follows:

$$\mathbf{m} = \hat{\chi}_m \mathbf{h}, \tag{11.39}$$

where the coefficients h_x, h_y in the right-hand side of Equation 11.38 form a **high-frequency magnetic susceptibility tensor** of the ferrite:

$$\hat{\chi}_m = \begin{pmatrix} \chi_m & i\chi_{ma} & 0 \\ -i\chi_{ma} & \chi_m & 0 \\ 0 & 0 & 0 \end{pmatrix}, \quad \chi_m = \frac{\omega_H \omega_M}{\omega_H^2 - \omega^2}, \quad \chi_{ma} = \frac{\omega\omega_M}{\omega_H^2 - \omega^2} \tag{11.40}$$

Knowing tensor $\hat{\chi}_m$, we find the **high-frequency magnetic permeability tensor** of the ferrite:

$$\hat{\kappa}_m = 1 + \hat{\chi}_m = \begin{pmatrix} \kappa_m & i\kappa_{ma} & 0 \\ -i\kappa_{ma} & \kappa_m & 0 \\ 0 & 0 & \kappa_{mz} \end{pmatrix}, \tag{11.41}$$

$$\kappa_m = 1 + \chi_m = 1 + \frac{\omega_H \omega_M}{\omega_H^2 - \omega^2}, \quad \kappa_{mz} = 1, \quad \kappa_{ma} = \chi_{ma} = \frac{\omega\omega_M}{\omega_H^2 - \omega^2}.$$

Tensor components κ_m and κ_{ma} have a resonant frequency dependence (the role of the resonant frequency is played by ω_H): κ_m, $\kappa_{ma} \to \infty$ if $\omega \to \omega_H$. The dependence of κ_m and κ_{ma} on the dc external field H_0 at a fixed frequency ω has the same character, and the resonance is achieved at the field value $H_{res} = \omega/\gamma\mu_0$.

If we solve the Landau–Lifshitz equation taking into account losses, the dependences of κ_m and κ_{ma} on ω will be in the form of smooth curves without singularities. In addition, κ_m and κ_{ma} will be complex numbers: $\kappa_m = \kappa'_m - i\kappa''_m$, $\kappa_{ma} = \kappa'_{ma} - i\kappa''_{ma}$. Field and frequency dependences of the imaginary parts κ''_m, κ''_{ma} have maxima in the region of resonance. Thus, resonant absorption of energy of the high-frequency field by the ferrite is observed. This phenomenon is called **ferromagnetic resonance**. The frequency and field dependences of the real and imaginary parts of κ_m and κ_{ma} are shown in Figure 11.2 (dashed line shows the path dependency in the absence of losses).

Exercise 11.3

The solution of the Landau–Lifshitz equation (this equation takes into account losses) gives the following dependence for the diagonal component of the magnetic permeability tensor on frequency:

$$\kappa_m = 1 + \frac{\omega_H \omega_M}{\omega_H^2 - \omega^2 + 2i\omega_r\omega},$$

where ω_r is the relaxation frequency. A ferrite sample has the following parameters: $\omega_H = 2 \times 10^{10}$ s^{-1}, $\omega_M = 3 \times 10^{11}$ s^{-1}, $\omega_r = 6 \times 10^8$ s^{-1}. Find the differences in the estimates of the real part of κ_m between the calculation that takes into account energy losses and the calculation that neglects losses. Calculate those differences for two frequencies: $\omega_1 = 0.5\omega_H$ and $\omega_2 = 0.9\omega_H$.

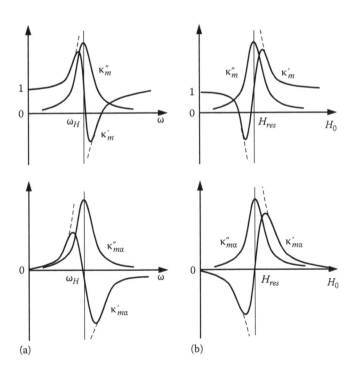

(a)   (b)

FIGURE 11.2 Frequency (a) and field (b) dependencies of real and imaginary parts of magnetic permeability in the vicinity of magnetic resonance (the dashed lines correspond to the case without energy dissipation).

Solution. Let us write separately the real and imaginary parts of $\kappa_m = \kappa_m' - i\kappa_m''$:

$$\kappa_m = 1 + \frac{\omega_H \omega_M (\omega_H^2 - \omega^2 - 2i\omega_r\omega)}{(\omega_H^2 - \omega^2)^2 + 4\omega_r^2\omega^2}$$

$$= 1 + \frac{\omega_H \omega_M (\omega_H^2 - \omega^2)}{(\omega_H^2 - \omega^2)^2 + 4\omega_r^2\omega^2} - i \frac{2\omega_H \omega_M \omega_r\omega}{(\omega_H^2 - \omega^2)^2 + 4\omega_r^2\omega^2}.$$

Thus, for $\kappa_m'(\omega)$, we get the following expression:

$$\kappa_m'(\omega) = 1 + \frac{\omega_H \omega_M \left(\omega_H^2 - \omega^2\right)}{\left(\omega_H^2 - \omega^2\right)^2 + 4\omega_r^2\omega^2}.$$

By substituting into this expression the given data and frequencies $\omega_1 = 0.5\omega_H$ and $\omega_2 = 0.9\omega_H$, we get $\kappa_m'(\omega_1) = 20.97$ and $\kappa_m'(\omega_2) \approx 74$. Substitution of the frequencies mentioned earlier into the expression for magnetic permeability, which does not take into account losses,

$$\kappa_m(\omega) = 1 + \frac{\omega_H \omega_M}{\omega_H^2 - \omega^2}$$

gives the following values of κ_m: $\kappa_m(\omega_1) = 21$, $\kappa_m(\omega_2) = 79.95$. Thus,

$$\Delta\kappa_m(\omega_1) = \kappa_m(\omega_1) - \kappa_m'(\omega_1) = 0.03,$$

$$\Delta\kappa_m(\omega_2) = \kappa_m(\omega_2) - \kappa_m'(\omega_2) = 5.95.$$

11.4 WAVES IN A TRANSVERSELY MAGNETIZED FERRITE

Consider a plane monochromatic wave propagating in a ferrite perpendicular to the direction of an external magnetic field (e.g., $\mathbf{E}(y) = \mathbf{E}\exp(-iky)$, $\mathbf{H}(y) = \mathbf{H}\exp(-iky)$, wave propagating in the direction of the y-axis). The ferrite is characterized by a scalar dielectric permittivity κ and a magnetic permeability tensor $\hat{\kappa}_m$ is given by Equation 11.41. We write Maxwell's equations for the complex amplitudes of the wave field vectors $\mathbf{E}$ and $\mathbf{H}$ as

$$\nabla \times \mathbf{E} = -i\omega\mu_0\hat{\kappa}_m\mathbf{H},$$

$$\nabla \times \mathbf{H} = i\omega\varepsilon_0\kappa\mathbf{E} \tag{11.42}$$

and project them on the axes of the chosen Cartesian coordinate system. In the left-hand side of the resulting expressions, we take into account that for the plane wave under consideration, the partial derivatives with respect to x and z are equal to zero. In the projections of the first equation of Equations 11.42, it is necessary to consider the tensor nature of magnetic permeability. The result is a system of six scalar equations, which is divided into two independent subsystems. The first subsystem contains the field components E_z, H_x, H_y:

$$\frac{\partial E_z}{\partial y} = -i\omega\mu_0(\kappa_m H_x + i\kappa_{ma}H_y),$$

$$0 = -i\omega\mu_0(-i\kappa_{ma}H_x + \kappa_m H_y), \tag{11.43}$$

$$-\frac{\partial H_x}{\partial y} = i\omega\kappa\varepsilon_0 E_z.$$

The second subsystem includes other components E_x, E_y, H_z:

$$\frac{\partial H_z}{\partial y} = i\omega\kappa\varepsilon_0 E_x,$$

$$0 = i\omega\kappa\varepsilon_0 E_y, \qquad (11.44)$$

$$\frac{\partial E_x}{\partial y} = i\omega\mu_0\kappa_{mz}H_z.$$

The two subsystems of the equations correspond to two independent (normal) waves of transversely magnetized ferrite.

The solution of subsystem of Equations 11.44 is the first normal wave. This wave is analogous to the wave in an isotropic dielectric. Its fields are linearly polarized and have only transverse components E_x and H_z, but $E_y = 0$. The wave characteristics of the *ordinary wave*, k_1 and Z_1, do not possess frequency and field dispersion:

$$k_1 = k_0\sqrt{\kappa\kappa_{mz}}, \quad Z_1 = \sqrt{\frac{\mu_0\kappa_{mz}}{\varepsilon_0\kappa}}, \qquad (11.45)$$

$$E_x(y) = E_x\exp(-ik_1 y), \quad H_z(y) = H_z\exp(-ik_1 y).$$

The solution of Equation 11.43 is the so-called *extraordinary wave*. It has only one (transverse) component of the electric field, E_z, which is parallel to the magnetizing field, and two components of a magnetic field: transverse component, H_x, and longitudinal component, H_y. The magnetic field components are related by the equation

$$H_y = i\frac{\kappa_{ma}}{\kappa_m}H_x, \qquad (11.46)$$

which follows from the second equation of subsystem (11.43). Thus, this normal wave is transverse and linearly polarized only concerning the electric field. The magnetic field of this wave is elliptically polarized in the (x, y) plane. The factor i in Equation 11.46 indicates that the phase shift of oscillations of components H_x and H_y is $\pi/2$, and the ratio of amplitudes of the magnetic field components H_y/H_x is equal to κ_{ma}/κ_m. Note that the longitudinal component of the magnetic field in the wave is directly related to the gyrotropy of the ferrite: if $\kappa_{ma} = 0$, then, according to Equation 11.43, $H_y = 0$.

If we eliminate from the subsystem of Equation 11.43 the component H_y, we arrive at the following two equations:

$$\frac{\partial E_z}{\partial y} = -i\omega\mu_0\kappa_{m\perp}H_x, \quad \frac{\partial H_x}{\partial y} = -i\omega\varepsilon_0\kappa E_z, \qquad (11.47)$$

where the ferrite effective magnetic permeability, $\kappa_{m\perp} = \kappa_m - \kappa_{ma}^2/\kappa_m$, has been introduced. By analogy with the solution of Equation 11.26, the solution of Equation 11.47 can be written as a monochromatic plane wave, for which the wave number k_2 and impedance Z_2 are

$$k_2 = k_0\sqrt{\kappa\kappa_{m\perp}}, \quad Z_2 = \sqrt{\frac{\mu_0\kappa_{m\perp}}{\varepsilon_0\kappa}}, \qquad (11.48)$$

$$E_z(y) = E_z\exp(-ik_2 y), \quad H_{x,y}(y) = H_{x,y}\exp(-ik_2 y).$$

As in the case of the components of high-frequency magnetic permeability tensor, κ_m and κ_{ma}, the effective magnetic permeability also has a strong dispersion: it is characterized by a resonance dependence on angular frequency ω and the external field H_0.

Consider a linearly polarized wave incident normally on a flat layer of a transversely magnetized ferrite (Figure 11.3). If the electric field of the incident wave is polarized at an angle φ_0 to the field $\mathbf{H}_0$, ($\varphi_0 \neq 0$, $\pi/2$), two normal waves are excited in the ferrite layer. The electric field component, parallel to the field $\mathbf{H}_0$, excites "extraordinary" wave, while the component orthogonal to the field $\mathbf{H}_0$ excites "ordinary" wave in the magnetized ferrite. These waves propagate in the ferrite with different phase velocities, $v_{ph1} = \omega/k_1$ and $v_{ph2} = \omega/k_2$. Therefore, at the output from the ferrite layer with a thickness l, these waves acquire a phase shift:

$$\psi(l) = (k_2 - k_1)l = \left(\sqrt{\kappa \kappa_{m\perp}} - \sqrt{\kappa \kappa_{mz}}\right)k_0 l, \tag{11.49}$$

which results in the transmitted wave being elliptically polarized (compare with Equation 11.24). The parameters of the polarization ellipse depend on the layer thickness, the external field, and the frequency ω. The effect of transforming the wave polarization by a layer of transversely magnetized ferrite is called the ***Cotton–Mouton effect***. This effect and also the resonance properties of "extraordinary" wave in a transversely magnetized ferrite are widely used in microwave technology to create devices controlled by an external magnetic field (e.g., valves, polarizers, and modulators).

Exercise 11.4

A linearly polarized electromagnetic wave is incident normally on a ferrite plate. The plate with thickness l is placed in an external magnetic field $\mathbf{H}_0$, which lies in the plane of the plate and magnetizes it uniformly. The polarization plane of the incident wave is at an angle $\varphi_0 = 30°$ with respect to the direction of the external field. Find the parameters of the polarization ellipse of the wave at the plate's surface through which the wave travels.

Solution. Let us assume that the electromagnetic wave is propagating along the y-axis and that the magnetic field is directed along the z-axis (see Figure 11.3). Two normal waves are generated inside the ferrite plate. Since the plane of polarization of the incident wave is at an angle of 30°

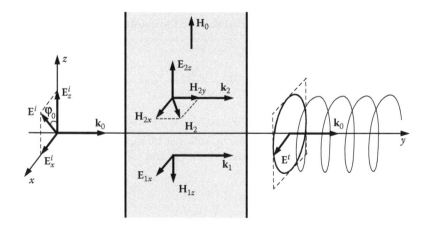

FIGURE 11.3 Illustration of wave propagation through the ferrite slab.

with respect to the external field, then the electric field vector inside of the plate has two vector components:

$$E_x = E_m \sin\left(\frac{\pi}{6}\right)\sin(\omega t - k_1 y) = \left(\frac{E_m}{2}\right)\sin(\omega t - k_1 y),$$

$$E_z = E_m \cos\left(\frac{\pi}{6}\right)\sin(\omega t - k_2 y) = \left(\frac{\sqrt{3}E_m}{2}\right)\sin(\omega t - k_2 y).$$

The component $E_x(t, y)$ of the wave field, which is orthogonal to the field $\mathbf{H}_0$, excites ordinary wave and the component, which is parallel to field, $E_z(t, y)$, excites extraordinary wave. These waves have two different phase velocities:

$$v_{ph1} = \frac{\omega}{k_1} = \frac{\omega}{k_0\sqrt{\kappa\kappa_{mz}}} \quad \text{and} \quad v_{ph2} = \frac{\omega}{k_2} = \frac{\omega}{k_0\sqrt{\kappa\kappa_{m\perp}}}.$$

An elliptically polarized wave is formed after the ordinary and extraordinary waves are added. If there is no absorption in the layer, then after transmission through the plate, the vector components of the wave field have the form

$$E_x = \left(\frac{E_m}{2}\right)\sin(\omega t - k_1 l),$$

$$E_z = \left(\frac{\sqrt{3}E_m}{2}\right)\sin(\omega t - k_2 l).$$

As it is seen, these two waves have a phase difference equal to

$$\psi(l) = (k_2 - k_1)l = \left(\sqrt{\kappa\kappa_{m\perp}} - \sqrt{\kappa\kappa_{mz}}\right)k_0 l.$$

Adding the two orthogonally polarized waves results in an elliptically polarized wave. The parameters of the polarization ellipse depend on the plate thickness l, the external field H_0, and the frequency ω. In order to find these parameters, let us rewrite the equations given earlier for E_x and E_z in the form

$$\frac{2E_x}{E_m} = \sin\omega t \cos k_1 l - \cos\omega t \sin k_1 l,$$

$$\frac{2E_z}{\sqrt{3}E_m} = \sin\omega t \cos k_2 l - \cos\omega t \sin k_2 l.$$

Let us multiply the first equation by $\cos k_2 l$ and the second equation by $\cos k_1 l$, and after this, subtract one from another. In a similar fashion, let us multiply the first equation by $\sin k_2 l$ and the second equation by $\sin k_1 l$, and after this, subtract one from another. As a result, we obtain a system of two equations:

$$\cos k_2 l \cdot \left(\frac{2E_x}{E_m}\right) - \cos k_1 l \cdot \left(\frac{2E_z}{\sqrt{3}E_m}\right) = \cos\omega t \cdot \sin(k_2 l - k_1 l),$$

$$\sin k_2 l \cdot \left(\frac{2E_x}{E_m}\right) - \sin k_1 l \cdot \left(\frac{2E_z}{\sqrt{3}E_m}\right) = \sin\omega t \cdot \sin(k_2 l - k_1 l).$$

Let us square each of these equations and add them to each other. As a result, we get the equation for ellipse with minor and major semiaxes $a = E_m/2$ and $b = \sqrt{3}E_m/2$:

$$\left(\frac{E_x}{a}\right)^2 + \left(\frac{E_z}{b}\right)^2 - 2\left(\frac{E_x}{a}\right)\left(\frac{E_z}{b}\right)\cos\psi(l) = \sin^2\psi(l).$$

It can be shown that the major axis of this ellipse is turned with respect to z-axis by an angle $\varphi(l)$, which is given by the relationship

$$\tan\left[2\varphi(l)\right] = \frac{2ab}{b^2 - a^2}\cos\psi(l) = \sqrt{3}\cos\psi(l),$$

$$\varphi(l) = \frac{1}{2}\arctan\left[\sqrt{3}\cos\psi(l)\right].$$

11.5 WAVES IN A LONGITUDINALLY MAGNETIZED FERRITE

Let us now consider the propagation of a plane monochromatic wave in z-direction as it is defined by Equation 11.15 ($\mathbf{E}(z) = \mathbf{E}\exp(-ikz)$, $\mathbf{H}(z) = \mathbf{H}\exp(-ikz)$) in a ferrite magnetized to saturation in the direction of a magnetic field $\mathbf{H}_0 = H_0\mathbf{z}_0$. We write down the projections of Equation 11.42 on the x- and y-axes. Taking into account that the magnetic permittivity of the ferrite is a tensor of the form in Equation 11.41 and also that for a plane wave the derivatives of the field vectors with respect to coordinates x and y are equal to zero, we get

$$\frac{\partial E_y}{\partial z} = i\omega\mu_0(\kappa_m H_x + i\kappa_{ma}H_y),$$

$$\frac{\partial E_x}{\partial z} = i\omega\mu_0(i\kappa_{ma}H_x - \kappa_m H_y),$$

$$\frac{\partial H_y}{\partial z} = -i\omega\varepsilon_0\kappa E_x,$$ (11.50)

$$\frac{\partial H_x}{\partial z} = i\omega\varepsilon_0\varepsilon E_y.$$

The projections on the z-axis are $E_z = 0$, $H_z = 0$, that is, the wave is transverse with respect to both the electric and the magnetic fields.

We now introduce the new vectors $\mathbf{E}_\pm$ and $\mathbf{H}_\pm$, which are related to $\mathbf{E}$ and $\mathbf{H}$ by the equations

$$E_\pm = E_x \pm iE_y, \quad H_\pm = H_x \pm iH_y.$$ (11.51)

For this purpose, we sum the second and fourth equations from the system of Equations 11.50 with the first and the third, multiplied by i or $-i$. As a result, we end up with two subsystems: the first one, for components E_+ and H_+,

$$\frac{\partial E_+}{\partial z} = -\omega\mu_0\left(\kappa_m + \kappa_{ma}\right)H_+,$$ (11.52)

$$\frac{\partial H_+}{\partial z} = \omega\varepsilon_0\kappa E_+,$$

and the second subsystem, for components E_- and H_-,

$$\frac{\partial E_-}{\partial z} = \omega\mu_0\left(\kappa_m - \kappa_{ma}\right)H_-,$$

(11.53)

$$\frac{\partial H_-}{\partial z} = -\omega\varepsilon_0\kappa E_-.$$

The solutions of these subsystems correspond to the two normal waves of a longitudinally magnetized ferrite:

$$H_\pm(z) = A_1 \exp(-ik_\pm z),$$

(11.54)

$$E_\pm(z) = Z_\pm A_1 \exp(-ik_\pm z),$$

where $k_\pm = k_0\sqrt{\kappa\left(\kappa_m \pm \kappa_{ma}\right)}$, $Z_\pm = \sqrt{\mu_0\kappa_{m\pm}/\varepsilon_0\kappa}$. These waves are transverse; they are circularly polarized with opposite directions of rotation. If we look in the direction opposite to the wave propagation, the end of field vector $\mathbf{E}_+$ or $\mathbf{H}_+$ rotating in the counterclockwise direction. This is right circular polarized wave. The wave with fields $\mathbf{E}_-$ and $\mathbf{H}_-$ exhibits a left circular polarization instead, that is, it is opposite to E_+ and H_+ (Figure 11.4).

To confirm what has been described earlier, let us remember that for a circularly polarized wave the mutually orthogonal components of the field vectors have the same amplitudes and a phase shift $\pm\pi/2$. Therefore, they must be related as $H_y = \mp iH_x$ and $E_y = \mp iE_x$, where the upper sign corresponds to the right and the bottom one to the left circular polarization. For the magnetic field of a wave with right circular polarization (phase shift is equal to $-\pi/2$), one gets

$$H_+ = H_x + iH_y = 2iH_y = 2H_x \neq 0,$$

(11.55)

$$H_- = H_x - iH_y = 0.$$

For a wave with left circular polarization (phase shift is equal to $+\pi/2$),

$$H_+ = H_x + iH_y = 0$$

(11.56)

$$H_- = H_x - iH_y = -2iH_y = 2H_x \neq 0.$$

Normal waves of a longitudinally magnetized ferrite, $\kappa_{m+} = \kappa_m + \kappa_{ma}$ and $\kappa_{m-} = \kappa_m - \kappa_{ma}$, have different wave characteristics—k_+, Z_+ and k_-, Z_-—due to the difference of their effective magnetic permeability. The dependence of κ_{m+} on angular frequency ω and external field H_0 has a resonant character: in the absence of losses, $\kappa_{m+} \to \pm\infty$ when $\omega \to \omega_H$ or $H_0 \to H_{res} = \omega/\gamma\mu_0$. Thus, for a wave with right circular polarization, the ferromagnetic resonance can be observed. On the contrary, the dependence of κ_{m-} is monotonic, that is, ferromagnetic resonance for a wave with left circular polarization is absent (Figure 11.5). The curves are plotted with losses taken into account that removes

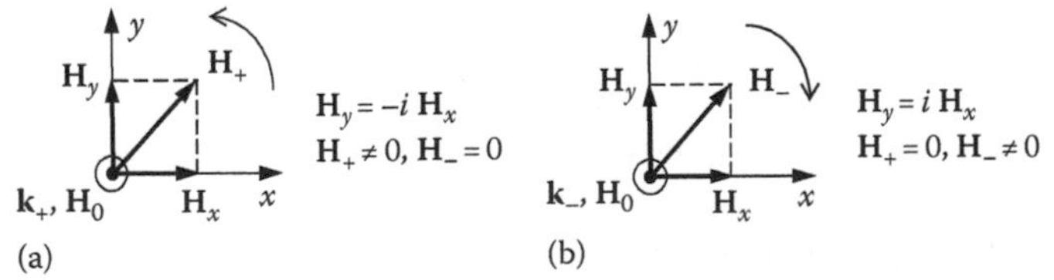

$H_y = -i H_x$
$H_+ \neq 0, H_- = 0$

(a)

$H_y = i H_x$
$H_+ = 0, H_- \neq 0$

(b)

FIGURE 11.4 Directions of rotation of the magnetic field vectors for (a) left and (b) right circular polarizations.

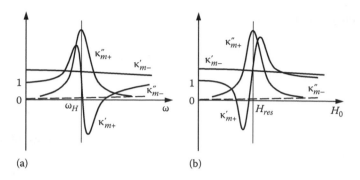

FIGURE 11.5 Frequency (a) and field (b) dependencies of real and imaginary parts of magnetic permeability in the vicinity of magnetic resonance.

$+\infty$ and $-\infty$ for κ_{m+}, and in addition, κ_{m+} and κ_{m-} become complex numbers and their imaginary parts are determined by losses (compare with Figure 11.2).

Consider a linearly polarized wave, whose magnetic field is oriented along the x-axis, incident normally from a vacuum on a longitudinally magnetized ferrite. A linearly polarized wave, which travels in the medium along the z-direction, can be represented as a superposition of two circularly polarized waves with the same amplitudes and opposite directions of rotation:

$$H_+ + H_- = 2H_x = A\left(e^{-ik_+z} + e^{-ik_-z}\right),$$

$$H_+ - H_- = 2iH_y = A\left(e^{-ik_+z} - e^{-ik_-z}\right). \tag{11.57}$$

Introducing the notations $k = (k_+ + k_-)/2$ and $\Delta k = (k_+ - k_-)/2$, we will write down Equation 11.57 in the form

$$H_x = 0.5 \cdot Ae^{-ikz}\left(e^{-i\Delta kz} + e^{i\Delta kz}\right) = Ae^{-ikz}\cos(\Delta kz),$$

$$H_y = 0.5 \cdot iAe^{-ikz}\left(e^{-i\Delta kz} - e^{i\Delta kz}\right) = Ae^{-ikz}\sin(\Delta kz). \tag{11.58}$$

The angle θ_F between the polarization plane of the wave and the x-axis after traveling a distance z is given by the relations

$$\theta_F(z) = \arctan\left(\frac{H_y}{H_x}\right) = \arctan\left(\tan(\Delta kz)\right) = \Delta kz, \tag{11.59}$$

$$\Delta k = k_0\left(\sqrt{\kappa\kappa_{m+}} - \sqrt{\kappa\kappa_{m-}}\right). \tag{11.60}$$

From Equation 11.59, it is seen that the plane of polarization of the wave is rotated (Figure 11.6). If the direction of the propagation is along the external magnetic field, the rotation is clockwise (Figure 11.6). If the wave propagates along the direction opposite to $\mathbf{H}_0$, the rotation of the polarization plane is reversed.

Thus, if the wave passes through the ferrite first along the direction of $\mathbf{H}_0$ and then in the opposite direction by the same distance, its polarization plane does not return to its initial position but is instead rotated by an angle $2\Delta kz$ (Figure 11.6).

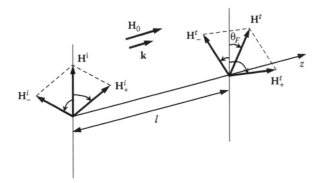

FIGURE 11.6 Rotation of the polarization plane for the wave propagating along the magnetic field in homogeneously magnetized ferrite.

Note that in a real ferrite, because of losses, both the magnetic and electric fields of the transmitted wave acquire some ellipticity. This is due to the fact that the various components of the field vectors experience unequal absorption, and, therefore, their amplitudes are different.

Exercise 11.5

A linearly polarized electromagnetic wave is propagating in a uniformly magnetized ferrite sample in the form of a slab along the direction of an external magnetic field. Find the smallest path length that the wave has to travel so that after transmission through the sample its polarization plane is rotated by an angle of $\pi/4$. The external field is perpendicular to both faces of the sample. Ferrite parameters are $\kappa = 9.80$, $\omega_M = 3.00 \times 10^{10}$ s^{-1}, $\omega_H = 4.00 \times 10^{11}$ s^{-1}, and wave frequency $\omega = 10^{10}$ s^{-1}.

Solution. For the propagating electromagnetic wave in a magnetized ferrite along the external field, the specific Faraday rotation can be found from the expression

$$\Delta k = k_0 \left(\sqrt{\kappa\kappa_{m+}} - \sqrt{\kappa\kappa_{m-}} \right) = \frac{\omega\sqrt{\kappa}}{c} \left(\sqrt{\kappa_{m+}} - \sqrt{\kappa_{m-}} \right).$$

The total rotation angle for a sample with thickness l is $\theta_F = \Delta k l$. Therefore, to rotate the wave's polarization plane by an angle of $\pi/4$, it is necessary that the thickness of the sample be

$$l = \frac{\pi}{4\Delta k} = \frac{\pi c}{4\omega\sqrt{\kappa}} \cdot \frac{1}{\sqrt{\kappa_{m+}} - \sqrt{\kappa_{m-}}} = \frac{\pi c}{4\omega\sqrt{\kappa}} \cdot \frac{1}{\sqrt{\kappa_m + \kappa_{ma}} - \sqrt{\kappa_m - \kappa_{ma}}}$$

$$= \frac{\pi c}{4\omega\sqrt{\kappa\kappa_m}} \cdot \frac{1}{\sqrt{1 + \kappa_{ma}/\kappa_m} - \sqrt{1 - \kappa_{ma}/\kappa_m}},$$

where the components of magnetic permeability tensor of ferrite have the form

$$\kappa_m = 1 + \frac{\omega_H\omega_M}{\omega_H^2 - \omega^2}, \quad \kappa_{ma} = \frac{\omega\omega_M}{\omega_H^2 - \omega^2}.$$

Let us take into account that for the given angular frequency and chosen parameters, the ratio

$$\frac{\kappa_{ma}}{\kappa_m} = \frac{\omega\omega_M}{\omega_H(\omega_H + \omega_M) - \omega^2} \ll 1.$$

Taking into account this inequality, we get

$$\sqrt{1 \pm \frac{\kappa_{ma}}{\kappa_m}} \simeq 1 \pm \frac{\kappa_{ma}}{2\kappa_m},$$

$$\sqrt{1 + \frac{\kappa_{ma}}{\kappa_m}} - \sqrt{1 - \frac{\kappa_{ma}}{\kappa_m}} \simeq 1 + \frac{\kappa_{ma}}{2\kappa_m} - \left(1 - \frac{\kappa_{ma}}{2\kappa_m}\right) = \frac{\kappa_{ma}}{\kappa_m}.$$

Thus,

$$l = \frac{\pi c}{4\omega\sqrt{\kappa\kappa_m}} \cdot \frac{1}{\sqrt{1 + \kappa_{ma}/\kappa_m} - \sqrt{1 - \kappa_{ma}/\kappa_m}} \simeq \frac{\pi c}{4\omega\sqrt{\kappa\kappa_m}} \cdot \frac{\kappa_m}{\kappa_{ma}}.$$

By substituting in the last expression numerical values, we get $l \simeq 5$ cm.

PROBLEMS

11.1 The concentration of electrons, N, in the plasma of the ionosphere is of the order 10^{14} m^{-3}. Calculate the plasma frequency and the value of the static magnetic field B_0 and the magnetic field intensity H_0 for which the cyclotron frequency is equal to the plasma frequency. (*Answer*: $\omega_p = 5.60 \times 10^8$ s^{-1}, $B_0 = 3.20 \times 10^{-3}$ T, $H_0 = 2.55 \times 10^3$ A/m.)

11.2 The relative dielectric permittivity tensor in a magnetic field is determined by Equation 11.11. Find tensor $\hat{\eta}$, which then allows us to determine the electric field vector from the electric displacement: $\mathbf{E} = (\varepsilon_0 \hat{\kappa})^{-1} \mathbf{D} = \varepsilon_0^{-1} \hat{\eta} \mathbf{D}$. Use the following two methods: (a) use the matrix relation $\hat{\eta}\hat{\kappa} = \hat{\kappa}\hat{\eta} = \hat{I}$, where $\hat{I}$ is the unity diagonal matrix and (b) solve Equation 11.10 for the electric field components assuming that the components of the electric displacement are known. ($Answer$: The components of tensor $\hat{\eta}$ have the form

$$\eta_{xx} = \frac{\kappa_{yy}}{\kappa_{xx}\kappa_{yy} - \kappa_{xy}\kappa_{yx}} = \frac{\kappa}{\kappa^2 - \kappa_a^2}, \quad \eta_{yy} = \frac{\kappa_{xx}}{\kappa_{xx}\kappa_{yy} - \kappa_{xy}\kappa_{yx}} = \frac{\kappa}{\kappa^2 - \kappa_a^2},$$

$$\eta_{xy} = -\frac{\kappa_{xy}}{\kappa_{xx}\kappa_{yy} - \kappa_{xy}\kappa_{yx}} = -\frac{i\kappa_a}{\kappa^2 - \kappa_a^2}, \quad \eta_{yx} = -\frac{\kappa_{yx}}{\kappa_{xx}\kappa_{yy} - \kappa_{xy}\kappa_{yx}} = \frac{i\kappa_a}{\kappa^2 - \kappa_a^2}, \quad \eta_{zz} = \kappa_{zz}^{-1}.)$$

11.3 An electrically neutral plasma consists of electrons and positive ions with charge $+e$. The electron and ion masses are m and M, respectively, and their concentration per unit volume is $N_e = N_i = N$. The plasma is placed in the magnetic field B_0. Write (a) the dielectric tensor of the plasma and (b) the expressions for the components of this tensor taking into account both types of charges. Find the ratio of the plasma and cyclotron frequencies for electrons and ions in the case of a helium plasma (mass of helium ion is $M = 4 \times 1.67 \times 10^{-27}$ kg).

$$Answer: \hat{\kappa} = \begin{pmatrix} \kappa & i\kappa_a & 0 \\ -i\kappa_a & \kappa & 0 \\ 0 & 0 & \kappa_z \end{pmatrix}, \quad \kappa = 1 - \frac{\omega_{pe}^2}{\omega^2 - \omega_{ce}^2} - \frac{\omega_{pi}^2}{\omega^2 - \omega_{ci}^2}, \quad \kappa_a = \frac{\omega_{pe}^2\omega_{ce}}{\omega(\omega^2 - \omega_{ce}^2)} - \frac{\omega_{pi}^2\omega_{ci}}{\omega(\omega^2 - \omega_{ci}^2)},$$

$$\kappa_z = 1 - \frac{\omega_{pe}^2}{\omega^2} - \frac{\omega_{pi}^2}{\omega^2}, \frac{\omega_{pe}}{\omega_{pi}} = 7.34 \times 10^3.$$

11.4 Find the angle of rotation of the polarization plane (Faraday rotation) of a linearly polarized wave in a collisionless gas plasma in a magnetic field $B_0 = 0.50$ T. Assume that the electron concentration, N, in the plasma is equal to 10^{18} m^{-3}, the wavelength is $\lambda = 1.00$ cm,

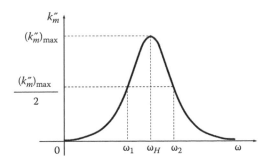

FIGURE 11.7 The frequency dependence of the imaginary part of the magnetic permeability κ_m''.

and the wave propagation direction coincides with the direction of the magnetic field. The wave path length $l = 10.0$ cm. (*Answer*: The angle of rotation of the polarization plane for $l = 10.0$ cm is $\psi = 1.6$ rad $\approx 92°$.)

11.5 A linearly polarized plane wave of amplitude E_0 and wavelength $\lambda = 3.00$ cm is incident on an area $y \geq 0$ filled with plasma with electron concentration $N = 10^{16}$ m^{-3}. The wave polarization plane makes an angle of $\pi/4$ with the y-axis. The plasma is placed in a magnetic field $B_0 = 0.20$ T, with field direction perpendicular to the direction of the wave propagation. Find the minimum nonzero distance that the wave must travel in the plasma in order to have the following polarizations: (a) circular and then (b) linear. (*Answer*: $y_{circular} \approx 0.19$ m, $y_{linear} \approx 0.38$ m.)

11.6 A sample of yttrium iron garnet with parameters $\omega_M = 3.10 \times 10^{11}$ s^{-1} and relaxation frequency $\omega_r = 6.00 \times 10^8$ s^{-1} is placed in an external magnetic field $B_0 = 0.10$ T. Find the half width of the curve $\kappa_m''(\omega)$, that is, the width of the curve, $\Delta\omega$, at $\kappa_m''(\omega) = (\kappa_m'')_{max}/2$ (Figure 11.7). (*Answer*: $\Delta\omega = 1.20 \times 10^9$ s^{-1}.)

11.7 Two orthogonally polarized waves with a frequency $f = 0.30$ GHz propagate through a ferrite sample with thickness $l = 2.00$ cm in a direction perpendicular to an external magnetic field $B_0 = 0.05$ T (and the magnetization). Find the saturation magnetization M_0 for the case when a phase difference equal to $\pi/12$ exists between the two output waves. The permittivity of the ferrite does not depend on frequency and is $\varepsilon = \kappa\varepsilon_0 = 5.25 \times \varepsilon_0$. (*Answer*: $M_0 = 9.80 \times 10^4$ A/m.)

11.8 A linearly polarized wave with amplitude E_0 is incident on a ferrite sample of thickness l, which is placed in an external magnetic field. A wave propagates in the ferrite along the magnetic field (and therefore the magnetization). This wave can be decomposed into a right circularly polarized and a left circularly polarized wave. Find the ratio of the amplitudes of waves of these circularly polarized waves when they exit the sample. $\left(Answer\text{: } \dfrac{A_+(l)}{A_-(l)} = \exp\left[(k_-'' - k_+'')l\right]. \right)$

12 Electromagnetic Waves in Amplifying Media

The analysis of the phenomenon of refraction of electromagnetic waves at the interface between two media suggests that their velocity when passing from one medium to another changes. It has previously been shown that the phase velocity of a wave in a medium is determined by the expression $v_{ph} = c/n$, where n is the medium refractive index. Numerous measurements indicate a dependence of the propagation velocity and the refractive index on the frequency (or wavelength). The wave velocity (or the refractive index) dependence on frequency is called **dispersion**. This phenomenon is present in all media except vacuum, and we considered several specific cases of dispersion in Chapters 6, 7, and 11. The phenomenon of light dispersion can also be observed when a beam of white light passes through a glass prism. One can observe a pattern of multicolor bands on a screen placed behind the prism. This pattern is a result of the decomposition of white light into its spectral components that have different colors and hence different wavelengths.

The dispersion in any medium is caused by absorption. An electromagnetic wave loses part of its energy in order to excite atoms and nuclei of the medium from a lower to a higher energy state. Deexcitation of atoms and nuclei produces secondary emission. In an ideal homogeneous nonconducting medium, the secondary reradiated waves fully return the absorbed fraction of the energy used on the excitation of the medium. In the real medium, not all the energy of the atoms and nuclei is returned in the form of secondary electromagnetic waves. Part of it is converted into other forms of energy, primarily into heat.

In addition to the absorption under certain conditions, some media can amplify optical radiation.

In this chapter, we describe the basic principles of amplification and discuss the main features of wave propagation in **amplifying media**. Amplifying media are defined to be media that can amplify the optical radiation at a particular frequency that corresponds to a quantum transition. Amplification occurs due to the induced coherent emission by excited atoms under the influence of the field of the electromagnetic wave.

12.1 DISPERSION OF ELECTROMAGNETIC WAVES

1. Maxwell's equations do not contain any atomic and molecular parameters, so they cannot explain the phenomenon of dispersion caused by absorption. This requires knowledge of the properties of matter at the atomic level. Dispersion of light occurs as a result of forced oscillations of atomic electrons and ions caused by the time-dependent electric field of an electromagnetic wave. When an electromagnetic wave propagates in a medium, the electric component of the wave field causes an oscillatory motion of electrons in the atoms and molecules of the medium. Oscillating electrons themselves become sources of secondary electromagnetic waves of the same frequency. These secondary waves are superimposed on each other and, together with the incident wave, form the net resulting field in the medium. The change of the phase velocity of an electromagnetic wave in a medium is a consequence of the superposition of the incident wave with the secondary waves. The change of the phase velocity results in a corresponding change of the refractive index.

Along with these effects, any medium (except vacuum) exhibits absorption of electromagnetic wave. The absorption mechanism can be explained as follows. The electromagnetic wave uses part of its energy to excite oscillations of the electrons. In an ideal homogeneous nonconducting medium, the reradiated secondary waves completely give back all the absorbed energy used for the excitation of oscillations. In a real medium, not all the energy of the oscillating electrons is returned in the form of secondary electromagnetic waves. Part of it is transformed into other forms of energy, mainly into heat in the medium.

Let us consider the interaction of electromagnetic waves with matter, using an idealized model, that of a gaseous medium. This model is characterized by its simplicity because, to first approximation, it is possible to ignore interactions between atoms or molecules. In addition, we can assume that the field acting on a single atom coincides with the average field in the medium. Under such assumptions, it is sufficient to consider the action of the electromagnetic wave field on an isolated atom to find the equations that describe the system.

Electrons in an atom can be divided into two groups: external electrons and electrons of the inner shells. Note that only the external electrons interact with the incoming radiation. Frequencies that correspond to the oscillations of the inner electrons are so high that the field of waves up to the light frequencies does not interact with them. The consideration of these electrons becomes essential only for X-ray radiation. The displacement of an external electron from its position of equilibrium (in the classical representation) results in the appearance of a restoring force that tends to return the electron to its equilibrium position. Therefore, under the influence of the electric field of an electromagnetic wave, the electron will perform forced oscillations.

2. In the case of propagation of a linearly polarized monochromatic wave with frequency ω in a medium, the wave electric field is

$$\mathbf{E}(t, \mathbf{r}) = \mathbf{E}_m \exp\left[i(\omega t - \mathbf{k} \cdot \mathbf{r})\right]. \tag{12.1}$$

Since the atomic dimension $a \approx 2 \times 10^{-10}$ m is much smaller than the wavelength ($\lambda \sim 10^{-8} - 10^{-5}$ m), it is possible to neglect the change in the wave phase across atomic distances: $\mathbf{k} \cdot \mathbf{r} \approx 2\pi a/\lambda \ll 1$. Therefore, the electric field of an electromagnetic wave acting on an atom can be assumed to depend only on time, that is, $\mathbf{E}(t) = \mathbf{E}_m \exp(i\omega t)$. This field causes a displacement of an external electron in an atom from its equilibrium position.

The excited atoms emit secondary monochromatic waves with frequencies that are specific for each substance. Therefore, we can assume that the coupling strength of such electron with its atom is quasielastic and the force dependence on the electron displacement is determined by $\mathbf{F}_u = -\beta\mathbf{r}$, where β is the coupling constant. The energy losses caused by the emission of the secondary waves and electron deceleration can be approximately taken into account, if we introduce a damping force $\mathbf{F}_r = -b(d\mathbf{r}/dt)$, where $d\mathbf{r}/dt$ is the electron velocity and b is the coefficient that characterizes damping of the electron motion. Along with these forces, the electron is affected by a driving force that according to Equation 12.1 has the form

$$\mathbf{F}(t) = -e\mathbf{E}(t) = -e\mathbf{E}_m \exp(i\omega t). \tag{12.2}$$

Using Newton's second law, $m(d^2\mathbf{r}/dt^2) = \mathbf{F}(t) + \mathbf{F}_r + \mathbf{F}_u$, we obtain

$$\frac{d^2\mathbf{r}}{dt^2} + 2\gamma \frac{d\mathbf{r}}{dt} + \omega_0^2 \mathbf{r} = -\left(\frac{e}{m}\right) \mathbf{E}_m \exp(i\omega t), \tag{12.3}$$

where
  $\gamma = b/2m$ is the damping coefficient
  $\omega_0 = \sqrt{\beta/m}$ is the intrinsic frequency of the electron oscillations

Assuming that the displacement in Equation 12.3 is harmonic ($\mathbf{r}(t) = \mathbf{r}_m \exp(i\omega t)$), after substituting it into Equation 12.3, we get what is known as a "forced oscillation" solution:

$$\mathbf{r}(t) = -\frac{(e/m)}{\left(\omega_0^2 - \omega^2\right) + 2i\gamma\omega} \mathbf{E}(t). \tag{12.4}$$

With the displacement of electrons in the atoms known, the polarization of the medium is given by the equation $\mathbf{P} = N\mathbf{p}$. Here, N is the number of atoms in unit volume, and $\mathbf{p}$ is the dipole moment of atom.

For atoms with one external electron, the dipole momentum is $\mathbf{p} = -e\mathbf{r}$, where e is the elementary charge, and $\mathbf{r}$ is the electron displacement under the influence of an electric field. In this case, the polarization vector is $\mathbf{P} = -Ne\mathbf{r}$.

In Chapter 1, we showed that the polarization vector is related to the electric field vector by $\mathbf{P} = \varepsilon_0\chi\mathbf{E}$, where the dielectric susceptibility χ of the medium is related to the permittivity by relation $\kappa = 1 + \chi$. Using these relations, we obtain

$$\mathbf{P} = -Ne\mathbf{r} = \frac{(Ne^2/m)}{\left(\omega_0^2 - \omega^2\right) + 2i\gamma\omega}\mathbf{E} = \varepsilon_0\chi\mathbf{E}, \tag{12.5}$$

from where we find an expression for the relative permittivity:

$$\tilde{\kappa} = 1 + \chi = 1 + \frac{\omega_p^2}{\left(\omega_0^2 - \omega^2\right) + 2i\gamma\omega}, \tag{12.6}$$

where $\omega_p = \sqrt{Ne^2/\varepsilon_0 m}$ is a plasma frequency. The permittivity generally is a complex number, and therefore, it can be presented in the form $\tilde{\kappa} = \kappa' - i\kappa''$, where the real and imaginary parts are

$$\kappa'(\omega) = 1 + \frac{\omega_p^2\left(\omega_0^2 - \omega^2\right)}{\left(\omega_0^2 - \omega^2\right)^2 + 4\gamma^2\omega^2}, \quad \kappa''(\omega) = \frac{2\gamma\omega\omega_p^2}{\left(\omega_0^2 - \omega^2\right)^2 + 4\gamma^2\omega^2}. \tag{12.7}$$

At very low frequency, as ω tends to zero, one gets from these equations $\kappa'(\omega) \approx 1 + \omega_p^2/\omega_0^2$, $\kappa''(\omega) \approx 0$ and at the very high frequency (ω tends to infinity) $\kappa'(\omega) \approx 1$, $\kappa''(\omega) \approx 0$ (see Figure 12.1). The imaginary part, $\kappa''(\omega)$, reaches its maximum value at $\omega = \omega_0$.

3. The complex refractive index, which is related to the complex permittivity, in its general form is given by

$$\tilde{n} = \sqrt{\tilde{\kappa}\kappa_m} = n - i\kappa, \tag{12.8}$$

where
  κ_m is the magnetic permeability
  n is the real part of the refractive index that in the literature is named simply refractive index in
    contrast to the complex refractive index, $\tilde{n}$
  κ is the imaginary part also known as the extinction coefficient

It is important that the phase velocity is determined by the refractive index: $v_{ph} = c/n$. In this section, we will consider a nonmagnetic medium, for which with a high degree of accuracy, $\kappa_m = 1$. Let us now square the expression $\sqrt{\tilde{\kappa}} = n - i\kappa$, and as a result, we obtain

$$n^2 - \kappa^2 = \kappa', \quad 2n\kappa = \kappa''. \tag{12.9}$$

The solution of this system of equations leads to the following expressions for the refractive index and the extinction coefficient:

$$n = \frac{1}{\sqrt{2}}\left(\sqrt{(\kappa')^2 + (\kappa'')^2} + \kappa'\right)^{1/2}, \quad \kappa = \frac{1}{\sqrt{2}}\left(\sqrt{(\kappa')^2 + (\kappa'')^2} - \kappa'\right)^{1/2}. \tag{12.10}$$

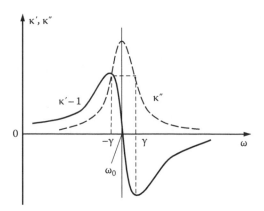

FIGURE 12.1 Frequency dependence of the real part $\kappa'(\omega)$ (solid line) and imaginary part $\kappa''(\omega)$ (dashed line) of a complex relative permittivity in the vicinity of the intrinsic absorption frequency ω_0.

Using relationships (12.7) for the real and imaginary parts of the dielectric permittivity, it is possible to determine, from Equation 12.10, the frequency dependence of the refractive index and the extinction coefficient. The dependence of the refractive index on frequency, $n(\omega)$, is known as **dispersion**. The dependence of the extinction coefficient on frequency, $\kappa(\omega)$, determines the spectral line shape of the absorption.

In the model described earlier, the real part of the refractive index, $n(\omega)$, behaves like $\kappa'(\omega)$ in Figure 12.1, that is, for frequencies $\omega < \omega_0 - \gamma$ increases with increasing frequency, reaching a maximum value in the vicinity of $\omega = \omega_0 - \gamma$, and then decreases with increasing frequency. The refractive index, $n(\omega)$, becomes equal to 1 at ω_0 if $\gamma = 0$; otherwise, it is larger than 1 at $\omega = \omega_0$. It becomes less than unity for $\omega > \omega_0$. The minimum of function $n(\omega)$ occurs in the vicinity of the frequency $\omega = \omega_0 + \gamma$. At higher frequencies, the real part of the refractive index increases and asymptotically approaches unity. The extinction coefficient $\kappa(\omega)$ behaves like $\kappa''(\omega)$ in Figure 12.1, that is, it has its maximum value at frequency $\omega = \omega_0$; the width of the resonance curve (i.e., the spectral absorption line) at half maximum is equal to 2γ.

The phenomenon of the refractive index decrease with increasing frequency, which takes place within the width of the spectral absorption line, is called **anomalous dispersion**. In this frequency range, $dn(\omega)/d\omega < 0$.

At frequencies far from the intrinsic frequency of electrons' oscillations, where the condition $|\omega - \omega_0| \gg 2\gamma$ is satisfied, the imaginary part in Equation 12.6 can be neglected. In this case, we obtain the following approximate expression for the frequency dependence of refractive index:

$$n^2(\omega) = \kappa(\omega) = 1 - \frac{\omega_p^2}{\omega^2 - \omega_0^2}. \tag{12.11}$$

At those frequencies for which the expression given earlier is applicable, the refractive index increases with increasing frequency, that is, $dn/d\omega > 0$. Such dependence of $n(\omega)$ is called **normal dispersion**.

For low frequencies ($\omega < \omega_0$), the refractive index is greater than unity, that is, the phase velocity $v_{ph} = c/n$ of the wave in a medium is smaller than the velocity of light in vacuum. This means that the phase of a wave in the medium is retarded compared to its propagation in a vacuum. Such a situation occurs, for example, during the propagation of visible radiation through transparent substances such as glass or quartz. The intrinsic optical absorption frequencies of the electrons in the atoms of these substances are in the ultraviolet spectral range and for visible light $\omega < \omega_0$.

If the frequency of the radiation is greater than the intrinsic frequency ($\omega > \omega_0$), then $n(\omega) < 1$ and the phase velocity of the wave in the medium is greater than the velocity of light in a vacuum, that is, the wave modified by the medium in such a way that its phase advances with respect to the wave propagation in vacuum. This fact is not in contradiction with the theory of relativity, which states that the velocity of signal transmission cannot exceed the speed of light in vacuum. The concepts of refractive index and phase velocity are applicable only to simple harmonic (monochromatic) waves that have an infinite extent in space and time. Thus, a monochromatic wave cannot be used for signal transmission.

Equation 12.11 is true for a wide frequency range, and it is substantially simplified for high frequencies, $\omega \gg \omega_0$. The high-frequency approximation describes well the dispersion of X-rays in glass as the X-ray radiation has frequencies that are several thousand times higher than the frequency of visible light and frequency ω_0. Here $n(\omega) < 1$, although not very different from 1, since in this case the frequency is high. A similar result is obtained when describing the propagation of radio waves in the ionosphere. The ionosphere consists of gaseous plasma in which electrons are practically free. Thus, in Equation 12.11 it is possible to put $\omega_0 = 0$, and for such electrons the condition $\omega \gg \omega_0$ will always be satisfied even in the radio frequency range. In this case, Equation 12.11 takes the form

$$n^2(\omega) = 1 - \frac{\omega_p^2}{\omega^2}. \tag{12.12}$$

The discussed theoretical considerations are valid not only for electrons but also for ions for which a classical description is more appropriate as their masses are substantially higher than the electron mass. It should be added that as a rule, in all substances, even in monoatomic gases with one optically active electron (i.e., only one free electron per one ion), there is not one but several absorption bands. To account for this fact, we assume that any substance is made of particles of various types—electrons and ions, which behave as damped harmonic oscillators with different intrinsic frequencies and damping coefficients. In this case, Equation 12.6 can be generalized as follows:

$$\tilde{\kappa}(\omega) = 1 - \sum_l \frac{f_l \omega_{pl}^2}{\left(\omega^2 - \omega_{0l}^2\right) - 2i\gamma_l \omega}, \tag{12.13}$$

where the factor f_l is called the ***oscillator strength*** (for each type of oscillator) and must satisfy the condition

$$\sum_l f_l = 1. \tag{12.14}$$

According to Equation 12.13 to each intrinsic frequency, there corresponds an absorption band, near which the index of refraction changes anomalously. The frequency dependence of the refractive index is shown schematically in Figure 12.2.

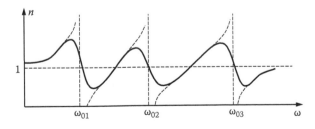

FIGURE 12.2 Frequency dependence of refractive index of a medium in the presence of several intrinsic frequencies ω_{0l}.

Exercise 12.1

A medium is described by a complex dielectric permittivity whose frequency dependence is given by Equation 12.6. Find the frequency dependence of the real part of the medium's conductivity.

Solution. The dielectric and conducting properties of a medium may be described either by medium's complex dielectric permittivity $\tilde{\kappa} = \kappa' - i\kappa''$ or by its complex conductivity $\tilde{\sigma} = \sigma' - i\sigma''$. In order to demonstrate this, let us rewrite Ampere's law as

$$\nabla \times \mathbf{H} = \mathbf{j} + \frac{\partial \mathbf{D}}{\partial t}$$

and the electric displacement and current density for the medium are $\mathbf{D} = \varepsilon_0 \kappa \mathbf{E}$ and $\mathbf{j} = \sigma \mathbf{E}$, where κ and σ are assumed to be real. For a monochromatic wave with frequency ω,

$$\frac{\partial \mathbf{D}(t)}{\partial t} = i\omega \mathbf{D}(t) = i\omega \varepsilon_0 \kappa \mathbf{E}(t).$$

Let us substitute these last relationships into Ampere's law:

$$\nabla \times \mathbf{H} = \mathbf{j} + \frac{\partial \mathbf{D}}{\partial t} = \sigma \mathbf{E} + i\omega \varepsilon_0 \kappa \mathbf{E} = (\sigma + i\omega \varepsilon_0 \kappa)\mathbf{E} = i\omega \varepsilon_0 \tilde{\kappa} \mathbf{E} = \tilde{\sigma} \mathbf{E}.$$

Here,

$$i\omega \varepsilon_0 \tilde{\kappa} = i\omega \varepsilon_0 (\kappa' - i\kappa'') = \sigma + i\omega \varepsilon_0 \kappa, \quad \kappa = \kappa', \quad \sigma = \sigma' = \omega \varepsilon_0 \kappa''.$$

In order to find medium's conductivity, let us use the expression for the complex dielectric permittivity from Equation 12.7:

$$\tilde{\kappa} = \kappa' - i\kappa'' = 1 - \frac{\omega_p^2 \left(\omega^2 - \omega_0^2 \right)}{\left(\omega^2 - \omega_0^2 \right)^2 + 4\gamma^2 \omega^2} - i \frac{2\gamma \omega \omega_p^2}{\left(\omega^2 - \omega_0^2 \right)^2 + 4\gamma^2 \omega^2}.$$

Taking into account the expression for the imaginary part of the complex dielectric permittivity, we get

$$\sigma = \sigma' = \omega \varepsilon_0 \kappa'' = \omega \varepsilon_0 \frac{2\gamma \omega \omega_p^2}{\left(\omega^2 - \omega_0^2 \right)^2 + 4\gamma^2 \omega^2} = \frac{2\varepsilon_0 \gamma \omega^2 \omega_p^2}{\left(\omega^2 - \omega_0^2 \right)^2 + 4\gamma^2 \omega^2}.$$

This expression gives the conductivity of a medium with bound electrons, that is, the conductivity of a dielectric. For a medium with free electrons, $\omega_0 = 0$. Therefore, for the conductivity of a metal, we get the following frequency dependence:

$$\sigma = \sigma' = \frac{2\varepsilon_0 \gamma \omega_p^2}{\omega^2 + 4\gamma^2}.$$

12.2 ATTENUATION OF WAVES IN AN ABSORBING MEDIUM: BOUGUER'S LAW

An electromagnetic wave is attenuated as it passes through a medium. The main reason for this attenuation is absorption—the phenomenon that consists of the decrease in the intensity of an electromagnetic wave propagating in matter due to the interaction of the wave with the constituent particles (atoms and molecules) of matter. As a result of such interaction, part of the wave intensity is transformed into heat of the medium. The law of attenuation of light in a medium has been experimentally investigated by the French scientist P. Bouguer.

Consider a beam of parallel rays propagating in an absorbing medium along the y-axis. The beam's initial intensity (at $y = 0$) is equal to I_0. After traveling through the medium a distance y, the light is attenuated, due to absorption, and its intensity becomes smaller than I_0.

Let us consider a layer of thickness dy. The intensity of light at a distance $y + dy$ will differ by dI from the intensity $I(y)$ that was at a distance y, that is, it is equal to $I + dI$. The value dI represents the flux of light energy absorbed by thickness dy and is therefore negative. This value is proportional to the thickness dy of this layer and the intensity $I(y)$ of the light incident on this layer, that is,

$$dI = -\alpha I(y) dy. \tag{12.15}$$

The coefficient of proportionality α is called the **coefficient of absorption** of the electromagnetic wave. If we separate the variables in Equation 12.15, we get

$$\frac{dI}{I} = -\alpha\, dy. \tag{12.16}$$

We then integrate this expression in the range from I_0 for $y = 0$ to $I(y)$. Thus, we get

$$\int_{I_0}^{I} \frac{dI}{I} = -\alpha \int_{0}^{y} dy, \quad I(y) = I_0 \exp(-\alpha y). \tag{12.17}$$

This expression is called Bouguer's law. The physical meaning of the absorption coefficient is easy to establish by transforming Equation 12.17 as follows:

$$\alpha = y^{-1} \ln \frac{I_0}{I(y)}. \tag{12.18}$$

If the layer thickness to be chosen is such that $I_0/I(y) = e$, then $\ln(I_0/I(y)) = 1$ and the thickness of the layer is equal to $y = \alpha^{-1}$. Thus, the absorption coefficient has the dimension of reciprocal length (m^{-1}) and the reciprocal value of the absorption coefficient is numerically equal to the distance in a medium, at which the intensity of light is decreased by a factor e.

The exponential dependence of the wave intensity on distance y is shown in Figure 12.3. For a plane monochromatic wave propagating in a metal, Equation 7.11 yields the following expressions for the amplitudes of the electric and magnetic fields of a wave that propagates in y-direction:

$$\mathbf{E}(y) = \mathbf{E}_0 \exp\left(-\frac{y}{\delta}\right), \quad \mathbf{H}(y) = \mathbf{H}_0 \exp\left(-\frac{y}{\delta}\right), \tag{12.19}$$

where parameter δ is the penetration depth of the wave in the metal. This parameter is related with the extinction coefficient through the expression $\delta = 1/k_0\kappa = c/\omega\kappa$. If we take into account that the

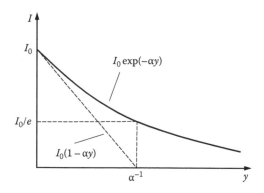

FIGURE 12.3   Dependence of intensity on a depth of wave penetration into a medium.

intensity of a wave is proportional to the square of the amplitude of the electric field, that is, $I \sim |\mathbf{E}|^2$, we get $\alpha = 2/\delta = 2\omega\kappa/c$.

Generally, the absorption coefficient depends on the frequency of the incident radiation and also on the absorbing material. For example, monoatomic gases and vapors of metals that have large interatomic distances from each other have absorption coefficients close to zero. Sharp absorption maxima occur only for very narrow spectral ranges in the vicinity of the frequencies of intrinsic oscillations of the electrons in the constituent atoms. This is known as a *line absorption spectrum*.

The absorption spectrum of molecules is determined by atomic oscillations in molecules is characterized by wider absorption bands. The absorption coefficient for dielectrics in a wide spectral range is insignificant ($\alpha \sim 10^{-3}$–10^{-1} m^{-1}). The coefficient α sharply increases in particular wavelength ranges (absorption bands), which form the *continuous absorption spectrum*. This is due to the fact that there are no free electrons in dielectrics, and light absorption is caused by quantum transitions between energy bands, which are characteristic for each dielectric material.

The absorption coefficient for a metal has high values ($\alpha \sim 10^5$–10^7 m^{-1}), and for this reason metals are opaque to light. In metals, free electrons moving under the influence of the electric field of the light wave generate fast alternating currents, which results in the generation of Joule heating. Therefore, the energy of a light wave decreases rapidly with distance, and it is converted into thermal energy of the metal. The higher the metal conductivity, the greater the light absorption rate.

Wave attenuation is not necessarily associated with the absorption of electromagnetic energy. If the dielectric permittivity is real but negative (i.e., $\kappa'' = 0$, $\kappa' < 0$), the refractive index $\tilde{n} = i\sqrt{|\kappa'|}$ is imaginary, so the radiation cannot penetrate into the material, and as a result, total reflection of the incident wave takes place.

The dependence of the absorption coefficient on wavelength explains the color of absorbing bodies. For example, glasses that absorb weakly the red and orange part of the spectrum but strongly absorb green and blue wavelengths, when illuminated with white light, will appear reddish. This phenomenon is used for the fabrication of *light filters*, which depending on chemical composition transmit light of only certain wavelengths while absorbing others.

Bouguer's law holds for a wide range of electromagnetic radiation—from radio waves to X-rays and gamma rays. However, it must be kept in mind that under certain conditions it is only an approximation. Thus, for linear wave processes, which take place at low radiation intensity, the absorption coefficient does not depend on the intensity. For sufficiently large values of the intensity, the absorption coefficient of a medium begins to depend on intensity and Bouguer's law is violated.

As a rule, an increase in the intensity of incident radiation results in a decrease of the absorption coefficient. This fact cannot be explained within the framework of classical physics but finds an explanation if quantum mechanics is used instead. During light absorption, a fraction of the atoms and molecules in the absorbing medium appear in their excited states. These atoms or molecules can no longer participate in further absorption until they return to their ground state. Bouguer's law holds true only if the number of excited atoms and molecules is a small fraction of the total number of atoms and molecules in the medium.

Exercise 12.2

A plane wave $E = E_0 \exp[i(\omega t - k_0 x)]$ is incident normally on an absorbing layer. Here, ω, c, and $k_0 = \omega/c$ are the frequency, speed, and wave number of the electromagnetic wave in vacuum, respectively. The layer thickness is equal to l, and its complex dielectric permittivity is equal to $\tilde{\kappa} = \kappa' - i\kappa''$. Find the amplitude of the electric field of the electromagnetic wave after it exits the absorbing layer, and write an expression for the wave's field for incident wave (reflected waves must be ignored), wave in the absorbing layer, and wave transmitted through the layer.

Solution. Let us assume that the surface on which the wave is incident is defined by the coordinate $x = 0$. Therefore, the field inside the layer is described by the expression

$$E(t, x) = E_1 \exp\left[i(\omega t - k_1 x)\right],$$

where the amplitude $E_1 = E_0$ from the continuity of the tangential components of the electric field at the interface of the two media. The magnitude of the wave vector is determined by the equation $k_1 = k_0(n - i\kappa)$, where n and κ are real. The refraction index, n, and extinction coefficient, κ, are the real and imaginary parts of complex refractive index, $\tilde{n}$:

$$\tilde{n} = n - i\kappa = \sqrt{\kappa' - i\kappa''}.$$

By the substitution of n and κ into the expression for the wave's field, we get the following expression for the field in the layer:

$$\begin{aligned} E(t, x) &= E_0 \exp\left[i(\omega t - k_0(n - i\kappa)x)\right] \\ &= E_0 \exp(-k_0 \kappa x) \exp\left[i(\omega t - k_0 n x)\right] = E_m(x) \exp\left[i(\omega t - k_0 n x)\right], \quad 0 \le x \le l, \end{aligned}$$

where the amplitude of the wave's field inside of the layer is equal to

$$E_m(x) = E_0 \exp(-k_0 \kappa x) = E_0 \exp\left(-\frac{x}{\delta}\right), \quad \delta = \frac{1}{k_0 \kappa}, \quad \alpha = \frac{2}{\delta} = 2k_0 \kappa.$$

The amplitude of the wave's field after transmission through the film is equal to

$$E_m(l) = E_0 \exp(-k_0 \kappa l).$$

The expressions for n and κ can be found if we will square the complex refraction index $\tilde{n}$:

$$n^2 - \kappa^2 - 2in\kappa = \kappa' - i\kappa'',$$
$$n^2 - \kappa^2 = \kappa', \quad 2n\kappa = \kappa''.$$

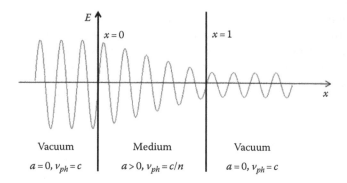

FIGURE 12.4 Distribution of the wave's electric field inside and outside of the plate.

The solution of these equations leads to the following relationships:

$$n = \sqrt{\frac{\kappa'}{2}\left(\sqrt{1+\left(\frac{\kappa'}{\kappa''}\right)}+1\right)}^{1/2},$$

$$\kappa = \sqrt{\frac{\kappa'}{2}\left(\sqrt{1+\left(\frac{\kappa'}{\kappa''}\right)}-1\right)}^{1/2}.$$

The wave's field after transmission through the plate is described by the expression

$$E(t,x) = E_2 \exp\left[i(\omega t - k_0 x)\right], \quad x \geq l,$$

where $E_2 = E_m(l) = E_0 \exp(-k_0 \kappa l)$. The phase velocity of the wave in vacuum is equal to $v_{ph} = c$ and in the medium to $v_{ph} = c/n$.

Figure 12.4 shows the electric field of a wave that is incident on a plate, traveling through the plate, and after transmission through the plate.

12.3 AMPLIFICATION OF ELECTROMAGNETIC WAVES IN A MEDIUM

1. To amplify a light beam, the absorbing medium must be in a state for which the absorption coefficient is negative. In this case, according to Bouguer's law, the intensity of light wave as it propagates in the medium will increase. This state can be achieved if the population of excited atoms and molecules exceeds that of those in the lower energy state (the lowest energy state is called ground state). Such a situation is known as ***population inversion*** and the medium itself can amplify an optical signal.

Gases, vapors, different liquid solutions, crystalline or amorphous materials, or some complex composites can be used as amplifying media. The components of the amplifying media must be chosen so that atoms of the media have higher energy levels/states where electron can be excited. The electron remains in the excited level a finite time before it goes to the lower state this is why the higher energy levels are named ***metastable levels***. Note that in the stable state, an electron can stay infinitely long, while in the metastable state, the lifetime of an electron is finite; this is mostly due to the existence of an empty state with a lower energy. As the electron goes from higher energy state to lower energy state, it emits quanta of radiation that are named photon. So an electron emits photon as it goes from higher energy state to lower energy state, and it absorbs photon as it goes from lower energy state to higher energy state. In the ordinary (passive) medium, the radiation of separate atoms occurs spontaneously—independently from each other at different times and in different directions.

In a passive medium, the number of atoms in the ground state is always larger than the number populating the exited states. An amplifying medium, prepared for light amplification, must have most of the atoms in their excited state. In order to realize this condition, electrons must continuously gain energy from an external source so that they can populate a metastable energy level.

An electron in a long-lived metastable level can make stimulated transition to the ground state and emit a photon. The stimulated transition is caused by a photon (quantum of electromagnetic radiation) of the same frequency as the emitted one. The photon emitted during such a forced transition has the same phase as the original photon that induced the transition. After each interaction, known as *stimulated emission*, the number of photons is doubled, and an avalanche of stimulated (induced) emission appears in the medium. This phenomenon—amplification of a light beam by an amplifying medium—is called *negative absorption*.

If the absorption coefficient is negative (i.e., $\alpha < 0$), the intensity of the radiation propagating in the medium increases exponentially (Equation 12.17):

$$I(y) = I_0 \exp(-\alpha y) = I_0 \exp(|\alpha| y). \tag{12.20}$$

Radiation amplification underlies the operation of quantum amplifiers and generators of microwave and optical radiation (masers and lasers). The principle of their operation has a quantum mechanical origin and is related with *stimulated* (or *induced*) emission of radiation.

In the framework of quantum mechanics, it can be described as follows: A photon passing through an amplifying inverted medium induces in an excited atom a transition to its ground state. As a result, an additional photon that has exactly the same energy and momentum as the original photon is generated. These two photons, propagating in the medium, cause the generation of two more identical photons. Due to the induced transitions described earlier, we have an avalanche-like increase of the photon population, that is, an increase of the light beam intensity. Simultaneously, in addition to the stimulated transitions, accompanied by emission of energy, we also have radiation absorption, caused by transitions of the medium's atoms from the ground to their excited state. These transitions result in a decrease of the light intensity. Therefore, the net intensity of a light beam passing through an amplifying medium is determined by the prevailing process. If absorption dominates, the beam intensity decreases; if on the other hand stimulated emission dominates, the beam intensity increases.

In order to better understand the principle of amplification of the light wave in the amplifying medium, it is necessary to examine more closely the processes of absorption and emission of photons by atoms. Atom can exist in different energy states with energies E_1, E_2, ..., E_n. In the simple picture of Bohr's atom, these states are stable states. An atom, in the absence of external perturbations, can occupy state of the lowest energy indefinitely. This is called the ground state. All other states have a finite lifetime. An excited atom can exist in these states only for a very short time (about 10^{-8} s) after which it spontaneously makes a transition to one of the lower states, emitting a quantum of light (photon). An electron occupying a metastable energy level of an atom can exist in that state for a much longer time—of the order of 10^{-3} s.

The transition of an atom to a higher energy state can occur only during resonant absorption of a photon whose energy is equal to the difference between the energies of the atom in the initial and final states, that is, $\hbar\omega = |E_1 - E_2|$, where $\hbar = 1.05 \times 10^{-34}$ J·s is Planck's constant and ω is the frequency of the absorbed photon. It should also be noted that the transitions between the energy levels of the atom are not necessarily related to the absorption or emission of photons. Atoms can absorb or emit energy and move to a different quantum state as a result of *nonradiative transition*, that is, as a result of interaction with other atoms or collisions with electrons.

Transition of an electron in an atom from the upper to the lower energy level can occur under the influence of an external electromagnetic field whose frequency is equal to the frequency of the transition. The resulting stimulated emission is significantly different from the spontaneous emission. This is due to the fact that, as a result of interaction of photon with excited atom, the atom emits a photon of the same frequency that propagates in the same direction. In terms of the wave theory, this means that the atom emits an electromagnetic wave whose frequency, phase, polarization, and direction of propagation are the same as in the initial wave. As a result of the stimulated emission of photons, the amplitude of the propagating wave in the medium increases.

2. Let us consider a layer of transparent material whose atoms can populate only two states with energy E_1 and $E_2 > E_1$. Such system is called the ***two-level system***. Let the radiation with the resonant frequency of the transition $\omega = (E_2 - E_1)/\hbar$ propagate in the layer. According to the Boltzmann distribution, at thermodynamic equilibrium, there are more atoms in the lowest energy state. Some of the atoms will be in the upper energy state receiving the necessary energy in collisions with other atoms. The number of atoms in the state with energy E_1 is equal to n_1 while the number of atoms in the state with energy E_2 is equal to n_2. These numbers are called the ***populations*** of the lower and upper levels, and $n_2 < n_1$ when $E_2 > E_1$. During the propagation of resonant radiation in such medium, three processes occur as shown in Figure 12.5. Einstein showed that the process of resonant absorption of a photon by the unexcited atom (a) and the process of induced emission of a photon by an excited atom (c) have the same probabilities. Since in the equilibrium system $n_2 < n_1$, the photon absorption will occur more frequently than the induced emission. As a result of passing through the material layer, the radiation intensity decreases.

Spontaneous transitions without external influence occur if electrons are at the energy state E_2 above the available energy state E_1. The reason for these transitions is that the system tends to minimize energy and eventually turns into an unexcited state. The probability of spontaneous transition of an electron with emitted energy $\hbar\omega = (E_2 - E_1)$ is determined by the Einstein coefficient A_{21}, that is, $W_{21} = A_{21}$ (s^{-1}). The power of spontaneous emission is given by

$$P_{21}^{spont} = \hbar\omega A_{21}n_2 \ [(\text{J/s}) \ \text{m}^{-3}]. \tag{12.21}$$

Radiation produced as a result of spontaneous transitions is incoherent and propagates in all directions.

Induced transitions can occur with an increase and decrease in the energy, and these transitions correspond to the absorption and emission, respectively. For induced transitions, the following statements are valid:

- The photons with energy $\hbar\omega$ of the induced transitions of the electromagnetic field are identical to the photons that caused this transition.

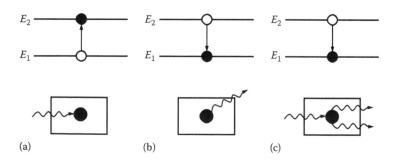

(a)   (b)   (c)

FIGURE 12.5 Energy diagram and schematic representation of the absorption (a), spontaneous (b), and induced (c) emissions of a photon.

- The probability of induced transitions is different from zero only for the external field that has resonant frequency $\omega = (E_2 - E_1)/\hbar$, and the probability of transition is proportional to the spectral energy density (with the dimension J s/m^3) at this frequency

$$W_{12}^{ind} = B_{12}u(\omega), \quad W_{21}^{ind} = B_{21}u(\omega). \tag{12.22}$$

Here, B_{12} and B_{21} are the Einstein coefficients that determine the probability of direct and inverse induced transitions. The spectral energy density is related with the volume energy density by the equation

$$u(\omega) = \frac{dw}{d\omega}, \quad w = \int u(\omega)d\omega, \tag{12.23}$$

where the integration has to be carried out over a frequency range over which the emission occurs. Generally, the spectral interval of stimulated emission $\Delta\omega_l$ (width of the emission line) is much smaller than the average frequency at which the radiation ω_0 occurs (the carrier frequency of the wave packet), that is, $\Delta\omega_l \ll \omega_0$. Accordingly, the volume energy density and spectral density can be expressed as $w = u(\omega)\Delta\omega_l$.

Let us write the expression for the power of stimulated absorption and emission:

$$P_{12}^{ind} = \hbar\omega B_{12}u(\omega)n_1, \quad P_{21}^{ind} = \hbar\omega B_{21}u(\omega)n_2. \tag{12.24}$$

Under conditions of thermodynamic equilibrium, the power absorbed by a medium is equal to the total radiated power, that is,

$$P_{12}^{ind} = P_{21}^{spont} + P_{21}^{ind},$$

$$\hbar\omega B_{12}u(\omega)n_1 = \hbar\omega A_{21}n_2 + \hbar\omega B_{21}u(\omega)n_2. \tag{12.25}$$

From this equation, we obtain the following expression for the spectral density:

$$u(\omega) = \frac{A_{21}n_2}{B_{12}n_1 - B_{21}n_2} = \frac{A_{21}}{B_{21}} \cdot \frac{1}{B_{12}n_1/B_{21}n_2 - 1}. \tag{12.26}$$

In accordance with statistical mechanics, the number of atoms, which may be in the state of energy E_l, is equal to

$$n_l = n_0 \exp\left(\frac{-E_l}{k_B T}\right), \tag{12.27}$$

where
k_B is the Boltzmann constant
T is the absolute temperature

According to this equation, the ratio of the number of atoms in the ground and excited states is

$$\frac{n_1}{n_2} = \exp\left(\frac{E_2 - E_1}{k_B T}\right). \tag{12.28}$$

From this relation, we obtain the following expression:

$$u(\omega) = \frac{A_{21}}{B_{21}} \left[\frac{B_{12}}{B_{21}} \exp\left(\frac{E_2 - E_1}{k_B T} \right) - 1 \right]^{-1} = \frac{A_{21}}{B_{21}} \frac{1}{\exp\left(\hbar\omega / k_B T \right) - 1}. \tag{12.29}$$

Here, we have introduced the photon energy emitted by the atom, $E_2 - E_1 = \hbar\omega$, and took into account that the probabilities of direct and inverse induced transitions are the same, that is, $B_{12} = B_{21}$. From a comparison of this expression with Planck's equation for the spectral energy density of blackbody radiation,

$$u(\omega) = \frac{\hbar\omega^3}{\pi^2 c^3} \frac{1}{\exp(\hbar\omega / k_B T) - 1}. \tag{12.30}$$

We find the relation of probability of spontaneous and induced transitions as

$$A_{21} = \frac{\hbar\omega^3}{\pi^2 c^3} B_{21}. \tag{12.31}$$

Let us now write expression (12.25) for the total radiated power in the following form:

$$P_{21} = \hbar\omega A_{21} n_2 + \hbar\omega B_{21} u n_2 = \hbar\omega \left(1 + \frac{\pi^2 c^3}{\hbar\omega^3} u \right) A_{21} n_2 = \hbar\omega \left(\frac{\hbar\omega^3}{\pi^2 c^3} + u \right) B_{21} n_2. \tag{12.32}$$

This equation emphasizes two components in radiation—stimulated emission and spontaneous emission.

3. Let us consider now the interaction of electromagnetic waves with a medium, which is an ensemble of atoms in one of two possible energy states. Assume that the radiation propagates in the medium in the form of a traveling wave in the direction of the y-axis with speed c. The wave intensity I (W/m^2) decreases as a result of the passage through an absorbing layer with thickness dy in accordance with Equation 12.15:

$$dI = -\alpha I \, dy,$$

where the absorption coefficient α is generally a function of frequency.

The intensity of the wave is associated with the wave energy density w by the relation $I = cw$, and therefore,

$$\alpha = -\frac{1}{I} \frac{dI}{dy} = -\frac{1}{cw} \frac{dw}{dt} = -\frac{1}{cw} P, \tag{12.33}$$

where we took into account that the power density is associated with the energy density as $dw/dt = P$.

The change of the wave power is equal to the difference between the absorption and emission in the media:

$$P = \hbar\omega A_{21} n_2 + \hbar\omega B_{21} u(\omega) n_2 - \hbar\omega B_{12} u(\omega) n_1. \tag{12.34}$$

The probability of spontaneous transition is usually small compared to the probability of an induced transition, so it can be neglected, wherein

$$P = -\hbar\omega B_{12} u(\omega)(n_1 - n_2) = -\frac{\hbar\omega B_{12} w(n_1 - n_2)}{\Delta\omega_l}. \qquad (12.35)$$

As a result, we arrive at the following expression for the absorption coefficient:

$$\alpha(\omega) = \frac{\hbar\omega B_{12}(n_1 - n_2)}{c\Delta\omega_l}. \qquad (12.36)$$

Thus, the solution of Equation 12.15 in this case is a function $I(y) = I_0\exp(-\alpha y)$ that takes the form of Bouguer's law (12.20). From this law, it follows that $\alpha > 0$ for $n_1 > n_2$. In this case, the situation corresponds to absorption. Amplification is possible only in the case in which the excited state is occupied with more particles than the ground state, that is, $n_1 < n_2$ and the coefficient $\alpha < 0$. The condition with the positive parameter $n_{21} = n_2 - n_1$ is called **inversion**, and the corresponding states of the medium are called **states with inverted population**.

In two-level system, it is impossible to achieve such a state. If we move the electrons by pumping from the lower to the upper level, then the equilibrium state n_2 can only get closer to n_1, that is, $n_{21} \to 0$. In this case, the medium is bleached, but the states with an inverted population are not achieved. So in order to achieve amplification, it is necessary to create the conditions under which population inversion is always achieved, that is, $n_2 > n_1$. To achieve that, it is often necessary to use a system with three or more energy levels.

Exercise 12.3

Compare the illumination from a point source (bulb) with a power of 100 W at a distance of $R = 10.0$ m from the source and from a helium–neon laser with a power of 1 mW at a distance of $R = 10.0$ m. The wavelength of the laser is $\lambda = 0.632$ μm and at the exit from the laser, the beam radius is $r_0 = 0.50$ mm.

Solution. Illumination is defined by $E = \Phi/A$, where Φ is the power of the source (the light flux emitted by the source) and A is an area of the illuminated surface. Lamp illuminates in all directions, so at the distance R from the lamp the surface of the space of radius R is equal to $4\pi R^2$ and illumination is equal to

$$E_{lamp} = \frac{\Phi_{lamp}}{4\pi R^2} = \frac{100}{4\times 3.14\times 100} = 0.08 \text{ lx}.$$

For a laser illuminated surface, area is determined by the section of the beam whose area at a distance R from the source can be approximately expressed via the angle θ of divergency of the laser beam:

$$A_b = \pi r^2 = \pi\left(\frac{\theta R}{2}\right)^2,$$

where
  r is the radius of the beam at the distance R
  $\theta = 2\lambda/\pi r_0$

so

$$A_b = \frac{\pi R^2}{4}\left(\frac{2\lambda}{\pi r_0}\right)^2 = \frac{3.14\times10^2}{4}\left(\frac{2\times0.632\times10^{-6}}{3.14\times0.5\times10^{-3}}\right)^2 = 5.00\times10^{-5} \text{ m}^2.$$

Illumination caused by laser at $R = 10$ m is

$$E_{las} = \frac{\Phi_{las}}{A_b} = \frac{10^{-3}}{5\times10^{-5}} = 20.0 \text{ lx}$$

and the ratio of two illuminations is given by

$$\frac{E_{las}}{E_{lamp}} \approx 253.$$

12.4   OPTICAL QUANTUM GENERATORS (LASERS)

1. *The laser* is a light source with properties that are markedly different from other sources such as incandescent lamps, fluorescent lamps, and natural light—the sun and stars. The name laser is the acronym of the English phrase "Light Amplification by Stimulated Emission of Radiation." Laser radiation possesses a number of unique properties: its spectral width is much narrower than that of other light sources, the intensity of laser radiation is larger than the intensity of the most intense light sources, and a laser beam possesses very small degree of divergence and therefore can reach the Moon with a spot diameter of only several hundred meters. The laser consists of three main components: the amplifying medium, in which the realization of inverse population of atomic levels is possible; the *pumping system*, which creates the population inversion of the amplifying medium; and the optical *resonator*, which creates *positive feedback*. Depending on the type of the amplifying medium used, we classify lasers as gas lasers, solid-state lasers, semiconductor lasers, and high-energy liquid laser. Each of them has its own characteristics associated with the design and the method of excitation. A special class of system is a combination of a laser and a quantum amplifier. The amplifier consists of an amplifying medium, but it has no resonator. The amplifier is placed at the laser output, and the laser pulse causes induced transitions in the amplifying medium of the amplifier. So the output from the amplifier has substantially higher intensity than the output from the laser. Such systems are working predominantly in pulse mode.

Lasers can operate in the continuous mode or in the pulsed mode. Gas lasers emit continuous radiation, solid-state lasers have a pulsed output, and semiconductor and liquid lasers have both types of output. The pulse generation mode is usually caused by the pump operating in the pulse mode (flash lamp, laser flash, current impulse, etc.). To obtain single pulses of high intensity, an *optical shutter is placed* in front of one of the mirrors of its resonator. When the excess energy (from a fraction of a joule to several hundred joules) has already been accumulated by the amplifying medium in the resonator, the shutter is opened for a time of the order of 10^{-4}–10^{-10} s. At this moment, the laser radiates a very short pulse (duration up to 10^{-15} c) with a power that reaches the order of 10^9 W.

2. An amplifying medium can emit radiation if it is placed inside a resonator. An elementary optical Fabry–Perot resonator consists of two parallel plane mirrors, one of which is semitransparent (Figure 12.6).

Any photon resulting from spontaneous emission and propagating perpendicular to the mirrors can start the amplification process and generate an avalanche of photons propagating in the same direction. A small fraction is absorbed by the mirrors, while another fraction escapes through the semitransparent mirror. The remaining part is reflected and causes a further increase of photon avalanche.

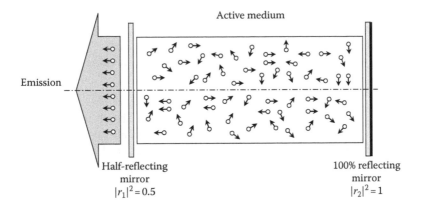

FIGURE 12.6   An elementary optical resonator.

Here, we consider conditions for light amplification in the resonator when its parallel mirrors are arranged at a distance l. We assume that the amplitude coefficients of reflection of each of the mirrors are r_1 and r_2 and that the effective (negative) absorption coefficient of the amplifying medium is equal to α_{ef}.

For convenience, we introduce the effective coefficient of amplification, γ. The equation of the primary plane wave, propagating along the optical axis of the system from the first mirror to the second in the positive direction of y-axis, is presented in the form

$$E(t,y) = E_0 \exp\left(\frac{\gamma y}{2}\right)\exp\left[i(\omega t - ky)\right], \tag{12.37}$$

where
  E_0 is the wave amplitude on the first mirror (i.e., at $y = 0$)
  $k = k_0 n$
  n is a refractive index of the amplifying medium

After passing through the amplifying medium from one to the other mirror and back, the light travels a total distance of $2l$. The amplitude of the wave field is given by

$$E(t,2l) = E_0 r_1 r_2 \exp(\gamma l)\exp\left[i(\omega t - 2kl)\right]. \tag{12.38}$$

The condition, at which the amplification compensates the losses appearing due to reflections from mirrors and absorption in the medium, is described by the inequality

$$r_1 r_2 \exp(\gamma l) > 1. \tag{12.39}$$

In addition to this condition, for amplification of a wave propagating in the resonator with an amplifying medium, it is necessary to satisfy a specific phase condition. An electromagnetic wave at a certain point of the amplifying medium after any number of reflections must have a phase that differs from the phase of the primary wave at the same point by a value multiple 2π. In turn, this condition imposes a relation that connects the wavelength and the distance between resonator mirrors:

$$2kl = 2\frac{2\pi}{\lambda}l = 2\pi m, \quad l = \frac{m\lambda}{2}, \tag{12.40}$$

where
  λ is the wavelength in the amplifying medium
  m is an integer

Thus, the resonator performs two important functions: (i) the reflection from the resonator mirrors causes the light wave to pass many times through the amplifying medium, significantly increasing its intensity, (ii) after reflection from the resonator mirrors, only those waves are amplified, for which the condition for the formation of standing waves (Equation 12.40) is satisfied.

3. Each atom of the amplifying medium possesses a discrete set of energy levels. Electrons of an atom in its ground state (the state with the lowest energy) can absorb light and be promoted to a higher energy level. A transition to a low energy level, accompanied by the emission of light, can occur spontaneously or under the influence of external radiation. Unlike photons of spontaneous radiation that are emitted in random directions, photons generated by stimulated emission are identical, that is, they have the same frequency and travel in the same direction as the original photon.

In order to have transitions with the radiation of energy, it is necessary to create a high concentration of excited atoms or molecules (i.e., to create population inversion). As was shown earlier, this should lead to the amplification of light traveling in the medium. The process of creating population inversion of levels is called pumping. There is a variety of pumping methods that depend on the laser type. The main task of the pumping process can be illustrated in the example of a three-level laser (Figure 12.7).

The lowest laser level with energy E_1 is the ground energy level of the system that is originally populated by all atoms of the amplifying medium. Pumping excites the atoms and thus promotes a fraction of a total number of atoms from the ground level to the level with energy E_3. The atoms occupying this level can return to the ground level with the emission of a photon or go to the upper laser level with energy E_2. In order to provide accumulation of excited atoms at this level, it is necessary to have a rapid transition from the level E_3 to the level E_2. The rate of this transition should exceed the decay rate of the upper laser level E_2. Thus, the created inverse population provides the necessary conditions for the amplification of radiation.

Since only photons, which propagate parallel to the resonator axis, participate in lasing, the efficiency of a laser does not usually exceed 1%. In certain cases, due to several improvements, the efficiency of lasers can be increased to 30%.

The pumping system is an external source of energy, which excites the atoms of the amplifying medium into an excited state. In gas lasers, pumping is carried out usually by an electric discharge, in solid-state lasers by a flash lamp, in liquid lasers by a laser light or flash lamp, and in semiconductor lasers by an electric current. The pumping system creates in the amplifying medium population inversion. Almost immediately, the excited atoms of the medium begin to spontaneously emit photons in random directions. Photons, emitted at an angle to the axis of the resonator, generate short cascades of stimulated radiation, which quickly leaves the medium. The photons, emitted along the resonator axis, are reflected by mirrors and repeatedly pass through the amplifying medium, inducing stimulated radiation. Generation of laser light starts when the increase in wave energy due to its amplification under each passage of the resonator exceeds the losses. The latter consists of internal

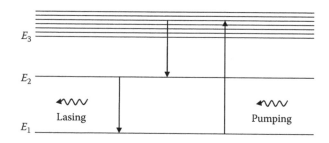

FIGURE 12.7   The scheme of a three-level laser.

losses (absorption and scattering of light in the amplifying medium, in resonator mirrors, and in other elements) and energy that exits the cavity through the output mirror.

Exercise 12.4

A spherical laser resonator consists of two spherical mirrors with radii R_1 and R_2, which are situated across each other and whose optical axis goes through the vertices of both mirrors. The distance between vertices is equal to L. Find the condition of stability of such resonator.

Solution. Whether the resonator is stable or not depends on whether the light ray, which propagates near the optical axis, stays inside the resonator or will after several reflections from the mirrors escape outside of the mirrors' limits. This condition defines if the lasing device will generate radiation or not. Using the three parameters R_1, R_2, and L, we must find if the resonator is stable or not.

If there are two points on the axis of the laser resonator such that the rays coming from one point (let us denote it as point 1) after their reflection from any of these mirrors are collected in a different point (let us denote it as point 2), then we have instability in the operation of laser resonator. Let us assume that the distances of these two points from the vertex of the left mirror are equal to f_1 and f_2 and the distances of these points from the vertex of the right mirror are equal to $L - f_1$ and $L - f_2$, correspondently (see Figure 12.8). For each of these mirrors, we can write the equation for a spherical mirror:

$$\frac{1}{f_1} + \frac{1}{f_2} = \frac{2}{R_1}, \quad \frac{1}{L - f_1} + \frac{1}{L - f_2} = \frac{2}{R_2}.$$

If this system of equations has solutions for f_1 and f_2, then such points do exist and the resonator will be unstable. If this system of equations does not have solutions, then the laser resonator is stable. Let us transform this system of equations to eliminate the denominators:

$$R_1(f_1 + f_2) - 2f_1f_2 = 0,$$

$$(2L - R_2)(f_1 + f_2) - 2f_1f_2 = 2L(L - R_2).$$

This is the system of two linear equations with two unknowns $f_1 + f_2$ and f_1f_2, which can be easily solved. As a result, we get

$$f_1 + f_2 = 2\frac{L(L - R_2)}{2L - R_1 - R_2}, \quad f_1f_2 = R_1\frac{L(L - R_2)}{2L - R_1 - R_2}.$$

The two roots x_1 and x_2 of equation $x^2 + px + q = 0$ satisfy the following conditions: $x_1 \cdot x_2 = q$, $x_1 + x_2 = -p$. By comparing these relationships with the solution mentioned earlier for f_1 and f_2, we conclude that f_1 and f_2 are the roots of quadratic equation. Therefore, the condition that quadratic equation

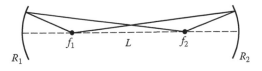

FIGURE 12.8 Schematics of the rays in spherical resonator when it is unstable: the rays come out from point f_1 (or f_2) and after reflection they cross at f_2 (or f_1).

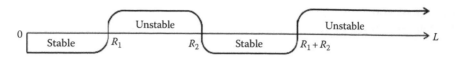

FIGURE 12.9   The regions of stability and instability of the laser resonator.

does not have roots is that the discriminant of the equation $p^2 - 4q$ must be negative. The discriminant can be expressed through $f_1 + f_2$ and $f_1 f_2$ as follows:

$$D = (f_1 + f_2)^2 - 4f_1 f_2 = 4 \frac{L(L - R_1)(L - R_2)(L - R_1 - R_2)}{(2L - R_1 - R_2)^2}.$$

Thus, the points f_1 and f_2 do exist if $D > 0$ and they do not exist if $D < 0$. From the expression for the discriminant, it follows that its sign coincides with the sign of numerator. This result allows us to formulate the following criterion for laser resonator stability:

- Resonator is stable if $(L - R_1)(L - R_2)(L - R_1 - R_2) < 0$.
- Resonator is unstable if $(L - R_1)(L - R_2)(L - R_1 - R_2) > 0$.

In order to make this result more transparent, let us show regions of stability and instability (that is regions where discriminant is positive and negative) as a function of L assuming that $R_1 \leq R_2$. We see that there are two regions of stability and two regions of instability and they are located as shown in Figure 12.9. In the case of symmetric resonator, when $R_1 = R_2 = R$, the criteria for stability and instability take the form $L < 2R$ and $L > 2R$, respectively.

12.5   MAIN FEATURES OF THE DIFFERENT TYPES OF LASERS

Solid-state lasers. The working substances of these lasers are crystals or glasses activated by impurity ions. Widely used solid-state lasers are based on ruby crystal aluminum oxide (Al_2O_3), in which about 0.05% of the aluminum atoms are replaced with chromium ions Cr^{3+}, or on the garnet YAG ($Y_3Al_5O_{12}$), doped with neodymium (Nd^{3+}), terbium (Tb^{3+}), ytterbium (Yb^{3+}), etc. More than 250 crystals and about 20 glasses emit stimulated radiation of various frequencies. Flash lamps are used for pumping. Solid-state lasers usually work in pulsed mode with a repetition rate of up to tens of megahertz. The energy of a single pulse can reach a few joules.

Gas lasers. In gases, the excited neutral atoms and molecules weakly ionized by electric discharge plasma serve as the sources of stimulated radiation. The number of electron-ion pairs appearing in the discharge column exactly compensates the loss of charged particles on the walls of a discharge tube. Therefore, the quantity of excited atoms is constant and their radiation is, as a rule, continuous. Since a gas medium is very uniform, the light beam is scattered in it weakly and the output beam diverges very little. Radiation power of gas lasers, depending on the type and structure, can vary from milliwatts to tens of kilowatts. This laser family has the largest number of lasers. They are as follows:

- *Lasers based on neutral atoms.* The lasers on a mixture of helium and neon (ratio 10:1) are the most widespread. They provide continuous radiation in the red region of the spectrum ($\lambda = 632.8$ nm).
- *Ion lasers.* The population inversion is created by electric discharge. Powerful radiation (tens of watts) is generated by ions Ar^{2+} ($\lambda = 488$ nm; 514.5 nm, in the blue–green range), Kr^{2+} ($\lambda = 561.2$ nm; 647.1 nm, yellow–red range), Kr^{3+}, and Ne^{2+} (UV range).

- *Molecular lasers.* They have high efficiency (up to 25%) and power (to tens of kW in the continuous mode and tens of kJ in pulse); they emit in the IR range. Population inversion is created by UV radiation or an electron beam. The lasers based on CO_2, H_2O, and N_2 are most widespread. Lasers based on vapor of dimer sulfur S_2 possess the following unique feature: due to the large number of metastable levels in this molecule, the laser emits simultaneously at 15 wavelengths in the visible range. Therefore, the laser beam on the S_2 appears white. The most powerful CO_2 lasers operate in the IR range ($\lambda = 10.6\ \mu m$). They generate radiation with power up to hundreds of kW in continuous mode emission.
- *Metal vapor lasers.* Ions and atoms of 27 metals possess the structures at energy levels, which are achieving population inversion. Cu vapor lasers emit at wavelengths of 510.4 nm (green light) and 578.2 nm (yellow light) with an average power of more than 40 W. Metal vapor lasers have very high amplification coefficient.
- *Excimer lasers.* They operate based on molecules in their excited state. Excimers are short-lived compounds of noble gases, with halogens or with oxygen (e.g., Ar_2, $KrCl$, XeO). These lasers have pulsed emission in the visible or UV spectral range with repetition rates up to 10^4 Hz and an average power of several tens of watts.

Currently, one of the most common gas lasers is laser using a mixture of helium and neon. The total pressure in the mixture is about 10^2 Pa with a ratio of He and Ne components of about 10:1. The amplifying gas is neon, wherein in the continuous mode lasing occurs at a wavelength of 632.8 nm (bright red). Helium—the buffer gas—is involved in the creation of a population inversion of one of the upper levels of neon. Emission of He–Ne laser is exceptionally unrivaled monochromatic. Calculations show that the width of the spectral lines generated by He–Ne laser is $\Delta f \approx 5 \times 10^{-4}$ Hz. This is a very small quantity. In this case, the **coherence time** of the radiation is of the order $\tau \approx 1/\Delta f \approx 2 \times 10^3$ s and the coherence length of $c\tau \approx 6 \times 10^{11}$ m, that is, greater than the diameter of Earth's orbit.

In practice, many technical reasons prevent the achievement of such a narrow spectral line of He–Ne laser. By careful stabilization of all parameters of the laser system, the relative width $\Delta f/f$ of the order of 10^{-14}–10^{-15} can be achieved, which is 3–4 orders of magnitude larger than the theoretical limit. But realistically achieved radiation monochromaticity of He–Ne laser makes this device absolutely indispensable in solving many scientific and engineering problems. Figure 12.10 shows a

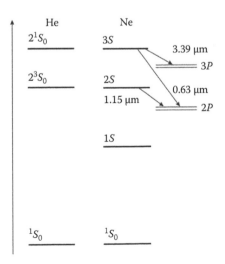

FIGURE 12.10 A diagram showing that light generation occurs for three possible transitions.

simplified energy level diagram of helium and neon and the mechanism of a population inversion of the laser transition. The arrows indicate the transitions, which can be generated. Bold lines represent groups of closely spaced levels. Generation corresponds to the transition between the individual levels, not between groups. For example, the lasing wavelength $\lambda = 632.8$ nm corresponds to the transition between levels $3S_2 \rightarrow 2P_4$.

As can be seen from the diagram, the generation is possible for three possible transitions. Maximum power (hundreds of milliwatts) is achieved in the generation of wavelength $\lambda = 3.39$ μm (transition $3S \rightarrow 3P$), but less power in the transition $2S \rightarrow 2P$ ($\lambda = 1.15$ μm). The lowest power is generated at the transition $3S \rightarrow 2P$ with the emission of the red line $\lambda = 0.63$ μm.

Dye lasers. The amplifying media are solutions of organic compounds, complex compounds of rare earth elements (Nd, Eu), and inorganic liquids. These materials combine to some extent the advantages of solid media (high density) and gases (high homogeneity). Population inversion is created by the irradiation of a cell filled with the dye by a laser or a gas-discharge lamp. Their emission power can reach tens of watts; the wavelength range can change ranging from 322 to 1260 nm by simply replacing the dye solution. Dye lasers can generate both continuous emission and sequences of ultrashort impulses with duration of up to 2×10^{-13} s.

Semiconductor lasers. The amplifying media of these lasers are semiconductor crystals GaAs, InSb, PbS, etc. In contrast to other amplifying media, in which the energy levels are discrete and therefore generate monochromatic radiation, semiconductors have rather broad energy bands. Therefore, the emitted radiation has a wide wavelength range and low coherence. Electrons and holes move in the amplifying medium and recombine in the vicinity of a p–n junction. In this case, electrical energy is directly transformed into radiation.

Semiconductor lasers have very high efficiency (up to 50%, and even about 100% for some models) and large amplification coefficient. Thanks to it, the size of the amplifying element is small (less than 1 mm). A wide range of semiconductor materials allow the generation of radiation in the 0.3–40 μm wavelength range. Lasers of different types work both in the continuous mode and in the pulsed mode, with powers ranging from a fraction of 1 mW to 1 MW per pulse. Semiconductor lasers are used as sights of portable weapons and pointers, in CD players, and as powerful light sources in beacons. Gas lasers are applied in range finders and theodolites, as frequency and time standards in metrology, and in recording holograms. Dye lasers are used for atmospheric measurements. High-power metal vapor lasers are used for cutting, welding, and processing of materials. Excimer lasers are applied in medicine for therapeutic procedures and in surgery. High-power lasers are used in thermonuclear fusion, in separation of isotopes, and in various physics and chemistry experiments.

Exercise 12.5

How will the frequency of the longitudinal mode change, if the length of the resonator L is increased by one wavelength λ?

Solution. An integer number of half waves for the longitudinal modes can fit the length of the original resonator, that is,

$$L = \frac{m\lambda}{2}.$$

The frequency of this mode is

$$f_1 = \frac{c}{\lambda} = \frac{mc}{2L}.$$

If we increase the resonator length by one wavelength, we have

$$L + \lambda = \frac{m\lambda}{2}, \quad \lambda = \frac{2L}{m-2}.$$

The frequency of the new longitudinal mode is

$$f_2 = \frac{c}{\lambda} = \frac{c(m-2)}{2L}.$$

The change of the frequency of the longitudinal mode is equal to

$$\Delta f = f_2 - f_1 = -\frac{c}{L}$$

PROBLEMS

12.1 A dilute electrically neutral plasma consists of electrons and positively singly ionized helium atoms, whose masses are $m = 9.1 \times 10^{-31}$ kg and $M = 4 \times 1.67 \times 10^{-27}$ kg, respectively. The electron and helium ion concentration is $N_e = N_i = N = 10^{14}$ m^{-3}. Find the phase velocity of an electromagnetic wave that propagates in this plasma at frequencies $\omega_1 = 10^{13}$ s^{-1} and $\omega_2 = 10^{14}$ s^{-1}. (*Answer:* At $\omega_1 = 10^{13}$ s^{-1} the phase velocity value is imaginary; at $\omega_2 = 10^{14}$ s^{-1} we get $v_{ph} = 3.63 \times 10^8$ m/s.)

12.2 A plane electromagnetic wave of frequency $f = 1.00$ GHz is propagating in a gaseous plasma. Find the dielectric permittivity ε and the conductivity σ of the plasma, if the complex wave number in the medium is equal to $k = (15 - 3i)$ m^{-1}. (*Answer:* $\varepsilon \approx 0.49\varepsilon_0$, $\sigma = 1.13 \times 10^{-2}$ ($\Omega \cdot$m)$^{-1}$.)

12.3 Find the concentration of free electrons in the ionosphere, if electromagnetic waves with frequencies below 0.50 GHz cannot propagate through the plasma as waves but instead are reflected by the plasma. (*Answer:* $N \geq 3.1 \times 10^{15}$ m^{-3}.)

12.4 In a p-type silicon sample, we have heavy and light holes. The concentration and the effective masses of heavy holes are $N_1 = 10^{23}$ m^{-3} and $m_{p1}^* = 0.5m$, and for the light holes, $N_2 = 1.7 \times 10^{22}$ m^{-3} and $m_{p2}^* = 0.2m$, where m is the free electron mass. Damping coefficients related to collisions of holes with other particles are $\gamma_1 = 9.2 \times 10^{12}$ s^{-1} for heavy holes and $\gamma_2 = 2.6 \times 10^{13}$ s^{-1} for light holes (compare with Equations 12.4 and 12.7). Find the conductivity of the silicon and the contribution to the dielectric constant of the holes for an electromagnetic wave with frequency $f = 3$ THz. $\left(Part\ of\ the\ answer;\ \kappa'' = \dfrac{2\omega_{p1}^2\gamma_1}{\omega\left(\omega^2 + 4\gamma_1^2\right)} + \dfrac{2\omega_{p2}^2\gamma_2}{\omega\left(\omega^2 + 4\gamma_2^2\right)}. \right)$

12.5 An electromagnetic wave propagates in a gaseous plasma that has a plasma frequency of $\omega_p = 5.00 \times 10^8$ s^{-1} and a damping parameter of $\gamma = 10^8$ s^{-1}. Find the frequency of the wave for which the conduction current density becomes equal to the density of the displacement current. (*Answer:* $\omega = 5.50 \times 10^8$ s^{-1}.)

12.6 A monochromatic plane wave with intensity I_0 is incident along the normal onto an inhomogeneously absorbing parallel plate with thickness L. The absorption coefficient of the material of the plate changes with position inside the plate in accordance to the equation $\alpha(z) = \alpha_0 + \gamma z^2$. Find the intensity of the wave after it emerges from the plate (reflection of the wave from the boundary surfaces of the plate may be neglected). (*Answer:* $I(L) = I_0 \exp[-(\alpha_0 L + \gamma L^3/3)]$.)

12.7   A monochromatic plane wave with intensity I_0 is incident along the normal onto an inhomogeneously absorbing parallel plate with thickness L. The absorption coefficient of the
material of the plate changes with position inside the plate in accordance with the equation $\alpha(z) = \alpha_0 + \gamma z^2$. The energy reflection coefficients of the wave from the input and output surfaces are R_1 and R_2. Find the intensity of the wave after it emerges from the plate
and the transmission coefficient T of the plate (secondary reflections can be neglected).
(*Answer:* $I_t = (1 - R_1)(1 - R_2)I_0\exp[-(\alpha_0 L + \gamma L^3/3)]$, $T = (1 - R_1)(1 - R_2)\exp[-(\alpha_0 L + \gamma L^3/3)]$.)

12.8   A point monochromatic source whose full luminous flux (power transferred by radiation
per unit area) is equal to Φ_0 is located at the center of a spherical shell of gaseous plasma
with inner and outer radii ρ_1 and ρ_2, plasma frequency ω_p, and damping parameter γ. Find
the frequency dependence of the ratio of the intensity of radiation that enters and the
intensity of the transmitted radiation (reflections at the boundary layers can be neglected).

$$\left(Answer: \frac{I_1}{I_2} = \frac{\rho_2^2}{\rho_1^2}\exp\left[k_0\kappa(\omega)(\rho_2 - \rho_1)\right]. \right)$$

12.9   Derive expressions for the phase and group velocity for a medium with one type of
oscillators with the intrinsic frequency ω_0 and plasma frequency ω_p. Absorption can be

neglected. $\left(Answer: v_{ph} = \dfrac{c}{\sqrt{1 - \dfrac{\omega_p^2}{\omega_0^2 - \omega^2}}}, v_{gr} = \dfrac{cn\left(\omega_0^2 - \omega^2\right)^2}{n^2\left(\omega_0^2 - \omega^2\right)^2 + \omega^2\omega_p^2}. \right)$

12.10 In a transparent medium, the group and phase velocities are related as follows: $v_{ph}v_{gr} = c^2$, where
c is the speed of light in a vacuum. Find the dependence on frequency of the wave's refractive
index in the medium. $\left(Answer: n(\omega) = \sqrt{1 + C/\omega^2}, \text{ where C is a constant of integration.} \right)$

13 Electromagnetic Waves in Media with Material Parameters That Are Complex Numbers

As we already discussed, electromagnetic waves propagate without attenuation only in vacuum or in a nonabsorbing medium. On the other hand, all real media to some extent absorb some of the energy of the propagating electromagnetic wave. The absorption depends strongly on the wave frequency. For some ranges of the electromagnetic wave spectrum, the absorption can be so weak that it can be neglected; for other ranges, absorption is important. Consider the features of the propagation of a plane monochromatic wave in a medium with absorption. Such a medium is generally described by a complex dielectric permittivity and magnetic permeability. The nature of the wave process in a medium significantly depends on the signs of the real and imaginary parts of these parameters. In this chapter, we will consider the features of wave propagation in media for which one of these complex parameters or both (dielectric and magnetic) have a nonzero imaginary part.

13.1 COMPLEX PERMITTIVITY, PERMEABILITY, AND IMPEDANCE OF A MEDIUM

Let us write down expressions for complex relative dielectric permittivity and magnetic permeability. Generally, these expressions can be represented in several forms:

$$\tilde{\kappa} = \kappa' - i\kappa'' = |\tilde{\kappa}|\exp(-i\delta_\kappa) = |\tilde{\kappa}|\cos\delta_\kappa - i|\tilde{\kappa}|\sin\delta_\kappa,$$
$$\tilde{\kappa}_m = \kappa'_m - i\kappa''_m = |\tilde{\kappa}_m|\exp(-i\delta_{\kappa_m}) = |\tilde{\kappa}_m|\cos\delta_{\kappa_m} - i|\tilde{\kappa}_m|\sin\delta_{\kappa_m},$$

(13.1)

where κ', κ'_m and κ'', κ''_m are the real and imaginary parts of the material parameters,

$$|\tilde{\kappa}| = \sqrt{(\kappa')^2 + (\kappa'')^2}, \quad |\tilde{\kappa}_m| = \sqrt{(\kappa'_m)^2 + (\kappa''_m)^2},$$

(13.2)

and $\delta_\kappa, \delta_{\kappa_m}$ are their arguments (phases). Phase angle δ_κ is called the ***angle of dielectric losses***. Analogously, we refer to δ_{κ_m} as the ***angle of magnetic losses***. In reference books, the magnitudes of tangents of the corresponding losses, that is, $\tan\delta_\kappa = \kappa''/\kappa'$ and $\tan\delta_{\kappa_m} = \kappa''_m/\kappa'_m$, are often listed.

The appearance of Maxwell's equations and their solutions does not change when we take into account the fact that parameters κ and κ_m are complex numbers. However, now in the analysis of the general dispersion and field relations, it is necessary to take into account that the refractive index $\tilde{n}$ (to emphasize the case of complex refractive index, we denote it by $\tilde{n}$), see also Equation 12.8, the wave propagation vector k, and the impedance (wave resistance) Z become complex:

$$\tilde{n} = \sqrt{\tilde{\kappa}\tilde{\kappa}_m} = n - i\kappa,$$
$$k = k_0\tilde{n} = k' - ik'' : \quad k' = k_0 n, \quad k'' = k_0\kappa,$$

(13.3)

$$Z = \sqrt{\frac{\mu_0 \tilde{\kappa}_m}{\varepsilon_0 \tilde{\kappa}}} = Z' - iZ'' = |Z|\exp(-i\zeta), \tag{13.4}$$

where

|Z| is the modulus
ζ is the phase of the complex impedance
we denote by n the real part and introduce κ for the imaginary part of $\tilde{n}$ (see Equation 12.8)

In an infinite medium with complex values for k and Z, the expressions for the components of the wave field of a plane linearly polarized wave, propagating along the positive z-axis, take the form

$$H_y(z,t) = H(z)\exp\left[i(\omega t - k'z + \alpha)\right],$$

$$E_x(z,t) = E(z)\exp\left[i(\omega t - k'z + \alpha - \zeta)\right], \tag{13.5}$$

where

$H(z)$ and $E(z)$ are the wave amplitudes
α is the initial phase of the magnetic field
ζ is the phase of the complex impedance

The analysis of expression (13.5) allows us to reach the following conclusions:

1. For a wave described by Equation 13.5, the role of the wave number is played by the real part of the complex wave number, which is related to the phase velocity and the wavelength in a medium, that is,

$$k' = \frac{\omega}{v_{ph}} = \frac{2\pi}{\lambda}. \tag{13.6}$$

2. The amplitudes of the magnetic and electric fields of a wave depend exponentially on z:

$$H(z) = H_0 \exp(-k''z),$$

$$E(z) = |Z|H(z) = |Z|H_0 \exp(-k''z). \tag{13.7}$$

Here, $H_0 = H(z = 0)$. For $k'' > 0$, the amplitude decreases in the direction of the z-axis (the wave is attenuated); for $k'' < 0$, the amplitude increases (the wave is amplified). Attenuation of a wave is characteristic of media with losses, and it is caused by the transformation of the wave energy to other forms of energy (usually internal energy of the medium). Such a medium is called **absorbing or passive**. Amplification of the wave is observed in lasers; this type of medium is called **amplifying** or **active**. The imaginary part of a wave number, k'', is called the **attenuation (amplification) coefficient**. Its value is inversely proportional to the distance z_e at which the wave amplitude decreases (increases) by a factor equal to e, that is, $k'' = 1/z_e$. Note, often in the literature the letter μ instead of k'' is used to denote the attenuation coefficient.

3. The ratio of the amplitudes of the electric and magnetic fields is equal at each point to the modulus of the complex impedance of the medium:

$$\frac{E(z)}{H(z)} = |Z| = \sqrt{(Z')^2 + (Z'')^2}. \tag{13.8}$$

4. If the imaginary part of the impedance Z'' is not zero, there is a phase shift between the oscillations of the electric and magnetic field vectors, which is equal to the phase of the impedance $\zeta = \arctan(Z''/Z')$. Its value is important as it determines the processes of energy transfer by an electromagnetic wave.

Next, we will consider only passive media in which there is attenuation of electromagnetic waves (in this case, it is not necessarily due to losses). As will be shown in the following text, for passive media, the angles of losses δ_κ, δ_{κ_m} must be positive to represent the complex values $\tilde{\kappa}$ and $\tilde{\kappa}_m$ with points in the complex plane. In Figure 13.1, the coordinate plane (κ', κ'') is shown. Each point corresponds to a specific value of the complex dielectric permittivity $\tilde{\kappa}$. This value can also be presented by a vector, which connects an origin of coordinates to the corresponding point of the (κ', κ'') plane.

The modulus of this vector is equal to $|\tilde{\kappa}|$ and the angle, which this vector forms with the horizontal axis, is equal to the loss angle δ_κ. In this figure, the upper half plane corresponds to passive media ($\kappa'' > 0$, $0 < \delta_\kappa < \pi$) while the lower half plane ($\kappa'' < 0$, $-\pi < \delta_\kappa < 0$) to active media.

Transparent media without losses (in this case—dielectrics) correspond to points with $\delta_\kappa = 0$, π (i.e., $\kappa'' = 0$). The coordinate plane for the relative magnetic permeability is analogous to the coordinate plane of the relative dielectric permittivity shown in Figure 13.1.

Exercise 13.1

Find the real and imaginary parts of the impedance of a medium with complex dielectric permittivity and magnetic permeability.

Solution. Using the definition of impedance, we get

$$Z = Z' - iZ'' = \sqrt{\frac{\mu_0 \tilde{\kappa}_m}{\varepsilon_0 \tilde{\kappa}}},$$

where $\tilde{\kappa} = \kappa' - i\kappa''$ and $\tilde{\kappa}_m = \kappa'_m - i\kappa''_m$. If we square the expression for impedance, we have

$$Z^2 = (Z')^2 - (Z'')^2 - 2iZ'Z'',$$

$$Z^2 = \frac{\mu_0 \tilde{\kappa}_m}{\varepsilon_0 \tilde{\kappa}} = \frac{\mu_0}{\varepsilon_0} \cdot \frac{\kappa'_m - i\kappa''_m}{\kappa' - i\kappa''} = \frac{\mu_0}{\varepsilon_0} \cdot \frac{(\kappa'_m - i\kappa''_m)(\kappa' + i\kappa'')}{(\kappa')^2 + (\kappa'')^2}$$

$$= \frac{\mu_0}{\varepsilon_0} \cdot \frac{\kappa'\kappa'_m + \kappa''\kappa''_m - i(\kappa'\kappa''_m - \kappa''\kappa'_m)}{(\kappa')^2 + (\kappa'')^2}.$$

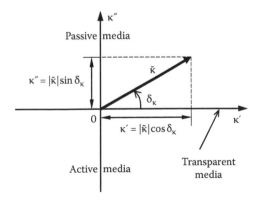

FIGURE 13.1   Representation of the complex permittivity on the coordinate plane (κ', κ'').

Thus,

$$(Z')^2 - (Z'')^2 = \frac{\mu_0}{\varepsilon_0} \cdot \frac{\kappa'\kappa'_m + \kappa''\kappa''_m}{(\kappa')^2 + (\kappa'')^2} = a, \quad 2Z'Z'' = \frac{\mu_0}{\varepsilon_0} \cdot \frac{\kappa'\kappa''_m - \kappa''\kappa'_m}{(\kappa')^2 + (\kappa'')^2} = b.$$

From the second equation, we get $Z'' = b/2Z'$, which we will substitute into the first equation. As a result, we get the equation for Z' and its solution:

$$(Z')^4 - a(Z')^2 - \frac{b}{2} = 0, \quad (Z')^2 = \frac{1}{2}\left(\sqrt{a^2 + b^2} + a\right).$$

In a similar fashion, we get the expression for $(Z'')^2$:

$$(Z'')^2 = \frac{1}{2}\left(\sqrt{a^2 + b^2} - a\right).$$

After taking the square root of these relationships, we get the following expressions for the real and imaginary parts of impedance:

$$\left.\begin{array}{c}Z'\\Z''\end{array}\right\} = \frac{1}{\sqrt{2}}\left(\sqrt{a^2 + b^2} \pm a\right)^{1/2}$$

$$= \sqrt{\frac{\mu_0}{2\varepsilon_0}} \frac{\left(\sqrt{(\kappa'\kappa'_m + \kappa''\kappa''_m)^2 + (\kappa'\kappa''_m - \kappa''\kappa'_m)^2} \pm (\kappa'\kappa'_m + \kappa''\kappa''_m)\right)^{1/2}}{\sqrt{(\kappa')^2 + (\kappa'')^2}}.$$

13.2 ENERGY FLOW IN A MEDIUM WITH COMPLEX MATERIAL PARAMETERS

1. Here we discuss the energy (or intensity) flow of an electromagnetic wave in a medium with complex permittivity κ and permeability κ_m. We assume that the wave is linearly polarized along the x-axis and that it propagates along the positive direction of the z-axis. We determine the average value of the energy flux over one period:

$$\langle S_z \rangle = \frac{1}{2}\operatorname{Re}(E_x H_y^*), \tag{13.9}$$

where the components of the wave field H_y and E_x are given by

$$H_y = H_0 \exp(-k''z)\exp\left[i(\omega t - k'z + \alpha)\right],$$
$$E_x = |Z| H_0 \exp(-k''z)\exp\left[i(\omega t - k'z + \alpha - \zeta)\right], \tag{13.10}$$

where
  H_0 and α are the amplitude and initial phase of the magnetic field of the wave, respectively
  ζ is the phase of the complex impedance

We write the complex conjugate of the expression for the y-component of the magnetic field:

$$H_y^* = H_0 \exp(-k''z)\exp\left[-i(\omega t - k'z + \alpha)\right],$$
(13.11)

and the product $E_x H_y^*$ is

$$E_x H_y^* = |Z| H_0^2 e^{-2k''z} e^{-i\zeta} = |Z| H_0^2 e^{-2k''z}(\cos\zeta - i\sin\zeta).$$

After substituting this relation into Equation 13.9, we obtain the following expression for the energy flux of the electromagnetic wave along the z-axis:

$$\langle S_z \rangle = \frac{1}{2} Z' H_0^2 \exp(-2k''z) = \frac{1}{2}|Z| H_0^2 e^{-2k''z} \cos\zeta.$$
(13.12)

From this expression, it follows that the energy flux of a single wave in an absorbing medium is proportional to the real part of the impedance, ReZ, and decays exponentially in the direction of propagation. If vectors **E** and **H** oscillate with phase shift $\zeta = \pm\pi/2$ (vector components of the field in this case, called *reactive*), the wave energy flux is zero and the transfer of electromagnetic energy by the wave in this case is impossible.

2. We now determine the total energy flow of the two counterpropagating waves of the same frequency in a medium with complex κ and κ_m. Let the waves propagate in the positive and negative directions of the z-axis and have the same linear polarization of the electric field along x-axis. The total field E_x and H_y of the two counterpropagating waves is determined by the expressions

$$H_y = A_1 e^{-k''z} e^{i(\omega t - k'z + \alpha_1)} - A_2 e^{k''z} e^{i(\omega t + k'z + \alpha_2)},$$

$$E_x = |Z| A_1 e^{-k''z} e^{i(\omega t - k'z + \alpha_1 - \zeta)} + |Z| A_2 e^{k''z} e^{i(\omega t + k'z + \alpha_2 - \zeta)}.$$
(13.13)

Let us write the complex conjugate of the expression for the y-component of the magnetic field:

$$H_y^* = A_1 e^{-k''z} e^{-i(\omega t - k'z + \alpha_1)} - A_2 e^{k''z} e^{-i(\omega t + k'z + \alpha_2)},$$
(13.14)

and the product $E_x H_y^*$ is

$$E_x H_y^* = |Z| A_1^2 e^{-2k''z} e^{-i\zeta} - |Z| A_2^2 e^{2k''z} e^{-i\zeta}$$

$$+ |Z| A_1 A_2 e^{-i\zeta} e^{i(2k'z + \alpha_2 - \alpha_1)} - |Z| A_1 A_2 e^{-i\zeta} e^{-i(2k'z + \alpha_2 - \alpha_1)}.$$
(13.15)

The last two terms in this equation can be combined into one taking into account that

$$e^{i\varphi} - e^{-i\varphi} = 2i\sin\varphi.$$

Then, Equation 13.15 takes the form

$$E_x H_y^* = |Z| A_1^2 e^{-2k''z} e^{-i\zeta} - |Z| A_2^2 e^{2k''z} e^{-i\zeta} +$$

$$+ 2i|Z| A_1 A_2 e^{-i\zeta} \sin(2k'z + \alpha_2 - \alpha_1).$$
(13.16)

Now, we will find the real part of this expression. We take into account that

$$\text{Re}(e^{-i\zeta}) = \cos\zeta, \quad |Z|\cos\zeta = Z',$$

$$\text{Re}(ie^{-i\zeta}) = \sin\zeta, \quad |Z|\sin\zeta = Z''.$$

As a result, we find that the total energy flux of the two waves can be expressed as the sum of three terms:

$$\langle S_z \rangle = S_1 + S_2 + S_{int}, \tag{13.17}$$

where the following notation is introduced:

$$S_1 = \frac{1}{2}Z'A_1^2 e^{-2k''z}, \quad S_2 = -\frac{1}{2}Z'A_2^2 e^{2k''z},$$

$$S_{int} = Z''A_1A_2 \sin(2k'z + \alpha_2 - \alpha_1).$$

The first two terms in Equation 13.17, S_1, S_2, have opposite signs. These represent the energy flow of each of the two counterpropagating waves. If there are losses, they are damped in the direction of the propagation of each wave (decay constant is $2k''$). The third term, S_{int}, is the interference energy flux of the counterpropagating waves, which is proportional to the product of their amplitudes (A_1A_2) and depends on the difference between their initial phases ($\alpha_2 - \alpha_1$).

The interference flux term is undamped, that is, it has a harmonic dependence on z. In the particular case of $k' = 0$, the flux S_{int} is constant. One more feature of the interference flow term is that the flows of individual waves, S_1 and S_2, are proportional to the real part of the impedance $Z' = |Z|\cos\zeta$, but S_{int} is proportional to the imaginary part $Z'' = |Z|\sin\zeta$.

If the vectors $\mathbf{E}$ and $\mathbf{H}$ of both waves oscillate with a phase shift $\zeta = \pm\pi/2$, the partial energy fluxes S_1 and S_2 are equal to zero. In this case, when there are two counterpropagating waves, transfer of electromagnetic energy is only possible due to the presence of the interference flux, S_{int}.

3. Suppose now that two waves of the same frequency generated by two independent sources propagate in the same direction along z-axis. In this case, the total energy flow in the direction of the wave propagation consists of three components (analogously to Equation 13.17):

$$\langle S_z \rangle = S_1 + S_2 + S_{int}, \tag{13.18}$$

where the following notations are used:

$$S_1 = \frac{1}{2}Z'A_1^2 e^{-2k''z}, \quad S_2 = \frac{1}{2}Z'A_2^2 e^{-2k''z},$$

$$S_{int} = Z'A_1A_2 e^{-2k''z} \cos(\alpha_2 - \alpha_1).$$

However, in the case of unidirectional waves, the properties of the interference flux S_{int} do not differ from those of the fluxes S_1 and S_2: the interference flux is proportional to a real part of an impedance Z' and decreases exponentially in the direction of propagation.

Exercise 13.2

Find the dependence of the phase ζ of the complex impedance Z on material parameters of the medium. What are the conditions that material parameters must satisfy for $\zeta = \pi/4$?

Solution. According to the definition of medium's complex impedance,

$$Z = Z' - iZ'' = |Z|\exp(-i\zeta) = |Z|(\cos\zeta - i\sin\zeta).$$

At the same time, the following relationships are valid:

$$\cos\zeta = \frac{Z'}{|Z|}, \quad \sin\zeta = \frac{Z''}{|Z|}, \quad \tan\zeta = \frac{Z''}{Z'}.$$

Let us substitute into the last relationship the expressions for real and imaginary parts of impedance:

$$\left.\begin{matrix} Z' \\ Z'' \end{matrix}\right\} = \sqrt{\frac{\mu_0/2\varepsilon_0}{(\kappa')^2 + (\kappa'')^2}}\left(\sqrt{(\kappa'\kappa_m' + \kappa''\kappa_m'')^2 + (\kappa'\kappa_m'' - \kappa''\kappa_m')^2} \pm (\kappa'\kappa_m' + \kappa''\kappa_m'')\right)^{1/2}.$$

As a result, we get the following expression:

$$\tan\zeta = \left(\frac{\sqrt{(\kappa'\kappa_m' + \kappa''\kappa_m'')^2 + (\kappa'\kappa_m'' - \kappa''\kappa_m')^2} - (\kappa'\kappa_m' + \kappa''\kappa_m'')}{\sqrt{(\kappa'\kappa_m' + \kappa''\kappa_m'')^2 + (\kappa'\kappa_m'' - \kappa''\kappa_m')^2} + (\kappa'\kappa_m' + \kappa''\kappa_m'')}\right)^{1/2}.$$

At $\zeta = \pi/4$, we get $\tan\zeta = 1$ and $\kappa'\kappa_m' + \kappa''\kappa_m'' = 0$.

13.3   RIGHT- AND LEFT-HANDED MEDIA

Here, we obtain and analyze the expressions for the real and imaginary parts of the complex refractive index n for the case when both material parameters κ and κ_m are complex. By representing these parameters in the exponential form, we get

$$\tilde{n}^2 = \tilde{\kappa}\tilde{\kappa}_m = |\tilde{\kappa}||\tilde{\kappa}_m|e^{-i(\delta_\kappa + \delta_{\kappa m})} = |\tilde{\kappa}||\tilde{\kappa}_m|\{\cos(\delta_\kappa + \delta_{\kappa_m}) - i\sin(\delta_\kappa + \delta_{\kappa_m})\}. \tag{13.19}$$

Taking the square root of the complex value $\tilde{n}^2$, for the refractive index, we obtain

$$\tilde{n} = \pm\sqrt{|\tilde{\kappa}||\tilde{\kappa}_m|}e^{-i(\delta_\kappa + \delta_{\kappa m})/2} = \pm|\tilde{n}|\left(\cos\delta_n - i\sin\delta_n\right), \tag{13.20}$$

where $|\tilde{n}| = \sqrt{|\tilde{\kappa}||\tilde{\kappa}_m|}$ and $\delta_n = (\delta_\kappa + \delta_{\kappa_m})/2$ are the modulus and phase angle of the complex refractive index, respectively. Equation 13.20 indicates that in the case under consideration, there is a problem in the choice of the sign before the square root. To resolve this problem, let us discuss the meaning of the real and imaginary parts of $\tilde{n}$ in Equation 13.3.

In a medium with a complex refractive index determined by Equation 13.3, the dependence of the field component of a monochromatic wave, given by Equation 13.10, on the z-coordinate and time can be represented as (here we put $\alpha = 0$)

$$\left.\begin{matrix} H_y(z,t) \\ E_x(z,t) \end{matrix}\right\} = \left\{\begin{matrix} H_0 \\ |Z|H_0\exp(-i\zeta) \end{matrix}\right\}\exp(-k_0\kappa z)\exp[i(\omega t - k_0 nz)]. \tag{13.21}$$

Here, the real part of the refractive index n determines the real part of the propagation vector $k' = k_0 n$ and the magnitude of the phase velocity $v_{ph} = c/n$. We assume that the positive direction of the z-axis coincides with the direction of the energy transfer by the wave, that is, with its Poynting

318   An Introduction to Applied Electromagnetics and Optics

vector $\mathbf{S} = \mathbf{E} \times \mathbf{H}$, which always forms a right-handed system with the directions of vectors $\mathbf{E}$ and $\mathbf{H}$. Then, for $n > 0$, the directions of vectors $\mathbf{k}$ and $\mathbf{S}$ coincide, and vector $\mathbf{k}$ forms with vectors $\mathbf{E}$ and $\mathbf{H}$ a right-handed system (Figure 13.2a). Analogously, for $n < 0$, vectors $\mathbf{k}$ and $\mathbf{S}$ have opposite directions (Figure 13.2b). This means that the phase and group velocities in such a medium are opposite to each other, that is, the directions of a phase propagation and energy transfer are opposite. Waves for which $\mathbf{v}_{ph} \uparrow\downarrow \mathbf{v}_{gr}$ and, therefore, $\mathbf{k} \uparrow\downarrow \mathbf{S}$ are called backward waves. Vectors $\mathbf{k}$, $\mathbf{E}$, and $\mathbf{H}$ of a backward wave form left-handed system, so for media with $n < 0$, the term ***left-handed materials*** is used. Therefore, it makes sense to call "traditional" media with $n' > 0$ as ***right-handed materials***.

According to Equation 13.21, the sign of the imaginary part κ determines the behavior of the wave amplitude in the propagation direction. For $\kappa > 0$, the amplitude decreases in the direction of energy transfer, that is, the wave decays (passive environment), while for $\kappa < 0$, the amplitude increases, that is, the wave is amplified (active medium). As a result, depending on the signs of n and κ, the four types of media can be distinguished:

1. Right-handed passive (type I): $n > 0$, $\kappa > 0$, $(0 < \delta < \pi/2)$
2. Left-handed passive (type II): $n < 0$, $\kappa > 0$, $(\pi/2 < \delta < \pi)$
3. Left-handed active (type III): $n < 0$, $\kappa < 0$, $(\pi < \delta < 3\pi/2$ or $-\pi < \delta < -\pi/2)$
4. Right-handed active (type IV): $n > 0$, $\kappa < 0$, $(3\pi/2 < \delta < 2\pi$ or $-\pi/2 < \delta < 0)$

Roman numerals denote the type of medium and correspond to the quadrant number of the coordinate plane (n, κ) shown in Figure 13.3. The range of values of the phase angle δ for each type of media is specified earlier in brackets.

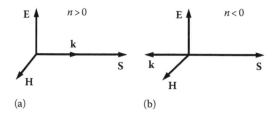

(a)   (b)

FIGURE 13.2 Relative orientation of vectors $\mathbf{k}$, $\mathbf{E}$, $\mathbf{H}$, and $\mathbf{S}$ in the right-handed (a) and left-handed (b) materials.

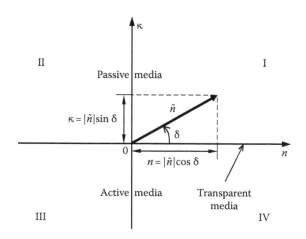

FIGURE 13.3 Representation of a complex refractive index $\tilde{n}$ on the coordinate plane (n, κ).

For passive media, $\kappa'', \kappa''_m > 0$ and $0 < \delta_\kappa, \delta_{\kappa_m} < \pi$, and, therefore, $0 < \delta < \pi$, that is, $\sin \delta > 0$. It means that in Equation 13.20, it is necessary to take the sign "+." In this case, $\kappa = |\tilde{n}| \sin \delta > 0$ corresponds to wave attenuation. The sign of a real part $n = |\tilde{n}| \cos \delta$ may be either positive or negative. The sign "−" in Equation 13.20 corresponds to counterpropagating wave transferring energy in the negative direction of z-axis.

Let us consider how the sign of n depends on the complex material parameters κ and κ_m. In the following, we analyze all possible special cases for a passive medium.

1. The real parts of the material constants are both positive ($\kappa' > 0, \kappa'_m > 0$). Thus, $0 \leq \delta_\kappa, \delta_{\kappa_m} < \pi/2$, therefore $0 \leq \delta < \pi/2$, and, respectively $n > 0$, that is, the right-handed passive medium (type I) is realized.
2. The values κ' and κ'_m are both negative ($\kappa' < 0, \kappa'_m < 0$). Thus, $\pi/2 < \delta_\kappa, \delta_{\kappa_m} \leq \pi, \pi/2 < \delta \leq \pi$, and $n < 0$, that is, it is the left-handed passive medium (type II).
3. The values κ', κ'_m have opposite signs (i.e., $\kappa' \kappa'_m < 0$). Suppose, for definiteness that $\kappa' < 0$, $\kappa'_m > 0$, that is, $\pi/2 < \delta_\kappa \leq \pi$, and $0 \leq \delta_{\kappa_m} < \pi/2$. Then, the phase angle δ of the refractive index lies in the interval $\pi/4 \leq \delta \leq 3\pi/4$. This means that the sign of n can be either positive or negative and the medium, depending on the ratio of the dielectric and magnetic losses, can be either right or left handed. Furthermore, the condition $n < \kappa$ is satisfied, that is, the wave in such medium strongly attenuates. If the losses are negligible, then $\delta_\kappa = \pi$, $\delta_{\kappa_m} = 0$, and $n \to 0$, that is, in such a medium, electromagnetic waves cannot propagate.

We also give a formula for the calculation of the real and imaginary parts of complex impedance of a medium:

$$Z = \sqrt{\frac{\mu_0 \tilde{\kappa}_m}{\varepsilon_0 \tilde{\kappa}}} = \sqrt{\frac{\mu_0 |\tilde{\kappa}_m|}{\varepsilon_0 |\tilde{\kappa}|}} e^{-i(\delta_{\kappa_m} - \delta_\kappa)} = \pm |Z| (\cos \zeta - i \sin \zeta), \tag{13.22}$$

where

$$|Z| = \sqrt{\tilde{\kappa}_m |\mu_0 / |\tilde{\kappa}| \varepsilon_0}$$

$$\zeta = (\delta_{\kappa_m} - \delta_\kappa)/2$$

The choice of sign in Equation 13.22 is based on the fact that the real part of the impedance, Z', is positive. It is easy to show that for passive media $-\pi/2 \leq \zeta \leq \pi/2$, that is, $\cos \zeta > 0$ always, and for the real part of the complex impedance, it is necessary to choose the sign "+." The sign of an imaginary part, $Z'' = |Z| \sin \zeta$, is determined by the sign of the difference of angles of magnetic and dielectric losses: $Z'' > 0$ for $\delta_\varepsilon < \delta_\mu$ and $Z'' < 0$ for $\delta_\varepsilon > \delta_\mu$.

Exercise 13.3

Determine the range of phases of complex electrical permittivity and magnetic permeability where a medium is "left handed" and absorbing (i.e., passive).

Solution. Let us consider a medium with complex relative dielectric permittivity κ and magnetic permeability κ_m:

$$\tilde{\kappa} = |\tilde{\kappa}| \exp(-i\delta_\kappa) = |\tilde{\kappa}| (\cos \delta_\kappa - i \sin \delta_\kappa),$$

$$\tilde{\kappa}_m = |\tilde{\kappa}_m| \exp(-i\delta_{\kappa_m}) = |\tilde{\kappa}_m|(\cos\delta_{\kappa_m} - i\sin\delta_{\kappa_m}),$$

where we introduced the moduli $|\tilde{\kappa}|$, $|\tilde{\kappa}_m|$ and the corresponding phases δ_κ, δ_{κ_m}. The complex refractive index is

$$\tilde{n} = \sqrt{\tilde{\kappa}\tilde{\kappa}_m} = \sqrt{|\tilde{\kappa}||\tilde{\kappa}_m|} \exp\left(-i\frac{\delta_\kappa + \delta_{\kappa_m}}{2}\right)$$

$$= \sqrt{|\tilde{\kappa}||\tilde{\kappa}_m|}\left[\cos\left(\frac{\delta_\kappa + \delta_{\kappa_m}}{2}\right) - i\sin\left(\frac{\delta_\kappa + \delta_{\kappa_m}}{2}\right)\right] = n - i\kappa.$$

In a passive medium with losses, the imaginary part of the refractive index should be positive: $\kappa > 0$. For the electric field of a plane wave with complex refractive index, the following relationships are valid:

$$E \sim \exp(-ik_0\tilde{n}x) \sim \exp(-ik_0nx)\exp(-k_0\kappa x).$$

Thus,

$$\sin\left(\frac{\delta_\kappa + \delta_{\kappa_m}}{2}\right) > 0 \quad \text{or} \quad 0 < \frac{\delta_\kappa + \delta_{\kappa_m}}{2} < \pi.$$

Let us take into account that the real components of $\tilde{\kappa}$ and $\tilde{\kappa}_m$ are negative. In this case, we find that

$$\cos\delta_\kappa < 0 \quad \text{or} \quad \frac{\pi}{2} < \delta_\kappa < \frac{3\pi}{2},$$

$$\cos\delta_{\kappa_m} < 0 \quad \text{or} \quad \frac{\pi}{2} < \delta_{\kappa_m} < \frac{3\pi}{2}.$$

Taking into account these relationships, we find that

$$\frac{\pi}{2} < \frac{\delta_\kappa + \delta_{\kappa_m}}{2} < \frac{3\pi}{2}.$$

Let us combine both of these conditions for $\delta_\kappa + \delta_{\kappa_m}$: $\kappa > 0$ and $\kappa' < 0$, $\kappa'_m < 0$. Thus, the medium is "left handed" and absorbing in the region where

$$\frac{\pi}{2} < \frac{\delta_\kappa + \delta_{\kappa_m}}{2} < \pi.$$

13.4 MEDIA WITH A NEGATIVE VALUE OF ONE OF THE MATERIAL CONSTANTS: ELECTROMAGNETIC TUNNELING

Let us consider features of propagation of electromagnetic waves in a medium with the negative value of one of its material constants. Consider the ideal case of absence of losses (i.e., $\kappa'' = \kappa''_m = 0$). For definiteness, we assume that $\kappa' = -|\kappa| < 0$, while $\kappa'_m = \kappa_m > 0$ (i.e., in Equation 13.1 $\delta_\kappa = \pi$, $\delta_{\kappa_m} = 0$). Calculation using Equation 13.20 gives the purely imaginary refractive index and wave number

$$\tilde{n} = -i\kappa = -i\sqrt{|\kappa|\kappa_m}, \quad k = -ik'' = -ik_0\sqrt{|\kappa|\kappa_m}, \tag{13.23}$$

where k_0 is the wave number in vacuum. According to Equation 13.22, the medium impedance will also be purely imaginary:

$$Z = \sqrt{-\frac{\mu_0 \kappa_m}{\varepsilon_0 |\kappa|}} = i\sqrt{\frac{\mu_0 \kappa_m}{\varepsilon_0 |\kappa|}} = iZ''. \tag{13.24}$$

Taking into account Equations 13.23 and 13.24, the expressions for the components of electric and magnetic fields of a wave, which propagates in the direction of z-axis, take the form

$$H_y = Ae^{-k''z}e^{i\varphi},$$
$$E_x = ZH_y = iZ''Ae^{-k''z}e^{i\varphi} = Z''Ae^{-k''z}e^{i(\varphi+\pi/2)}, \tag{13.25}$$

where $\varphi = \omega t + \alpha$. Let us calculate the real part of complex expressions (13.25):

$$H_y = \mathrm{Re}\, H_y = Ae^{-k''z}\cos\varphi,$$
$$E_x = \mathrm{Re}\, E_x = Z''Ae^{-k''z}\cos(\varphi+\pi/2) = -Z''Ae^{-k''z}\sin\varphi. \tag{13.26}$$

Analyzing these expressions, we come to the following conclusions:

1. Due to the fact that $k' = 0$, it is impossible to call this process a traveling wave. It is a harmonic electromagnetic oscillation with amplitude, which decreases exponentially in the positive z-direction.
2. The phase shift between the oscillations of the vectors $\mathbf{E}$ and $\mathbf{H}$ is $\pi/2$, that is, E_x and H_y are the reactive components of fields. Thus, the average energy (intensity) flux in the z-direction is equal to zero. This is confirmed by direct calculations:

$$S_z = E_x H_y = -\frac{1}{2}Z''A^2 e^{-2k''z}\sin 2\varphi, \tag{13.27}$$

   that is, the energy flow oscillates with frequency 2ω, and its average value over a period is equal to zero.
3. The energy densities of electric and magnetic fields are given by

$$u_e = -\frac{1}{2}|\kappa|\varepsilon_0 E_x^2 = -u_0(1-\cos 2\varphi),$$
$$u_m = \frac{1}{2}\kappa_m \mu_0 H_y^2 = u_0(1+\cos 2\varphi), \tag{13.28}$$

   where we introduced the average value of the energy density of each field component, and

$$u_0 = \frac{1}{4}|\kappa|\varepsilon_0 \left(Z''A\right)^2 e^{-2k''z} = \frac{1}{4}\kappa_m \mu_0 A^2 e^{-2k''z} \tag{13.29}$$

The values u_e and u_m oscillate with a frequency 2ω around their average values $-u_0$ and $+u_0$, respectively. This is accompanied by a periodic process of mutual transformation of the energy of the electric and magnetic fields. This process is analogous to the process in an ideal resonant circuit, which has a pure reactive resistance. The total energy density

$$u = u_e + u_m = 2u_0 \cos 2\varphi = 2u_0 \cos 2(\omega t + \alpha) \tag{13.30}$$

varies around the average value $\langle u \rangle = -u_0 + u_0 = 0$. From the point of view of energy relations, this process resembles a standing electromagnetic wave with the important difference that in such wave $\langle u \rangle \neq 0$.

The process described here can be realized near the planar interface between medium "1" (vacuum [air]), with $\kappa_1 = \kappa_{m1} = 1$, and medium "2," for which $\kappa_2 < 0$, $\kappa_{m2} > 0$. Consider a plane monochromatic wave incident normally from medium "1" onto the interface at $z = 0$. The electromagnetic field penetrates into the second medium and the decaying oscillations in the second medium are described by Equation 13.26. These decaying oscillations have electromagnetic energy that is determined by Equation 13.28. Also, there is an instantaneous energy flow (13.28) due to which energy enters into medium "2" during the first half of the period of oscillations, and then during the second half of the period of oscillations, it leaves medium "2," that is, it returns. Thus, the average flow through the boundary is equal to zero. This means that there is a total reflection of the incident wave (even at normal incidence), and a standing wave is formed in front of the boundary. The distribution of the electric field amplitude along the z-axis for this case is shown in Figure 13.4a. We note that the total reflection from the interface is "violated" (or "broken") if the wave is reflected from a thin enough layer of medium "2" (thickness of the medium must be of the order of $1/k''$). In this case, in the back surface layer, an additional oscillation decreasing in the negative z-direction is formed. In this case, $Z' = 0$, $Z'' \neq 0$. Among three components of the total energy flow of the counterpropagating waves, only the interference energy flow is not equal to zero. For $k' = 0$, it does not depend on z-coordinate:

$$S_{\text{int}} = Z'' |p||q| A^2 \sin(\alpha_p - \alpha_q), \tag{13.31}$$

where
   $|p|$ and $|q|$ are the moduli
   α_p and α_q are the phase angles of the amplitude coefficients of the two counterpropagating waves inside the layer (these values can be found from the boundary conditions)

Due to the existence of the interference that is described by Equation 13.31, the electromagnetic energy partially "leaks" or "tunnels" through the thin layer in which the solitary wave does not transfer energy. In this case, we say that ***broken total reflection*** or ***electromagnetic tunneling*** takes place (Figure 13.4b).

In a dielectric behind the thin layer (medium "3"), we have the propagation of a traveling wave. If we increase the layer thickness (medium "2"), then the amplitude of the wave in medium "3" asymptotically approaches to zero. The wave in front of the layer will not be purely standing as the value of the field at the nodes is not equal to zero.

In a real case, the medium experiences losses (κ'', $\kappa''_m \neq 0$) and the real parts of the propagation constant k' and impedance Z' differ from zero. Thus, the wave in medium "2" is traveling and transferring

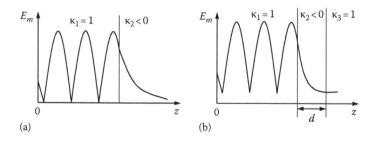

FIGURE 13.4 Distribution of the amplitude of electric field for the case of thick (a) and thin (b) layers "2" with negative κ_2.

energy that leads to violation of the total reflection even in the situation shown in Figure 13.4a. However, because of the relation $k' < k''$, the wave in the medium "2" will be strongly damped.

It should be noted that the properties of media with negative permittivity κ (or permeability κ_m) have been discussed here in the ideal case of a medium without dispersion. All real media (natural or artificial) with negative permittivity have frequency dispersion. Dielectrics in the vicinity of resonance absorption lines, metals in the visible region of the spectrum, magnetized ferrites, plasma, and many synthetic composite materials are examples of media that exhibit negative material constants (κ or κ_m) in a certain frequency range. A more rigorous analysis with taking into account the dispersion of $\kappa(\omega)$ and $\kappa_m(\omega)$ requires correction of the equations given earlier and especially of Equation 13.28 for the density of electromagnetic energy in a medium.

Exercise 13.4

A linearly polarized plane wave with electric field amplitude E_0 and frequency ω is incident perpendicularly from vacuum ($\kappa_1 = 1$ and $\kappa_{m1} = 1$) on a layer made of nonmagnetic material ($\kappa_{m2} = 1$) with real and negative electrical permittivity ($\kappa_2 < 0$). The thickness of the layer is equal to d. Find the intensity of the wave transmitted through the layer neglecting reflection from the back surface.

Solution. Let us assume that the wave is propagating in the positive z-direction. The surface on which the wave is incident is located at $z = 0$. Since $\kappa'' = \kappa_m'' = 0$, there are no energy losses inside the layer. Given that the dielectric permittivity of the layer is negative ($\kappa_2 < 0$), we can write $\kappa_2 = -|\kappa_2| < 0$, and this gives us a purely imaginary refractive index and wave number:

$$\tilde{n}_2 = \sqrt{-|\kappa_2|\,\kappa_{m2}} = -i\sqrt{|\kappa_2|}, \quad k_2 = k_0\tilde{n}_2 = -ik_0\sqrt{|\kappa_2|},$$

where k_0 is the wave number in vacuum. The impedance of the layer in this case is also a purely imaginary number in contrast to the impedance of vacuum, Z_0:

$$Z_2 = \sqrt{\frac{-\mu_0\kappa_{m2}}{\varepsilon_0\,|\kappa_2|}} = \frac{iZ_0}{\sqrt{|\kappa_2|}}.$$

Let us write the expressions for the components of the electric and magnetic fields of the wave propagating in the layer:

$$E_x(z) = E_0\exp\left(-\sqrt{|\kappa_2|}\,z\right)\exp(i\omega t),$$

$$H_y(z) = \frac{E_x(z)}{Z_2} = -\frac{i\sqrt{|\kappa_2|}}{Z_0}E_0\exp\left(-\sqrt{|\kappa_2|}\,z\right)\exp(i\omega t)$$

$$= \frac{\sqrt{|\kappa_2|}}{Z_0}E_0\exp\left(-\sqrt{|\kappa_2|}\,z\right)\exp(i(\omega t - \pi/2)).$$

The wave amplitude after transmission through the layer (at $z = d$) is

$$E_x(d) = E_0\exp\left(-\sqrt{|\kappa_2|}\,d\right)\exp(i\omega t),$$

$$H_y(d) = \frac{\sqrt{|\kappa_2|}}{Z_0}E_0\exp\left(-\sqrt{|\kappa_2|}\,d\right)\exp(i(\omega t - \pi/2)).$$

Let us write the real part of these complex expressions for the following fields:

$$\operatorname{Re} E_x(d) = E_0 \exp\left(-\sqrt{|\kappa_2|}d\right)\cos(\omega t),$$

$$\operatorname{Re} H_y(d) = \frac{\sqrt{|\kappa_2|}}{Z_0} E_0 \exp\left(-\sqrt{|\kappa_2|}d\right)\sin(\omega t).$$

The intensity of the wave is given by the modulus of the energy flux density, that is, by the Poynting vector. In this problem, only one of its components, S_z, is nonzero, and after transmission through the layer, it is equal to

$$S_z(d) = \operatorname{Re}\left(E_x(d)H_y(d)\right) = \frac{\sqrt{|\kappa_2|}E_0^2}{Z_0}\exp\left(-2\sqrt{|\kappa_2|}d\right)\cos(\omega t)\sin(\omega t) = S_{0d}\sin(2\omega t),$$

$$S_{0d} = \frac{\sqrt{|\kappa_2|}E_0^2}{2Z_0}\exp\left(-2\sqrt{|\kappa_2|}d\right).$$

From this expression, it follows that in this case, the energy flux density of the wave traveling through the layer has harmonic time dependence at a frequency twice the frequency of the wave. This means that during the first half period $\tau = \pi/2\omega$, the energy flux comes out from the layer and during the next half period $\tau = \pi/2\omega$, the energy flux enters the layer. The average flux during period 2τ is equal to zero as we neglected reflection from the back surface (see Equation 13.31 for comparison):

$$\langle S_z(t) \rangle = \frac{1}{2\tau}\int_0^{2\tau} S_z(t)dt = \frac{S_{0d}}{2\tau}\int_0^{2\tau} \sin(2\omega t)dt$$

$$= -\frac{S_{0d}}{4\omega\tau}\cos(2\omega t)\Big|_0^{2\tau} = \frac{S_{0d}}{4\omega\tau}\left(\cos(0)-\cos(2\pi)\right) = 0.$$

13.5 MEDIA WITH NEGATIVE VALUES OF BOTH MATERIAL CONSTANTS: NEGATIVE REFRACTION

The unusual electromagnetic properties of media with negative material parameters ($\kappa < 0$, $\kappa_m < 0$) were first considered in the theoretical works of V.G. Veselago (1967), who first introduced the concept of a negative refractive index. The first experimental report on artificial materials with such properties appeared in the beginning of 2000. These materials are composites consisting of a set of metal elements (electromagnetic particles) embedded into a dielectric matrix. The elements are arranged in a specific order to form a structure similar to a crystal lattice. In the lattice, elements of two types are alternated. The elements of first type are thin metal rods, representing antennas, which interact with the electric field of a wave propagating in a structure. The elements of second type are rings with cuts, interacting with the magnetic component of an electromagnetic wave. If the dimensions of these elements and the distance between them are much smaller than the wavelength of an incident radiation, this structure can be considered as a continuous medium, which is homogeneous and isotropic.

Consider a homogeneous isotropic nonabsorbing medium with negative material parameters: $\kappa < 0$ and $\kappa_m < 0$. Moreover, the parameters κ, κ_m are scalar and real, that is, $\kappa = -|\kappa|$, $\kappa_m = -|\kappa_m|$. In this case, Maxwell's equations that relate the vectors $\mathbf{k}$, $\mathbf{E}$, and $\mathbf{H}$ of a plane monochromatic wave are given by

$$\mathbf{k} \times \mathbf{E} = -\omega |\kappa_m| \mu_0 \mathbf{H},$$

$$\mathbf{k} \times \mathbf{H} = \omega |\kappa| \varepsilon_0 \mathbf{E}.$$

(13.32)

The analysis of these expressions shows that simultaneous change of signs of κ and κ_m converts the right-handed combination of three vectors $\mathbf{k}$, $\mathbf{E}$, and $\mathbf{H}$ into a left-handed combination (Figure 13.2). Consequently, the medium with $\kappa < 0$, $\kappa_m < 0$ is left handed with a negative refractive index, and the wave in such a medium is a backward wave: the directions of vectors $\mathbf{k}$ and $\mathbf{S}$ are opposite to each other.

In order to prove the negative sign of the refractive index, the negative parameters should be considered as complex quantities. Assuming $\delta_\kappa = \delta_{\kappa_m} = \pi$, we find that $\delta = \pi$, $n = -|\tilde{n}| < 0$, and $\kappa = 0$, that is,

$$n = -\sqrt{|\kappa||\kappa_m|} < 0.$$

(13.33)

An example of an electromagnetic wave in a material with a negative refractive index is given here. Consider the reflection and refraction of a plane monochromatic wave at the interface between medium 1 with $n_1 > 0$ and left-handed medium 2 with $n_2 < 0$. If we formally write Snell's law, taking into account the negative sign of n_2, we find that

$$\frac{\sin \theta_0}{\sin \theta_2} = \frac{n_2}{n_1}, \quad \sin \theta_2 = -\frac{n_1}{|n_2|} \sin \theta_0 < 0,$$

(13.34)

that is, the refraction angle θ_2 is negative. It means that unlike a "routine" case of reflection and refraction on interface between two "right-handed" media (Figure 13.5a), the incident and refracted beams lie on the same side of the interface normal at point of incidence (Figure 13.5b).

The Poynting vector $\mathbf{S}_2$ of the refracted wave in the left-handed medium is still directed into medium 2. Since the refracted wave in a left-handed medium is a backward wave, its wave vector $\mathbf{k}_2$ will be directed toward the interface.

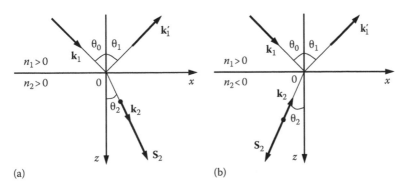

(a)   (b)

FIGURE 13.5 Reflection and refraction at the boundary of two right-handed materials (a) and right-handed and left-handed materials (b).

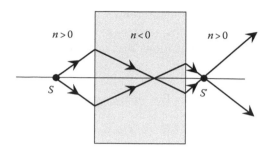

FIGURE 13.6 Path of the rays from point source S that are transmitted through a layer of the "left-handed" medium (Veselago lens).

In other words, in this case, the energy leaves from the boundary, whereas the phase front is moving toward the boundary. The anomalous direction of the refracted beam provides a realization of the equality $k_{1x} = k_{2x}$, necessary for the satisfaction of boundary conditions for vectors of an electromagnetic field. Left-handed media are also called as *media with negative refraction*.

Using this type of refraction, it is possible to fabricate an unusual convex lens using "left-handed" medium. A plane layer of a "left-handed" medium can be used to make such a lens. The ray path from a point source S outside and inside of the layer is shown in Figure 13.6. It is seen that a diverging light beam will concentrate inside the layer into a point (at a sufficient width of the layer) as a result of refraction. After this, the beam begins to diverge. After transmission through the layer, we observe the reverse refraction and the beam converges at point S'. Thus, Figure 13.6 demonstrates the example of a well-known Veselago lens consisting of a slab of left-handed material.

Exercise 13.5

Plot the ray path for oblique incidence of a plane wave on the interface between vacuum ($\kappa_1 = 1$ and $\kappa_{m1} = 1$) and a "left-handed" medium ($\kappa_2 < 0$ and $\kappa_{m2} < 0$) using the boundary conditions for the electric and magnetic fields. Consider the case of transmission through the plane layer of a "left-handed" material.

Solution. Let us consider transmission of the plane wave through the interface between two different media. The boundary conditions for the normal components of the electric and magnetic field, $D_n = \varepsilon_0 \kappa E_n$ and $B_n = \mu_0 \kappa_m H_n$, are

$$\kappa_1 E_{n1} = \kappa_2 E_{n2}, \quad \kappa_{m1} H_{n1} = \kappa_{m2} H_{n2}.$$

Since $\kappa_2 < 0$ and $\kappa_{m2} < 0$, then from these relationships it follows that during the transmission of wave from a "right-handed" medium to a "left-handed" medium, the following relationships are valid:

$$\frac{E_{n2}}{E_{n1}} = \frac{\kappa_1}{\kappa_2} < 0, \quad \frac{H_{n2}}{H_{n1}} = \frac{\kappa_{m1}}{\kappa_{m2}} < 0.$$

This indicates an unusual ray path for the refracted ray—the ray bends in the direction opposite to the direction of incidence. We get the same result if we use Snell's law:

$$\frac{\sin \theta_0}{\sin \theta_2} = \frac{n_2}{n_1} < 0, \quad \sin \theta_2 = \frac{n_1}{n_2} \sin \theta_0 < 0,$$

that is, the angle of refraction $\theta_2 < 0$ and the ray in the "left-handed" medium bends in the "negative" direction from the normal (Figure 13.7).

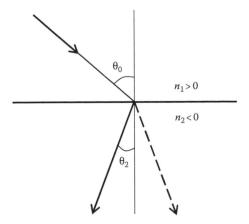

FIGURE 13.7   The ray paths of incident and refracted rays at the interface of "right-handed" and "left-handed" media.

Exercise 13.6

A plate of thickness d and negative refractive index $n_2 = -1$ is placed in the air (a medium with a positive refractive index $n_1 = 1$). A point source of monochromatic light is located at a distance l from the left face of the plate. Draw a diagram of the rays coming from the point source as they travel through the plate for the following cases: (a) $d = 3l/2$ and (b) $d = l/2$.

Solution. (a) $d = 3l/2$: When a light beam is incident on a plate with negative refractive index, the beam will refract inward at an angle determined by Snell's law (Figure 13.8a). Since the absolute value of the two indices of refraction is the same, the angle of refraction will be equal to the angle of incidence. The beams will converge at the point A at a distance l into the plate and then will diverge for the remaining distance of $l/2$. The beam will then refract inward again, at the same angle, upon leaving the plate and the beam will converge once again a distance $l/2$ away from the plate. Thus, the plate acts as a lens; for the observer, the image of point S will be located to the right of the plate at point S'.

(b) $d = l/2$: In this case, the incident beams refract inward at the air–plate interface; since the plate thickness is less than the distance l, the beam does not reconverge in the plate (Figure 13.8b). Upon exiting the plate, the beams refract again at the same angle and continue to diverge. Therefore, the plate acts as a lens and for the observer the image of point S will be located at point S' that is on the air–plate boundary. For a thinner plate, the image S' will move closer to S and as d tends to zero, S' coincides with S.

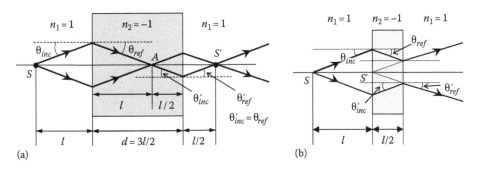

FIGURE 13.8   Path of the rays from point source S that are transmitted through a plate of the "left-handed" material of thickness $d = 3l/2$ (a) and $d = l/2$ (b).

PROBLEMS

13.1 Find the phase velocity and damping coefficient of waves propagating in a homogeneous isotropic medium with complex permittivity and permeability $\varepsilon = \varepsilon_0(\kappa' + i\kappa'')$ and $\mu = \mu_0(\kappa'_m + i\kappa''_m)$. (*Part of the answer:* $n(\omega) = \sqrt{1 + C/\omega^2}$.)

13.2 The dielectric permittivity and magnetic permeability of a medium depend on frequency as follows:

$$\varepsilon = \varepsilon_0 \kappa = \varepsilon_0 \left(1 - \frac{F_e}{\omega^2 + i\gamma_e \omega} \right), \quad \mu = \mu_0 \kappa_m = \mu_0 \left(1 - \frac{F_m}{\omega^2 + i\gamma_m \omega} \right),$$

where F_e, F_m and γ_e, γ_m are known constants. Find the dependence on frequency of the tangents of angles related to the dielectric and magnetic losses.

13.3 The dielectric permittivity and magnetic permeability of a medium depend on the frequency as follows:

$$\varepsilon = \varepsilon_0 \kappa = \varepsilon_0 \left(1 + \frac{F_e}{\omega_{0e}^2 - \omega^2} \right), \quad \mu = \mu_0 \kappa_m = \mu_0 \left(1 + \frac{F_m}{\omega_{0m}^2 - \omega^2} \right),$$

Find the frequency ranges where the material parameters ε and μ are negative, that is, where the medium can be considered as "left handed."

13.4 Find the phase impedance ζ for a medium with complex material parameters $\varepsilon = \varepsilon_0(\kappa' + i\kappa'')$, $\mu = \mu_0(\kappa'_m + i\kappa''_m)$.

13.5 A nonabsorbing medium with negative dielectric permittivity and positive magnetic permeability occupies the half-space $z \geq 0$. A linearly polarized plane wave is incident normally on the medium from vacuum. Find the energy reflection coefficient of the wave. (*Answer:* $R = 1$.)

13.6 A plane wave is incident from a homogeneous medium with a positive refractive index n_1 at the interface with a medium having a refractive index of n_2. Draw the paths of rays in the media with (a) $n_2 > 0$ and (b) $n_2 < 0$.

13.7 A plate with thickness d and with a negative refractive index $n_2 = -1$ is placed in a medium with a positive refractive index (air with $n_1 = 1$). A point source of monochromatic radiation is located at a distance l from one of the faces of the plate. Draw a diagram of the rays coming from the point source that travel through the plate. Find the distance at which these rays focus past the plate.

Appendix A

PHYSICAL CONSTANTS AND UNITS

TABLE A.1
SI Units

Quantity	Name	Symbol	Equivalent
Length	Meter, m	l	m
Mass	Kilogram, kg	M	kg
Time	Second, s	t	s
Speed, velocity		v	m s^{-1}
Acceleration		a	m s^{-2}
Angular velocity		ω	rad s^{-2}
Frequency	Hertz, Hz	f	s^{-1}
Force	Newton, N	F	kg m s^{-2}
Pressure, stress	Pascal, Pa	P	N m^{-2}
Work, energy	Joule, J	W	N m, kg m^2 s^{-2}
Impulse, momentum		p	N s, kg m s^{-1}
Power	Watt, W	N	J s^{-1}
Electric charge	Coulomb, C	q	A s
Electric potential	Volt, V	φ	J C^{-1}, W A^{-1}
Resistance	Ohm, Ω	R	V A^{-1}
Conductance	Siemens, S	σ	A V^{-1}, Ω^{-1}
Magnetic flux	Weber, Wb	Φ	V s
Inductance	Henry, H	L	Wb A^{-1}
Capacitance	Farad, F	C	C V^{-1}
Electric field strength		E	V m^{-1}, N C^{-1}
Magnetic flux density	Tesla, T	B	Wb m^{-2}, N A^{-1} m^{-1}
Electric displacement		D	C m^{-2}
Magnetic field strength		H	A m^{-1}
Temperature	Kelvin, K	T	K

TABLE A.2
Physical Constants

Constant	Symbol	Value	Units
Speed of light in a vacuum	c	$2.9979 \times 10^8 \approx 3 \times 10^8$	m s^{-1}
Elementary charge	e	1.602×10^{-19}	C
Electron mass	m_e	9.11×10^{-31}	kg
Electron charge to mass ratio	e/m_e	1.76×10^{11}	C kg^{-1}
Proton mass	m_p	1.67×10^{-27}	kg
Permittivity of free space	ε_0	$10^{-9}/(36\pi) \approx 8.854 \times 10^{-12}$	F m^{-1}
Permeability of free space	μ_0	$4\pi \times 10^{-7}$	H m^{-1}
Impedance of free space	$Z_0 = \sqrt{\mu_0/\varepsilon_0}$	$120\pi \approx 376.73$	Ω
Boltzmann's constant	k_B	1.38×10^{-23}	J K^{-1}
Planck's constant	h	6.626×10^{-34}	J s

TABLE A.3
Conversion of SI Units to Gaussian Units

Quantity	SI Unit	Gaussian Unit
Length	1 m	10^2 cm
Mass	1 kg	10^3 g
Force	1 N	10^5 dyne $= 10^5$ g cm s^{-2}
Energy	1 J	10^7 erg $= 10^7$ g cm^2 s^{-2}

$1 \text{ eV} = 1.602 \times 10^{-19} \text{ J} = 1.602 \times 10^{-12} \text{ erg}$

TABLE A.4
Standard Prefixes Used with SI Units

Prefix	Abbreviation	Meaning	Prefix	Abbreviation	Meaning
atto	a	10^{-18}	deka	da	10^1
femto	f	10^{-15}	hecto	h	10^2
pico	p	10^{-12}	kilo	k	10^3
nano	n	10^{-9}	mega	M	10^6
micro	μ	10^{-6}	giga	G	10^9
milli	m	10^{-3}	tera	T	10^{12}
centi	c	10^{-2}	peta	P	10^{15}
deci	d	10^{-1}	exa	E	10^{18}

Appendix B

SOME MATHEMATICAL RELATIONS AND IDENTITIES

The roots of a quadratic equation

$$x_{1,2} = \frac{-b \pm \sqrt{b^2 - 4ac}}{2a} \quad \text{for } ax^2 + bx + c = 0,$$

and

$$x_{1,2} = -\frac{p}{2} \pm \sqrt{\frac{p^2}{4} - q}, \quad x_1 + x_2 = -p, \quad x_1 \cdot x_2 = q \quad \text{for } x^2 + px + q = 0.$$

Algebraic formulas

$$(a \pm b)^2 = a^2 \pm 2ab + b^2,$$

$$(a \pm b)^3 = a^3 \pm 3a^2b + 3ab^2 \pm b^3,$$

$$a^2 - b^2 = (a - b)(a + b),$$

$$a^3 - b^3 = (a - b)(a^2 + ab + b^2),$$

$$a^3 + b^3 = (a + b)(a^2 - ab + b^2),$$

$$a^x a^y = a^{x+y}, \quad \sqrt[n]{ab} = \sqrt[n]{a}\sqrt[n]{b},$$

$$a^x b^x = (ab)^x, \quad \sqrt[n]{a/b} = \sqrt[n]{a}/\sqrt[n]{b},$$

$$\left(a^x\right)^y = a^{xy}, \quad \sqrt[n]{a^m} = \left(\sqrt[n]{a}\right)^m = a^{m/n},$$

$$\frac{a^x}{a^y} = a^{x-y}, \quad \sqrt[n]{\sqrt[m]{a}} = \sqrt[nm]{a} = a^{1/nm}.$$

Length of the circumference of radius R

$$l = 2\pi R, \quad l = \pi D, \quad D = 2R.$$

Area of a circle of radius R

$$A = \pi R^2, \quad A = \frac{\pi D^2}{4}.$$

Area of a sphere of radius R

$$A = 4\pi R^2, \quad A = \pi D^2.$$

Volume of a sphere of radius R

$$Vol = \frac{4}{3}\pi R^3, \quad Vol = \frac{1}{6}\pi D^3, \quad D = 2R.$$

Relations for a triangle

For $\gamma = \pi/2$ (see Figure B.1a)

$$a^2 + b^2 = c^2, \quad \sin\alpha = \frac{a}{c}, \quad \cos\alpha = \frac{b}{c}, \quad \tan\alpha = \frac{a}{b}, \quad \cot\alpha = \frac{b}{c}, \quad A = \frac{1}{2}ab.$$

For $\gamma \neq \pi/2$ (see Figure B.1b)

$$c^2 = a^2 + b^2 - 2ab\cos\gamma, \quad \frac{a}{\sin\alpha} = \frac{b}{\sin\beta} = \frac{c}{\sin\gamma},$$

$$A = \frac{1}{2}ab\sin\gamma = \frac{1}{2}ac\sin\beta = \frac{1}{2}bc\sin\alpha.$$

Values for some trigonometric functions

α	0°	30°	45°	60°	90°
	0	$\pi/6$	$\pi/4$	$\pi/3$	$\pi/2$
$\sin\alpha$	0	$\dfrac{1}{2}$	$\dfrac{\sqrt{2}}{2}$	$\dfrac{\sqrt{3}}{2}$	1
$\cos\alpha$	1	$\dfrac{\sqrt{3}}{2}$	$\dfrac{\sqrt{2}}{2}$	$\dfrac{1}{2}$	0
$\tan\alpha$	0	$\dfrac{1}{\sqrt{3}}$	1	$\sqrt{3}$	∞
$\cot\alpha$	∞	$\sqrt{3}$	1	$\dfrac{1}{\sqrt{3}}$	0

Trigonometric identities

$$\sin(-x) = -\sin x, \quad \sin x = \cos\left(\frac{\pi}{2} - x\right),$$

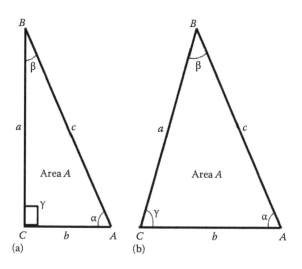

FIGURE B.1   Right (a) and arbitrary (b) triangles.

$$\cos(-x) = \cos x, \quad \cos x = \sin\left(\frac{\pi}{2} - x\right),$$

$$\sin^2 \alpha + \cos^2 \alpha = 1,$$

$$\sin(\alpha \pm \beta) = \sin \alpha \cos \beta \pm \cos \alpha \sin \beta,$$

$$\cos(\alpha \pm \beta) = \cos \alpha \cos \beta \mp \sin \alpha \sin \beta,$$

$$\sinh(\alpha \pm \beta) = \sinh \alpha \cosh \beta \pm \cosh \alpha \sinh \beta,$$

$$\cosh(\alpha \pm \beta) = \cosh \alpha \cosh \beta \pm \sinh \alpha \sinh \beta,$$

$$\cos 2\alpha = \cos^2 \alpha - \sin^2 \alpha,$$

$$\sin 2\alpha = 2 \sin \alpha \cos \alpha,$$

$$\sin 3\alpha = 3 \sin \alpha - 4 \sin^3 \alpha,$$

$$\cos 3\alpha = 4 \cos^3 \alpha - 3 \cos \alpha,$$

$$1 + \tan^2 \alpha = \frac{1}{\cos^2 \alpha} = \sec^2 \alpha, \quad 1 + \cot^2 \alpha = \frac{1}{\sin^2 \alpha} = \csc^2 \alpha,$$

$$\sin^2 \alpha = \frac{1}{2}(1 - \cos 2\alpha), \quad \cos^2 \alpha = \frac{1}{2}(1 + \cos 2\alpha),$$

$$\sin^3 \alpha = \frac{1}{4}(3 \sin \alpha - \sin 3\alpha), \quad \cos^3 \alpha = \frac{1}{4}(3 \cos \alpha + \cos 3\alpha),$$

$$2 \sin \alpha \cos \beta = \sin(\alpha + \beta) + \sin(\alpha - \beta),$$

$$2 \cos \alpha \cos \beta = \cos(\alpha + \beta) + \cos(\alpha - \beta),$$

$$2 \sin \alpha \sin \beta = \cos(\alpha - \beta) - \cos(\alpha + \beta),$$

$$\tan(\alpha + \beta) = \frac{\tan \alpha + \tan \beta}{1 - \tan \alpha \tan \beta}.$$

Logarithmic relations

$$\log_a x = y, \quad a^y = x, \quad a^{\log_a x} = x,$$

$$\log_a(xy) = \log_a x + \log_a y,$$

$$\log_a\left(\frac{x}{y}\right) = \log_a x - \log_a y,$$

$$\log_a x^k = k \log_a x, \quad \log_{a^n} x = \frac{1}{n} \log_a x,$$

$$\log_{10} x = \lg x, \quad \log_e x = \ln x, \quad e = 2,72,\ldots,$$

$$\log_a x = \frac{\log_b x}{\log_b a}, \quad \log_a b = \frac{1}{\log_b a},$$

$$\lg x = \frac{\ln x}{\ln 10} = (\ln x) \lg e.$$

Derivatives of some functions

Function	Derivative	Function	Derivative
$c = const$	0	$\log_a x$	$(1/\ln a)x^{-1}$
x^n	nx^{n-1}	$\sin x$	$\cos x$
e^x	e^x	$\cos x$	$-\sin x$
a^x	$a^x \ln a$	$\tan x$	$1/\cos^2 x$
$\ln x$	$1/x$	$\cot x$	$-1/\sin^2 x$

Integrals of some functions

Function	Integral	Function	Integral
x^n	$x^{n+1}/(n+1)$	$\sin x$	$-\cos x$
$1/x$	$\ln |x|$	$\cos x$	$\sin x$
e^x	e^x	$\tan x$	$-\ln|\cos x|$
a^x	$a^x/\ln a$	$\cot x$	$\ln|\sin x|$

Taylor series *for* $|x| < 1$

$$(1 \pm x)^n = 1 \pm nx + \frac{n(n-1)}{2!}x^2 \pm \frac{n(n-1)(n-2)}{3!}x^3 + \cdots,$$

$$\sin x = x - \frac{x^3}{3!} + \frac{x^5}{5!} - \cdots, \quad \cos x = 1 - \frac{x^2}{2!} + \frac{x^4}{4!} - \cdots,$$

$$\ln(1+x) = x - \frac{x^2}{2} + \frac{x^3}{3} - \cdots, \quad \ln(1-x) = -x - \frac{x^2}{2} - \frac{x^3}{3} - \cdots,$$

$$e^x = 1 + \frac{x}{1!} + \frac{x^2}{2!} + \frac{x^3}{3!} + \cdots, \quad n! = 1 \cdot 2 \cdot 3 \cdot 4 \cdots (n-1) \cdot n.$$

Trigonometric identities for a complex argument ($i = \sqrt{-1}$)

$$\cosh(ix) = \cos x, \quad \sinh(ix) = i \sin x,$$

$$\cos(ix) = \cosh x, \quad \sin(ix) = i \sinh x,$$

$$\sin(\alpha \pm i\beta) = \sin \alpha \cosh \beta \pm i \cos \alpha \sinh \beta,$$

$$\cos(\alpha \pm i\beta) = \cos \alpha \cosh \beta \mp i \sin \alpha \sinh \beta,$$

$$\sinh(\alpha \pm i\beta) = \sinh \alpha \cos \beta \pm i \cosh \alpha \sin \beta,$$

$$\cosh(\alpha \pm i\beta) = \cosh \alpha \cos \beta \pm i \sinh \alpha \sin \beta.$$

Euler's formula

$$e^{\pm ix} = \cos x \pm i \sin x,$$

$$\cos x = \frac{1}{2}(e^{ix} + e^{-ix}), \quad \sin x = \frac{1}{2i}(e^{ix} - e^{-ix}),$$

$$e^{\pm x} = \cosh x \pm \sinh x,$$

$$\cosh x = \frac{1}{2}(e^x + e^{-x}), \quad \sinh(x) = \frac{1}{2}(e^x - e^{-x}).$$

Appendix C

SOME RELATIONS FROM VECTOR AND TENSOR ALGEBRA

C.1 VECTOR ALGEBRA

Vector $\mathbf{A}$ in a Cartesian coordinate system has the form $\mathbf{A} = \mathbf{i}A_x + \mathbf{j}A_y + \mathbf{k}A_z$: its components are $A_x = A \cos \alpha$, $A_y = A \cos \beta$, and $A_z = A \cos \gamma$ (see Figure C.1). Here $\mathbf{i}$, $\mathbf{j}$, $\mathbf{k}$ are the unit vectors along the x, y, z axes, respectively (not shown in Figure C.1).

Analogous relations can be written for the position vector $\mathbf{r} = \mathbf{i}r_x + \mathbf{j}r_y + \mathbf{k}r_z$ with components $r_x = r \cos \alpha$, $r_y = r \cos \beta$, $r_z = r \cos \gamma$, or any other vector in the Cartesian coordinate system.

The magnitude of vector $\mathbf{r}$ is given by $r = \sqrt{r_x^2 + r_y^2 + r_z^2} = \sqrt{x^2 + y^2 + z^2}$.

The directional angles α, β, and γ of any vector satisfy the following identity:

$$\cos^2 \alpha + \cos^2 \beta + \cos^2 \gamma = 1.$$

ADDITION AND SUBTRACTION OF VECTORS

Vector $\mathbf{C} = \mathbf{A} \pm \mathbf{B}$ and its components are (Figure C.2)

$$\mathbf{C} = \mathbf{i}C_x + \mathbf{j}C_y + \mathbf{k}C_z = \mathbf{i}(A_x \pm B_x) + \mathbf{j}(A_y \pm B_y) + \mathbf{k}(A_z \pm B_z).$$

SCALAR (OR DOT) PRODUCT OF TWO VECTORS

$$\mathbf{A} \cdot \mathbf{B} = |\mathbf{A}| \times |\mathbf{B}| \cos \varphi = AB \cos \varphi = A_x B_x + A_y B_y + A_z B_z,$$

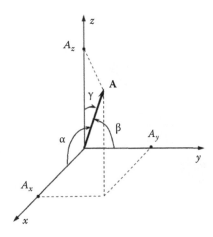

FIGURE C.1 Vector $\mathbf{A}$ in a Cartesian coordinate system.

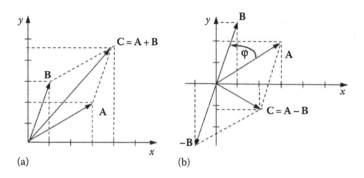

FIGURE C.2   Parallelogram rule for addition (a) and subtraction (b) of two vectors.

where

$$A = \sqrt{A_x^2 + A_y^2 + A_z^2}$$

$$B = \sqrt{B_x^2 + B_y^2 + B_z^2}$$

φ is the angle between vectors **A** and **B**

$$\cos \varphi = \frac{A_x B_x + A_y B_y + A_z B_z}{|\mathbf{A}| \cdot |\mathbf{B}|} = \frac{A_x B_x + A_y B_y + A_z B_z}{\sqrt{\left(A_x^2 + A_y^2 + A_z^2\right) \cdot \left(B_x^2 + B_y^2 + B_z^2\right)}}.$$

The following are the properties of scalar product:

$$\mathbf{A} \cdot \mathbf{B} = \mathbf{B} \cdot \mathbf{A},$$
$$\mathbf{A} \cdot (\mathbf{B} + \mathbf{C}) = \mathbf{A} \cdot \mathbf{B} + \mathbf{A} \cdot \mathbf{C},$$
$$\mathbf{A} \cdot \mathbf{A} = |\mathbf{A}|^2 = A^2,$$
$$\mathbf{i} \cdot \mathbf{j} = \mathbf{j} \cdot \mathbf{k} = \mathbf{k} \cdot \mathbf{i} = 0,$$
$$\mathbf{i} \cdot \mathbf{i} = \mathbf{j} \cdot \mathbf{j} = \mathbf{k} \cdot \mathbf{k} = 1.$$

It follows from these that two vectors are perpendicular to each other if their scalar product is equal to zero.

Vector (or Cross) Product of Two Vectors

The vector (or cross) product of two vectors **A** and **B** is a vector $\mathbf{C} = \mathbf{A} \times \mathbf{B}$ with the direction perpendicular to the plane formed by vectors **A** and **B**; the magnitude C is equal to the area of the parallelogram formed by vectors **A** and **B**. $C = AB \sin\varphi$, where φ is the angle between **A** and **B** (Figure C.3).

The right-hand rule must be applied as follows to determine the direction of $\mathbf{A} \times \mathbf{B}$. Rotate **A** in the plane defined by the vectors **A** and **B** along the shortest angle so that it coincides with **B**. Then curl the fingers of the right hand in the same direction; the direction of the thumb gives the direction of $\mathbf{A} \times \mathbf{B}$.

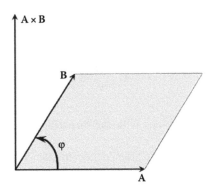

FIGURE C.3 Vector product of two vectors.

In the Cartesian coordinate system

$$\mathbf{A} \times \mathbf{B} = \begin{vmatrix} \mathbf{i} & \mathbf{j} & \mathbf{k} \\ A_x & A_y & A_z \\ B_x & B_y & B_z \end{vmatrix} = \mathbf{i}(A_y B_z - A_z B_y) + \mathbf{j}(A_z B_x - A_x B_z) + \mathbf{k}(A_x B_y - A_y B_x).$$

The following are some properties of the cross product:

$$\mathbf{A} \times \mathbf{B} = -\mathbf{B} \times \mathbf{A},$$

$$\mathbf{A} \times (\mathbf{B} \times \mathbf{C}) = \mathbf{B}(\mathbf{AC}) - \mathbf{C}(\mathbf{AB}),$$

$$\mathbf{A} \times (\mathbf{B} + \mathbf{C}) = \mathbf{A} \times \mathbf{B} + \mathbf{A} \times \mathbf{C},$$

$$\mathbf{A} \cdot (\mathbf{B} \times \mathbf{C}) = \begin{vmatrix} A_x & A_y & A_z \\ B_x & B_y & B_z \\ C_x & C_y & C_z \end{vmatrix}$$

$$= A_x(B_y C_z - B_z C_y) + A_y(B_z C_x - B_x C_z) + A_z(B_x C_y - B_y C_x).$$

$$\mathbf{A} \cdot (\mathbf{B} \times \mathbf{C}) = \mathbf{C} \cdot (\mathbf{A} \times \mathbf{B}) = \mathbf{B} \cdot (\mathbf{C} \times \mathbf{A}),$$

$$\mathbf{A} \times \mathbf{A} = 0, \quad \mathbf{i} \times \mathbf{j} = \mathbf{k}, \quad \mathbf{j} \times \mathbf{k} = \mathbf{i}, \quad \mathbf{k} \times \mathbf{i} = \mathbf{j}.$$

It follows from these that two vectors are parallel to each other if their cross product is equal to zero.
Del operator (in a Cartesian coordinate system)

$$\nabla = \mathbf{i} \frac{\partial}{\partial x} + \mathbf{j} \frac{\partial}{\partial y} + \mathbf{k} \frac{\partial}{\partial z}.$$

Laplace operator (in a Cartesian coordinate system)

$$\Delta = \nabla \cdot \nabla = \frac{\partial^2}{\partial x^2} + \frac{\partial^2}{\partial y^2} + \frac{\partial^2}{\partial z^2}.$$

Gradient of a scalar function $U(\mathbf{r})$ in a Cartesian coordinate system is determined by the following equation:

$$\text{grad } U(\mathbf{r}) \equiv \nabla U(\mathbf{r}) \equiv \mathbf{i}\frac{\partial U(\mathbf{r})}{\partial x} + \mathbf{j}\frac{\partial U(\mathbf{r})}{\partial y} + \mathbf{k}\frac{\partial U(\mathbf{r})}{\partial z}.$$

The direction of the vector ∇U coincides with the direction of the fastest increase of the scalar function $U(\mathbf{r})$.

Divergence of a vector field $\mathbf{F}(\mathbf{r})$ in a Cartesian coordinate system is determined by the following equation:

$$\text{div } \mathbf{F}(\mathbf{r}) \equiv \nabla \cdot \mathbf{F}(\mathbf{r}) = \frac{\partial F_x(\mathbf{r})}{\partial x} + \frac{\partial F_y(\mathbf{r})}{\partial y} + \frac{\partial F_z(\mathbf{r})}{\partial z}$$

This can be thought of as the scalar product of vectors operator ∇ with vector $\mathbf{F}$.

According to **Gauss's theorem**, the integral of the flux of vector $\mathbf{F}(\mathbf{r})$ over a closed surface A can be expressed as the integral over the volume enclosed by the surface A:

$$\oint_A \mathbf{F}(\mathbf{r}) \cdot d\mathbf{A} = \int_{Vol} \text{div } \mathbf{F}(\mathbf{r})dr^3,$$

where the elementary vector $d\mathbf{A} = \mathbf{n}dA$ and $\mathbf{n}$ is the unit vector that is outward normal to the surface $d\mathbf{A}$. If div $\mathbf{F}(\mathbf{r}) = 0$, the vector field $\mathbf{F}(\mathbf{r})$ is called "solenoidal" and in accordance with the preceding equation, the flux of the solenoidal field over any closed surface is equal to zero. Note that the flux $d\Phi$ of vector $\mathbf{F}$ through the area $d\mathbf{A}$ is determined by the equation $d\Phi = \mathbf{F}\cdot d\mathbf{A}$.

The curl of a vector field $\mathbf{F}(\mathbf{r})$ in the Cartesian coordinate system is determined by the following equation:

$$\text{curl } \mathbf{F}(\mathbf{r}) = \nabla \times \mathbf{F}(\mathbf{r}) = \begin{vmatrix} \mathbf{i} & \mathbf{j} & \mathbf{k} \\ \dfrac{\partial}{\partial x} & \dfrac{\partial}{\partial y} & \dfrac{\partial}{\partial z} \\ F_x & F_y & F_z \end{vmatrix},$$

that is, the curl of the vector field $\mathbf{F}(\mathbf{r})$ is

$$\nabla \times \mathbf{F}(\mathbf{r}) = \mathbf{i}\left(\frac{\partial F_z}{\partial y} - \frac{\partial F_y}{\partial z}\right) + \mathbf{j}\left(\frac{\partial F_x}{\partial z} - \frac{\partial F_z}{\partial x}\right) + \mathbf{k}\left(\frac{\partial F_y}{\partial x} - \frac{\partial F_x}{\partial y}\right).$$

According to **Stokes' theorem**, an integral over a closed path L of the vector $\mathbf{F}(\mathbf{r})$ can be transformed into an integral over the surface that has the path as its border:

$$\oint_L \mathbf{F}(\mathbf{r}) \cdot d\mathbf{l} = \int_A \nabla \times \mathbf{F}(\mathbf{r}) \cdot d\mathbf{A}.$$

If $\nabla \times \mathbf{F} = 0$, the vector field $\mathbf{F}(\mathbf{r})$ is called potential field and in accordance with the equation given earlier, the integral of $\mathbf{F}(\mathbf{r})\cdot d\mathbf{l}$ over any closed path is equal to zero.

For any scalar functions $U(\mathbf{r})$ and $V(\mathbf{r})$ or vector fields $\mathbf{F}(\mathbf{r})$ and $\mathbf{G}(\mathbf{r})$, the following relations are true:

$$\nabla\big(U(\mathbf{r})+V(\mathbf{r})\big) = \nabla U(\mathbf{r})+\nabla V(\mathbf{r}),$$

$$\nabla\cdot\big(\mathbf{F}(\mathbf{r})+\mathbf{G}(\mathbf{r})\big) = \nabla\cdot\mathbf{F}(\mathbf{r})+\nabla\cdot\mathbf{G}(\mathbf{r}),$$

$$\nabla\big(U(\mathbf{r})V(\mathbf{r})\big) = V(\mathbf{r})\nabla U(\mathbf{r})+U(\mathbf{r})\nabla V(\mathbf{r}),$$

$$\nabla\cdot\big(U(\mathbf{r})\mathbf{F}(\mathbf{r})\big) = \mathbf{F}(\mathbf{r})\cdot\nabla U(\mathbf{r})+U(\mathbf{r})\nabla\cdot\mathbf{F}(\mathbf{r}),$$

$$\nabla\times\big(U(\mathbf{r})\mathbf{F}(\mathbf{r})\big) = U(\mathbf{r})\big(\nabla\times\mathbf{F}(\mathbf{r})\big)+\big(\nabla U(\mathbf{r})\big)\times\mathbf{F}(\mathbf{r}),$$

$$\nabla\cdot\big(\mathbf{F}(\mathbf{r})\times\mathbf{G}(\mathbf{r})\big) = \mathbf{G}(\mathbf{r})\cdot\big(\nabla\times\mathbf{F}(\mathbf{r})\big)-\mathbf{F}(\mathbf{r})\cdot\big(\nabla\times\mathbf{G}(\mathbf{r})\big),$$

$$\text{div}(\text{curl }\mathbf{F}(\mathbf{r})) = \nabla(\nabla\times\mathbf{F}(\mathbf{r})) = 0,$$

$$\text{curl}(\text{grad }U(\mathbf{r})) = \nabla\times(\nabla U(\mathbf{r})) = (\nabla\times\nabla)U(\mathbf{r}) = 0,$$

$$\text{curl}(\text{curl }\mathbf{F}(\mathbf{r})) = \nabla\times(\nabla\times\mathbf{F}(\mathbf{r})) = \nabla(\nabla\cdot\mathbf{F}(\mathbf{r}))-(\nabla\cdot\nabla)\mathbf{F}(\mathbf{r}) = \text{grad}(\text{div }\mathbf{F}(\mathbf{r}))-\nabla^2\mathbf{F}(\mathbf{r}).$$

C.2   TENSOR ALGEBRA

The second-order tensor $\hat{T}$ is described by nine scalar components. In the Cartesian coordinates, they are denoted as $T_{xx},T_{xy},T_{xz},T_{yx},T_{yy},T_{yz},T_{zx},T_{zy},T_{zz}$ or $T_{11},T_{12},T_{13},T_{21},T_{22},T_{23},T_{31},T_{32},T_{33}$, respectively:

$$\hat{T} = \begin{pmatrix} T_{xx} & T_{xy} & T_{xz} \\ T_{yx} & T_{yy} & T_{yz} \\ T_{zx} & T_{zy} & T_{zz} \end{pmatrix} \quad\text{or}\quad \hat{T} = \begin{pmatrix} T_{11} & T_{12} & T_{13} \\ T_{21} & T_{22} & T_{23} \\ T_{31} & T_{32} & T_{33} \end{pmatrix}.$$

If $\mathbf{A}$ and $\mathbf{B}$ are vectors in a Cartesian coordinate system and a tensor $\hat{T}$ transforms vector $\mathbf{A}$ into the vector $\mathbf{B}$, the transformation can be defined by the following equation:

$$\mathbf{B} = \hat{T}\mathbf{A} \quad\text{or}\quad B_i = \sum_{j=1}^{3} T_{ij}A_j,$$

This transformation rotates vector $\mathbf{A}$ in space as well as stretches (or shrinks) its components A_i. Note that the equation given earlier can be written in the following matrix form:

$$\begin{pmatrix} B_1 \\ B_2 \\ B_3 \end{pmatrix} = \begin{pmatrix} T_{11} & T_{12} & T_{13} \\ T_{21} & T_{22} & T_{23} \\ T_{31} & T_{32} & T_{33} \end{pmatrix} \begin{pmatrix} A_1 \\ A_2 \\ A_3 \end{pmatrix}.$$

Addition of the second-rank tensors:

$$\begin{pmatrix} T_{11} & T_{12} & T_{13} \\ T_{21} & T_{22} & T_{23} \\ T_{31} & T_{32} & T_{33} \end{pmatrix} + \begin{pmatrix} D_{11} & D_{12} & D_{13} \\ D_{21} & D_{22} & D_{23} \\ D_{31} & D_{32} & D_{33} \end{pmatrix} = \begin{pmatrix} T_{11}+D_{11} & T_{12}+D_{12} & T_{13}+D_{13} \\ T_{21}+D_{21} & T_{22}+D_{22} & T_{23}+D_{23} \\ T_{31}+D_{31} & T_{32}+D_{32} & T_{33}+D_{33} \end{pmatrix},$$

$$T_{ij} + D_{ij} = (T+D)_{ij}.$$

Multiplication of tensor $\hat{T}$ by a scalar α:

$$\alpha \hat{T} = \alpha \begin{pmatrix} T_{11} & T_{12} & T_{13} \\ T_{21} & T_{22} & T_{23} \\ T_{31} & T_{32} & T_{33} \end{pmatrix} = \begin{pmatrix} \alpha T_{11} & \alpha T_{12} & \alpha T_{13} \\ \alpha T_{21} & \alpha T_{22} & \alpha T_{23} \\ \alpha T_{31} & \alpha T_{32} & \alpha T_{33} \end{pmatrix},$$

$$(\alpha \hat{T})_{ij} = \alpha T_{ij}.$$

Tensors are linear operators, so for arbitrary vectors **A** and **B** and scalars α and β, the following relation is true:

$$\hat{T}(\alpha \mathbf{A} + \beta \mathbf{B}) = \alpha \hat{T}\mathbf{A} + \beta \hat{T}\mathbf{B}.$$

Multiplication of tensors is performed as multiplication of matrices, that is,

$$\hat{T}\hat{D} = \hat{C}, \quad C_{ij} = \sum_{l} T_{il}D_{lj}.$$

The **identity tensor** has the following components:

$$I_{ij} = \delta_{ij} = \begin{cases} 1, & i = j, \\ 0, & i \neq j. \end{cases}$$

The identity tensor transforms any vector **A** into itself:

$$\hat{I}\mathbf{A} = \mathbf{A}, \quad \sum_{j=1}^{3} I_{ij}A_j = \sum_{j=1}^{3} \delta_{ij}A_j = A_i.$$

For any tensor $\hat{D}$ with components D_{ij}, there is a transpose tensor $\hat{D}^T$ with the components D_{ji}, so the transposition swaps the indices:

$$\text{if } \hat{D} = \begin{pmatrix} D_{11} & D_{12} & D_{13} \\ D_{21} & D_{22} & D_{23} \\ D_{31} & D_{32} & D_{33} \end{pmatrix}, \quad \text{then} \quad \hat{D}^T = \begin{pmatrix} D_{11} & D_{21} & D_{31} \\ D_{12} & D_{22} & D_{32} \\ D_{13} & D_{23} & D_{33} \end{pmatrix}.$$

The tensor is symmetric if $\hat{D} = \hat{D}^T$, that is, $D_{ij} = D_{ji}$, and the tensor is antisymmetric (or skew tensor) if $\hat{D} = -\hat{D}^T$, that is, $D_{ij} = -D_{ji}$.

The **determinant of the tensor** $\hat{D}$ is the determinant of its matrix:

$$\det \hat{D} = \det \begin{pmatrix} D_{11} & D_{12} & D_{13} \\ D_{21} & D_{22} & D_{23} \\ D_{31} & D_{32} & D_{33} \end{pmatrix}$$

$$= D_{11}(D_{22}D_{33} - D_{32}D_{23}) - D_{12}(D_{21}D_{33} - D_{31}D_{23}) + D_{13}(D_{21}D_{32} - D_{22}D_{31}).$$

The tensor $\hat{D}$ is called singular if its determinant is equal to zero.

For any nonsingular tensor $\hat{T}$, there is an **inverse tensor** $\hat{T}^{-1}$:

$$\hat{T}\hat{T}^{-1} = \hat{T}^{-1}\hat{T} = \hat{I}, \quad (\hat{T}^{-1})^{-1} = \hat{T}.$$

COORDINATE TRANSFORMATIONS

Let us consider two coordinate systems with basis vectors (e_1, e_2, e_3) and (e_1', e_2', e_3'), as shown in Figure C.4. In each Cartesian coordinate system, we have the following identities for the scalar products of the basis vectors:

$$e_i \cdot e_j = \delta_{ij} \quad \text{and} \quad e_i' \cdot e_j' = \delta_{ij} \quad \text{where } \delta_{ij} = \begin{cases} 1, & i = j \\ 0, & i \neq j \end{cases}.$$

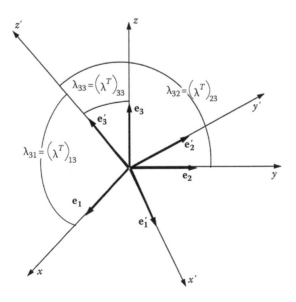

FIGURE C.4 Three out of nine components of tensor $\hat{\lambda}$ and $\hat{\lambda}^T$ are shown.

A vector **A** can be presented by its components in each of these two coordinate systems:

$$\mathbf{F} = F_1\mathbf{e}_1 + F_2\mathbf{e}_2 + F_3\mathbf{e}_3 = \sum_{j=1,2,3} F_j\mathbf{e}_j,$$

$$\mathbf{F} = F_1'\mathbf{e}_1' + F_2'\mathbf{e}_2' + F_3'\mathbf{e}_3' = \sum_{j=1,2,3} F_j'\mathbf{e}_j'.$$

The components F_i and F_i' are projections of the vector **F** on the corresponding basis vectors $\mathbf{e}_i$ и $\mathbf{e}_i'$:

$$F_i = \mathbf{e}_i \cdot \mathbf{F}, \quad F_i' = \mathbf{e}_i' \cdot \mathbf{F}.$$

Let us determine the relation between the components in these two coordinate systems:

$$F_i' = \mathbf{e}_i' \cdot \mathbf{F} = \left(\mathbf{e}_i' \cdot \sum_j F_j\mathbf{e}_j \right) = \sum_j (\mathbf{e}_i' \cdot \mathbf{e}_j)F_j = \sum_j \lambda_{ij}F_j,$$

where $\lambda_{ij} = \mathbf{e}_i' \cdot \mathbf{e}_j$ are the elements of the tensor $\hat{\lambda}$ that transforms unit vectors of coordinate system $(\mathbf{e}_1, \mathbf{e}_2, \mathbf{e}_3)$ into $(\mathbf{e}_1', \mathbf{e}_2', \mathbf{e}_3')$, $\mathbf{e}_k' = \sum_{i=1}^{3} \lambda_{ki}\mathbf{e}_i$. The transposed matrix $\hat{\lambda}^T$ performs the inverse transformation $\mathbf{e}_i = \sum_{k=1}^{3} (\lambda^{-1})_{ik}\mathbf{e}_k'$; here $(\lambda^{-1})_{ik} = (\lambda^T)_{ik} = \lambda_{ki}$ (see Figure C.4).

Let us write the components of a tensor $\hat{T}' = T_{ij}'$ in the basis $\mathbf{e}_k'$ knowing the components of $\hat{T} = T_{lk}$ in the given basis $\mathbf{e}_k$:

$$T_{ij}' = \sum_{k=1}^{3}\sum_{l=1}^{3} \lambda_{ik}T_{kl}(\lambda^T)_{lj}.$$

TENSOR DIAGONALIZATION

Transformation from one coordinate system to another changes the tensor components. So, a transformation can be found that diagonalizes any nonsingular tensor. As an example, let us consider the diagonalization of the relative dielectric permittivity tensor:

$$\hat{\kappa} = \begin{pmatrix} \kappa_{11} & \kappa_{12} & \kappa_{13} \\ \kappa_{21} & \kappa_{22} & \kappa_{23} \\ \kappa_{31} & \kappa_{32} & \kappa_{32} \end{pmatrix}.$$

We write a characteristic equation for the matrix of that tensor:

$$\begin{vmatrix} \kappa_{11} - \lambda & \kappa_{12} & \kappa_{13} \\ \kappa_{21} & \kappa_{22} - \lambda & \kappa_{23} \\ \kappa_{31} & \kappa_{32} & \kappa_{32} - \lambda \end{vmatrix} = 0.$$

This is a cubic equation and it will have three roots λ_1, λ_2, and λ_3. For each root λ_j (here $j = 1, 2, 3$), there is an eigenvector $\mathbf{a}_j(l_j, m_j, n_j)$ with components l_j, m_j, n_j that can be found by solving the system of equations:

$$(\kappa_{11} - \lambda_j)l_j + \kappa_{12}m_j + \kappa_{13}n_j = 0,$$

$$\kappa_{21}l_j + (\kappa_{22} - \lambda_j)m_j + \kappa_{23}n_j = 0,$$

$$\kappa_{31}l_j + \kappa_{32}m_j + (\kappa_{33} - \lambda_j)n_j = 0.$$

The vectors

$$\mathbf{a}_1(l_1, m_1, n_1), \quad \mathbf{a}_2(l_2, m_2, n_2), \quad \mathbf{a}_3(l_3, m_3, n_3)$$

can be normalized to get basis vectors of a new coordinate system

$$\mathbf{e}_1(l_1', m_1', n_1'), \quad \mathbf{e}_2(l_2', m_2', n_2'), \quad \mathbf{e}_3(l_3', m_3', n_3'),$$

where

$$l_j' \rightarrow \frac{l_j}{\sqrt{l_j^2 + m_j^2 + n_j^2}}$$

$$m_j' \rightarrow \frac{m_j}{\sqrt{l_j^2 + m_j^2 + n_j^2}}$$

$$n_j' \rightarrow \frac{n_j}{\sqrt{l_j^2 + m_j^2 + n_j^2}}$$

In the new coordinate system, the tensor $\hat{\kappa}$ is diagonal with components λ_1, λ_2, λ_3:

$$\hat{\kappa} = \begin{pmatrix} \lambda_1 & 0 & 0 \\ 0 & \lambda_2 & 0 \\ 0 & 0 & \lambda_3 \end{pmatrix}.$$

We complete discussion of diagonalization by considering numerical example. For the tensor

$$\hat{D} = \begin{pmatrix} 7 & -2 & 0 \\ -2 & 6 & -2 \\ 0 & -2 & 5 \end{pmatrix}.$$

The characteristic equation

$$\begin{vmatrix} 7-\lambda & -2 & 0 \\ -2 & 6-\lambda & -2 \\ 0 & -2 & 5-\lambda \end{vmatrix} = \lambda^3 - 18\lambda^2 + 99\lambda - 162 = 0$$

has three roots: $\lambda_1 = 3$, $\lambda_2 = 6$, and $\lambda_3 = 9$.

Solving the system of equations

$$(7-\lambda_j)l_j - 2m_j = 0,$$

$$-2l_j + (6-\lambda_j)m_j - 2n_j = 0,$$

$$-2m_j + (5-\lambda_j)n_j = 0$$

and performing normalization, we get three basis vectors of the coordinate system:

$$\mathbf{e}_1(l_1', m_1', n_1') = \mathbf{e}_1\left(\frac{1}{3}, \frac{2}{3}, \frac{2}{3}\right),$$

$$\mathbf{e}_2(l_2', m_2', n_2') = \mathbf{e}_2\left(\frac{2}{3}, \frac{1}{3}, -\frac{2}{3}\right),$$

$$\mathbf{e}_3(l_3', m_3', n_3') = \mathbf{e}_3\left(-\frac{2}{3}, \frac{2}{3}, -\frac{1}{3}\right),$$

in which the matrix of tensor $\hat{D}$ becomes diagonal with $\lambda_1 = 3$, $\lambda_2 = 6$, $\lambda_3 = 9$:

$$\hat{D} = \begin{pmatrix} 3 & 0 & 0 \\ 0 & 6 & 0 \\ 0 & 0 & 9 \end{pmatrix}.$$

Index

A

Absorbing media, 152, 153, 159, 223, 293–296, 312, 315
Alternating current, 59
Amperes' force, 29–30
Ampere's law, 34–36, 39, 42, 43, 108, 109, 168, 169, 292
Amplifying media, 312
    amplification in absorbing media, 296–302
    attenuation in absorbing media, 293–296
    dispersion, 287–292
    laser, 302–309
Amplitude array, 186
Amplitude–phase diffraction gratings, 186
Angle of dielectric losses, 311
Angle of magnetic losses, 311
Angular dispersion, 190
Anomalous dispersion, 290
Antenna
    aperture, 253
    closed oscillatory circuit, 252
    current and voltage, 252
    directional diagram
        antenna pattern, 249–250
        half-wave Hertz dipole, 250
        normalized, 248
        symmetric dipole antenna, 250–251
        toroid-shaped, 248–249
        unidirectional, 248–249
        width of diagram, 248–249
    feeder, 252
    ground-wave, 253
    horn
        combined, 256
        conical, 256
        phase distribution, aperture, 256–258
        pointed pyramidal, 256
        sectoral, 255–256
        wedge pyramidal, 256
    lattices, 253
    linear, 253
    open oscillatory circuit, 252
    operating height of a receiving, 253
    receiving antenna, 253
    transmitting, 251
Antinodes, standing waves, 93
Aperture
    antenna, 253
    lightguide, 235
Atomic number, 3
Attenuation (amplification) coefficient, 312

B

Bardeen, Cooper, and Schrieffer (BCS) theory, 167
Biaxial crystals, 133
Biot–Savart law, 32
Bloch wave number, 200
Bloch wave vector, 199
Bohr model, 172

Bose condensation, 167
Bouguer's law, 293–296, 301
Boundary conditions
    direct problems, 102
    electric field
        dielectric displacement vector, 104
        dielectric permittivities, 104
        direction change of vectors, 105–106
        electric displacement vector, Gauss's law, 103–104
        normal component of vector, 104
        tangential components of displacement, 105
        tangential components of vectors, 104
    magnetic field
        air gap of ferromagnetic core, 109–110
        dielectric permittivities, 107
        normal component of magnetic intensity, 107
        normal components of vectors, 107
        surface current, Ampere's law, 108
        tangential component of magnetic intensity, 108
Bragg's angle, 193
Bragg's diffraction, 193
Bragg's equation, 193
Bragg's reflection, 195
Brewster's angle, 115–116
Broken total internal reflection method, *see* Prism method
Broken total reflection, *see* Electromagnetic tunneling

C

Capacitance, 12–14
Chiral smectic LCs, 142
Cholesteric LCs, 142
Closed oscillatory circuit, 252
Coaxial transmission lines, 207–209, 230–231
Coherence length, 98
Coherence time, 97
Combined horn, 256
Complex permittivity and permeability; *see also*
        Electromagnetic tunneling; Negative refraction
    absorbing media, 312
    amplifying media, 312
    angle of dielectric losses, 311
    angle of magnetic losses, 311
    attenuation (amplification) coefficient, 312
    energy flow
        counterpropagating waves, 315–316
        impedance, 317
        linearly polarized wave, 314–315
        two independent sources, 314–315
    impedance real and imaginary parts, 313–314
    passive media, 312
    right- and left-handed media
        active, 318–319
        backward waves, 318
        impedance, 319
        passive, 318–320
        Poynting vector, 318–319
        refractive index, 317–320

Milton Keynes UK
Ingram Content Group UK Ltd.
UKHW052021071024
449327UK00027B/2362